T0321952

Modern Cellular Automata
Theory and Applications

ADVANCED APPLICATIONS IN PATTERN RECOGNITION
General editor: Morton Nadler

A STRUCTURAL ANALYSIS OF COMPLEX AERIAL PHOTOGRAPHS — Makoto Nagao and Takashi Matsuyama

PATTERN RECOGNITION WITH FUZZY OBJECTIVE FUNCTION ALGORITHMS — James C. Bezdek

COMPUTER MODELS OF SPEECH USING FUZZY ALGORITHMS — Renato de Mori

MODERN CELLULAR AUTOMATA: Theory and Applications — Kendall Preston, Jr., and Michael J. B. Duff

A Continuation Order Plan is available for this series. A continuation order will bring delivery of each new volume immediately upon publication Volumes are billed only upon actual shipment. For further information please contact the publisher

Modern Cellular Automata
Theory and Applications

Kendall Preston, Jr.
Carnegie-Mellon University
University of Pittsburgh
Pittsburgh, Pennsylvania
Kensal Consulting
Tuscon, Arizona

and

Michael J. B. Duff
University College London
London, England

Plenum Press • New York and London

Library of Congress Cataloging in Publication Data

Preston, Kendall, 1927–

Modern cellular automata.

(Advanced applications in pattern recognition)
Bibliography: p.
Includes indexes.
1. Cellular automata. I. Duff, M. J. B. II. Title. III. Series.
QA267.5.C45P74 1984 001.64 84-11672
ISBN 0-306-41737-5

A Division of Plenum Publishing Corporation
233 Spring Street, New York, N.Y. 10013

Printed in the United States of America

to Susan,
to Sally,
and especially
to *Paisan*

FOREWORD

It is with great pleasure that I present this fourth volume in the series "Advanced Applications in Pattern Recognition." It would be difficult to find two authors better versed in the design and application of parallel image processing systems, due to both their own many years of pioneering in the field and their encyclopedic knowledge of what is going on in university and industrial laboratories around the world.

The monograph is unique in its parallel presentation of orthogonal and hexagonal dissections, and the wealth of graphic illustration of algorithmic procedures for processing and analyzing images in the various known implementations of parallel image-processing architectures.

This volume should find a place on the bookshelf of every practitioner of pattern recognition, image processing, and computer graphics.

Morton Nadler
General Editor

PREFACE

This book endeavors to introduce the reader to the subject of cellular logic and cellular automata and is devoted particularly to those parts dealing with the manipulation of pictorial data. The study of cellular automata owes much to the pioneering work of John von Neumann during the 1950s. Von Neumann was interested in general problems in the behavior of computing structures and was immensely impressed by the complexity and performance of the human brain, which he felt must point towards successful designs for automatic computing machines. However, he also (correctly) realized that high complexity carries with it the need for high reliability and saw this fact as indicating the need for a self-repairing mechanism within the computing structure itself. This led to a detailed investigation of the theory of self-reproducing automata. Initially, von Neumann spoke in terms of actual mechanical systems having the physical property of self-reproduction, in which mechanical manipulators assembled duplicate machines, using a pool of spare parts as building blocks. As his studies proceeded into further depths, the concepts became more formalized and the physical aspects of the problem became submerged in an abstract representation.

The interval 1980-1984, during which this book was written, has been an era of great significance in the history of cellular automata. Early in 1980, the first major embodiment of the classic concept of von Neumann was made by a group of scientists and engineers working for one of us (Duff) at University College London. This cellular automaton is the Cellular Logic Image Processor number 4 (CLIP4) which is a 96×96 array of 9216 one-bit CPUs (Central Processing Units) fabricated as 1152 LSI (Large Scale Integration) circuit chips and comprised of about three million transistors. This extraordinary machine, when performing a simple operation such as the logical AND between two data arrays, operates at over one billion actual instructions per second. When conducting more complex nearest neighbor operations, its speed is equivalent to over ten billion instructions per second. This makes CLIP4 one of the fastest computers in the world, and, probably, the most cost effective when it is considered that it will be produced by commercial manufacturers in 1984 at a price of only 100 thousand dollars.

Later, in mid-1983, the MPP (Massively Parallel Processor) was completed by the Goodyear Aerospace Corp. (Akron, Ohio) under a 5 million dollar NASA (National Aeronautics and Space Administration) contract and installed at the NASA Goddard Laboratory in the United States. This machine, consisting of 16,384 active one-bit CPUs and employing 100 million transistors in its 128×128 cellular array, is now the world's fastest computer performing operations at a rate of approximately one thousand billion instructions per second.

In addition to this *tour de force*, the year 1983 saw the completion of over three hundred of the diff-series cellular logic image processors of Coulter Electronics, Inc. (Hialea, Florida). These machines are now in use in hematology laboratories worldwide and represent the major application of cellular operations in pictorial data processing. At the time of writing, they are used in the analysis of images of the human blood cells of some fifty thousand persons per day. They are based on the early work on the CELLSCAN cellular logic machine by one of us (Preston) at the Perkin-Elmer Corporation (Norwalk, Connecticut) in the 1960s.

The above accomplishments in both the design, fabrication, and utilization of cellular automata and cellular logic machines led the authors to produce this book. We were also influenced by the fact that no text was available which reviewed theoretical work or many important events which have taken place in the field. It is hoped that knowledge concerning cellular automata which heretofore has been scattered has now been successfully collected here. In order to achieve this goal, many aspects of both cellular automata theory and practice are contained in this book. Chapter 1 surveys the history of the field and makes every attempt to credit those pioneering workers who, in addition to von Neumann, contributed to the great progress made to date. Chapters 2 through 4 treat the theoretical aspects of the subject presenting the theory of both two-dimensional and multi-dimensional cellular logic transforms as well as describing two-dimensional numerical methods. Since the major application of these machines to date is in the field of image processing, further chapters are provided (Chapters 5 through 7) on some of the major image processing tasks which may be approached using the cellular methodology. These chapters cover image segmentation, skeletonization, and filtering. Specific applications in both science and biomedicine are treated in the next two chapters (Chapters 8 and 9).

Following this, Chapter 10 describes cellular logic machines, starting with CELLSCAN (built in 1961) and extending through a variety of more modern machines. Similarly, Chapter 11 covers the three major full array cellular automata now in existence (CLIP4, DAP, MPP). The next chapter (Chapter 12) is somewhat of a diversion from the work-a-day world and exposes the reader to games which can be played with the cellular automaton. Then, because of the importance of understanding cellular programming techniques and methods, Chapter 13 is devoted to the languages developed for CLIP4 and MPP to illustrate programming cellular arrays, while GLOL and TASIC are taken as examples of languages employed in programming cellular logic machines. This chapter is followed by the final chapter (Chapter 14) on fabrication methods wherein a description is provided of both the CLIP4 and the MPP chips.

The authors herewith gratefully acknowledge the benefit received from the work of both their staff and students in the Image Processing group of University College London, in the Department of Electrical and Computer Engineering and the Biomedical Engineering Program of Carnegie-Mellon University, and in the Department of Radiation Health of the University of Pittsburgh. It is in this latter department that the cellular logic computers are located which made possible the bulk of the illustrations contained herein.

In preparing this work, the authors have received significant assistance from many of their colleagues. In particular the authors would like to acknowledge those individuals who wrote certain portions of the manuscript. Judith Hilditch (formerly with University College London) produced that part of Chapter 9 which treats the application of the cellular automaton to chromosome image analysis. Terence Fountain (University College London) and Kenneth Batcher (Goodyear Aerospace Corp.) wrote the body of Chapter 14 on fabrication methods. In addition to these authors, certain text and figures have been employed from the works of others. In these cases, acknowledgements are made in the text and in the figure captions. Also, the authors have used figures which they have previously published elsewhere. In particular, we wish to acknowledge the use of material from Pattern Recognition and from several publications of the Institute of Electrical and Electronics Engineers

(IEEE Proceedings, Transactions on Acoustics Speech and Signal Processing, Transactions on Computers, Transactions on Pattern Analysis and Machine Intelligence). Acknowledgements of these figures are not contained in the captions themselves and are simply made herewith. It should be noted that certain trademarks are employed in this text which are not directly acknowledged. Rather, they are listed below.

Trademark	*Company*
diff3	Coulter Electronics
diff3-50	Coulter Electronics
diff4	Coulter Electronics
Imanco	Cambridge Instruments
Omnicon	Bausch & Lomb
Magiscan	Joyce Loebl
Microline	Perkin-Elmer
Unix	AT&T Bell Laboratories

Drafting and photography for this book was done primarily at Mellon Institute by Gary Thomas and his staff (Mike Westfall, Mark Martin, and Mary Adams) as well as Dublin Design (New York City) and by the photographic staff of University College London. The text itself was prepared by the staff of Executive Suite, Inc. (Tucson, Arizona), founded by the late Judith Tyree and now managed by Teresa Davenport. The Olivetti Venezia type font, using proportional spacing, was selected for the text; Esteem Elite and Esteem Pica for the equations. For this purpose, special daisy wheels were fabricated by Camwil Inc. (Honolulu, Hawaii). Rose Ballentine, of Executive Suite, was both responsible for this aspect of the work and prepared Chapters 2 through 4 and 8. Linda Daigle was responsible for Chapters 6, 7, and 10 through 12, while Jan Tweed prepared Chapters 1, 5, 9, 13 and 14. Ms. Tweed also was responsible for paging, contents, running heads, and the author and subject indices. Thanks also go to M. Nadler, General Editor of this series, who proofed the entire manuscript, and to L. S. Marchand (Senior Editor), J. Matzka (Managing Editor), and the other staff of Plenum Press who were involved in production.

Finally, the financial support of several government agencies, both for the preparation of the text and as regards the research reported herein, is gratefully acknowledged. These are the National Institutes of Health and the National Science Foundation of the United States and both the Medical and Science and Engineering Research Councils of Great Britain.

Kendall Preston Jr.
Tucson, Arizona, USA

Michael J. B. Duff
London, England

CONTENTS

3. CELLULAR LOGIC OPERATIONS IN N-SPACE

4.1 NUMERICAL OPERATIONS

5. SEGMENTATION

1. INTRODUCTION

1.1 EARLY HISTORY

Ironically, the name *von Neumann* is now strongly associated with the old-fashioned, single-CPU (Central Processing Unit) computer architecture. Many of today's computer scientists and engineers have completely forgotton that John von Neumann was also the major pioneer in parallel computing via his research on arrays of computers or *cellular automata* (Figures 1-1 and 1-2).

As stated in an essay by Thatcher (1970),

> "Von Neumann, as the originator of this area of study, was not interested in cellular automata as mathematical objects of study, but instead was attempting to find a manageable way of treating in detail the problem of how to make machines reproduce themselves ..."

In 1944, von Neumann was introduced to electronic computing via a description of the ENIAC (Electronic Numerical Integrator And Computer) by Goldstine (1972, p. 182). Shortly, he formed a group of scientists headed by himself, Howard Aiken (Harvard University), and Norbert Wiener (Massachusetts Institute of Technology) to work on problems in computers, communications, control, time-series analysis, and the "communication and control aspects of the nervous system" (Goldstine, 1972, p. 275). The last topic was included due to his great

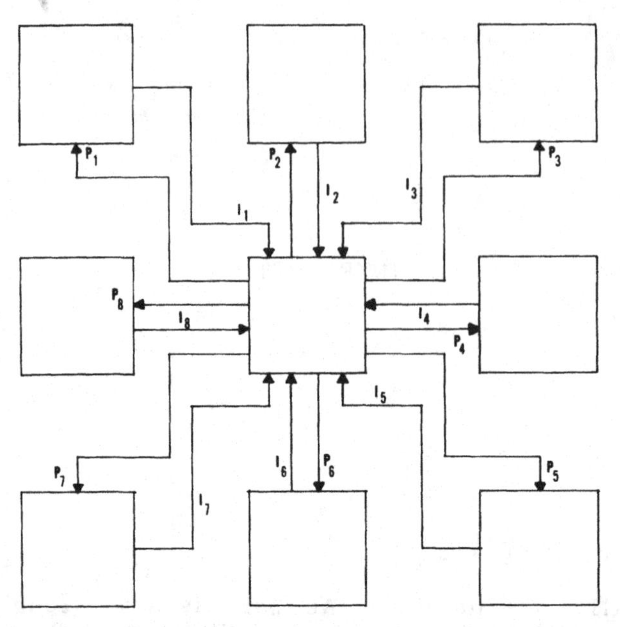

Fig. 1-1 The cellular automaton consists of an array of identical computers with each connected to its immediate neighbors.

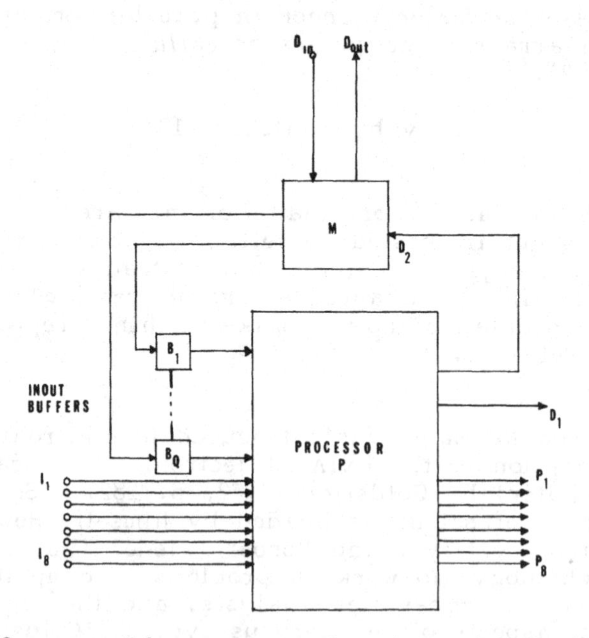

Fig. 1-2 One possible structure for the individual computer or processing element. This configuration is used in the University College London CLIP4.

interest in the work on neural networks of McCulloch and Pitts (1943). In 1946, von Neumann proceeded to design the EDVAC (Electronic Discrete Variable Computer) which was the first design of a stored-program machine. By 1947, under the influence of the ideas on automata developed by Post (1936) and Turing (1936), he had commenced studies on the complexity required for a device or system to be self-reproductive. These studies also included work on the problem of organizing a system from basically unreliable parts (a field of study which we now know as "fault tolerant computing"). At first, von Neumann investigated a continuous model of a self-reproducing automaton based on "a system of non-linear partial differential equations, essentially of the diffusion type" (letter, von Neumann to Goldstine, 28 October 1952). He also pursued the idea of a kinematic automaton which could, using a description of itself, proceed to mechanically assemble a duplicate from available piece parts (von Neumann, 1951).

When von Neumann found it difficult to provide the rigorous and explicit rules and instructions needed to realize such an automaton and when it became evident that the value of such an automaton would be moot, he redirected his efforts towards a model of self-reproduction using an array of computing elements. Both Burks (1970) and Goldstine (1972) confirm that the idea of such an array was suggested to von Neumann by Stanislaw Ulam. Von Neumann was also attracted to this idea of using parallelism because he saw that it would eventually lead to high computational speed. By 1952 he had put his ideas in writing and in 1953 described them more fully in his Vanuxem lectures at Princeton University as summarized by Kemeny (1955). Unfortunately, his premature death in 1957 prevented him from completely achieving his goals. Thus, it can be said that, in the early 1950s, von Neumann conceived of the cellular automaton.

1.2 EARLY APPLICATIONS

At the same time, in a completely unrelated vein, Jacob Bronowski convened a meeting of the National Coal Board (Great Britain) in 1951 to investigate the use of a combination of the electronic computer and the television scanner for "the possibility of making a machine to replace the human observer" (Walton, 1952). The immediate goal was to quantitate airborne particulate matter (particularly coal dust) by counting and sizing dust particles when precipitated on microscope slides. The real leadership in this effort came from biomedical engineers in the Department of Clinical Pathology (Radcliffe Infirmary, Oxford) and the Department of Anatomy (University College London) and was funded by the Medical Research Council. By the mid-1950s, Causley and Young (1955) had created a flying-spot microscope. Using simple neighborhood logic circuits, it was employed in counting and sizing microscopic objects at the rate of one field-of-view every few seconds.

The connection between this work and the work of von Neumann and his colleagues appears tenuous. However, it is now recognized that the logic circuits associated with the first computerized television microscopes were essentially sequential mechanisms for emulating the action of a cellular array. Further refinements along these lines were investigated by Moore (1966) who used a scanner coupled to a general purpose computer, namely, the SEAC (Standards Electronic Analyzer and Computer) at the United States National Bureau of Standards. He studied the application of cellular computing to images of metallurgical specimens. Kirsch (1957), also at the National Bureau of Standards, built on the work of Moore and Selfridge (1955) at the Massachusetts Institute of Technology performed similar research. The results of some of the early experiments of Kirsch are shown in Figure 1-3. As can be seen, the images dealt with were bilevel (sometimes called "binary"), i.e., composed of groups of 1-elements on a background of 0-elements. Neighborhood operations performed on such images became known as "cellular logic transforms" because their mathematics was that of Boolean logic. The special purpose machines which executed these transforms were called "cellular logic machines." (See Chapters 2, 3, and 10.)

At the same time, there was a growing interest in industry in the development of automatic machinery for reading printed characters. Dinneen (1955) was the first to use cellular logic transforms for the purpose of pre-processing images of individual printed characters for the purpose of noise removal (Figure 1-4). Unger (1959) reported on more elaborate cellular logic methods for feature extraction and character recognition and presented the design of a cellular automaton to execute his algorithms. Although not recognized as such at the time, it is now understood that the cellular logic transform of Dinneen is the bilevel ranking transform (over a 5×5 cell). Dinneen demonstrated not only the median filter (rank 13) but also showed both image augmentation and reduction by employing ranks other than the median. (See Chapter 7.)

This work was followed in 1960 by that of one of the authors (Preston) who constructed the world's first dedicated cellular logic machine (CELLSCAN). (See Chapter 10.) This machine, based on Golay's original cellular logic patent (Figure 1-5), was capable of not only performing the 3×3 binary ranking transform but also of executing this transform using connectivity-preserving logic so as to permit what is now known

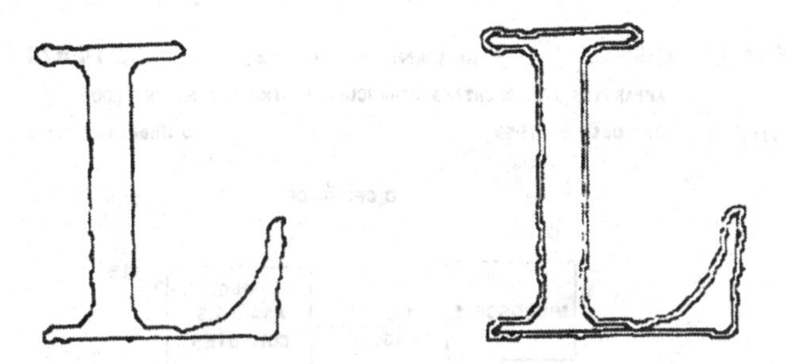

Fig. 1-3 Example of the work of Kirsch (1957) demonstrating the cellular logic operation called "custering" by Moore (1966) which finds the edges of the edges displayed on the left picture to give the result shown on the right. (See Chapters 2 and 12 for details.)

Fig. 1-4 Dinneen (1955), working with Selfridge (1955), was the first to investigate the ranking transform using a 5×5 kernel. The illustrations above show (left-right, top-bottom) original, rank 5, rank 10, rank 13 (now called the "median filter").

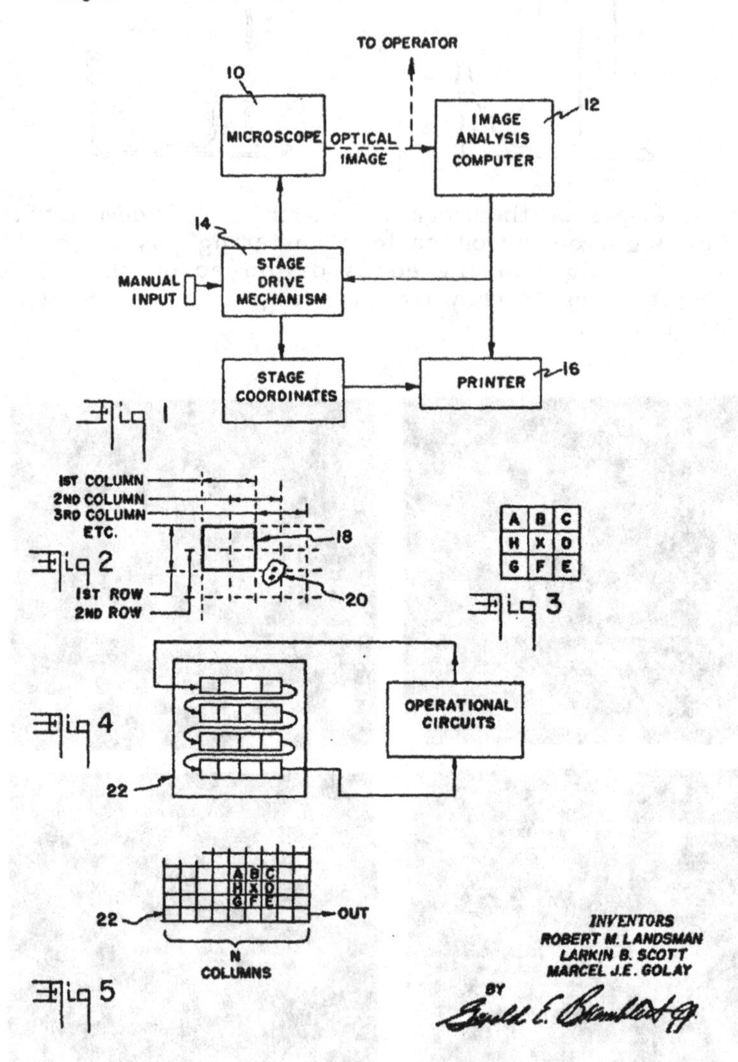

Fig. 1-5 Golay (1969), along with Landsman and Scott of Perkin-Elmer, was the first to patent a cellular logic machine. This machine, reduced to practice as CELLSCAN (Preston, 1961), performed iterative operations on a bilevel image array using a 3×3 kernel. It was capable of performing a *connected components analysis* and generating the *residue histogram*. (See Chapters 2 and 10.)

as "skeletonization." (See Chapter 6.) An illustration of the iterative application of a connectivity-preserving ranking transform using CELLSCAN is given in Figure 1-6.

1.3 LATER WORK

The pioneering work on cellular automata by von Neumann over the interval 1947-1955 and the work in cellular logic for picture processing and pattern recognition (1951-1961) led to many theoretical and practical studies and developments in the field of cellular automata in the ensuing years. Some of these developments are reviewed in this section.

1.3.1 Self-Reproducing Machines

Von Neumann's original construct for a self-reproducing cellular array machine, i.e., cellular automaton, required that each computer in the array support a set of 29 states. The array itself required some 200,000 computers. This degree of complexity was needed since von Neumann sought to design his automaton as a universal computing system or *Turing machine*, i.e., a construct capable of performing any desired calculation. In the 1960's other researchers, in particular those at the Massachusetts Institute of Technology, found self-reproducing constructs, although not universal, of an extremely simple nature. In particular, a cellular automaton consisting of two-state computing elements with each connected only to its four nearest neighbors in the square tessellation was found to self-reproduce *any* initial configuration when any element with an odd number of neighboring elements in the 1-state is set to the 1-state; an even number (including zero), to the 0-state. This discovery was made by Fredkin (see Gardner, 1971).

Ulam and Schrandt (1967) also investigated the performance of these simple cellular arrays of two-state computers, and, in fact, extended their research by studying three-dimensional cellular automata. (See Chapter 3.) Their work, along with the work of such researchers as Thatcher, Moore, Myhill, Stein, and Holland, are collected in the book *Essays on Cellular Automata* by Burks (1970). Subsequently, Banks (1970) (a student of Fredkin) proved the existence of a von Neumann-type self-reproducing cellular automaton using four-state computing elements. Yamada and Amoroso (1971), at the University of Pennsylvania and at Fort Monmouth, studied cellular automata in other than the square tessellation. Smith (1970, 1971), at the Polytechnic Institute of Brooklyn, and Akers (1972), at General Electric, expanded on these activities. The first for-

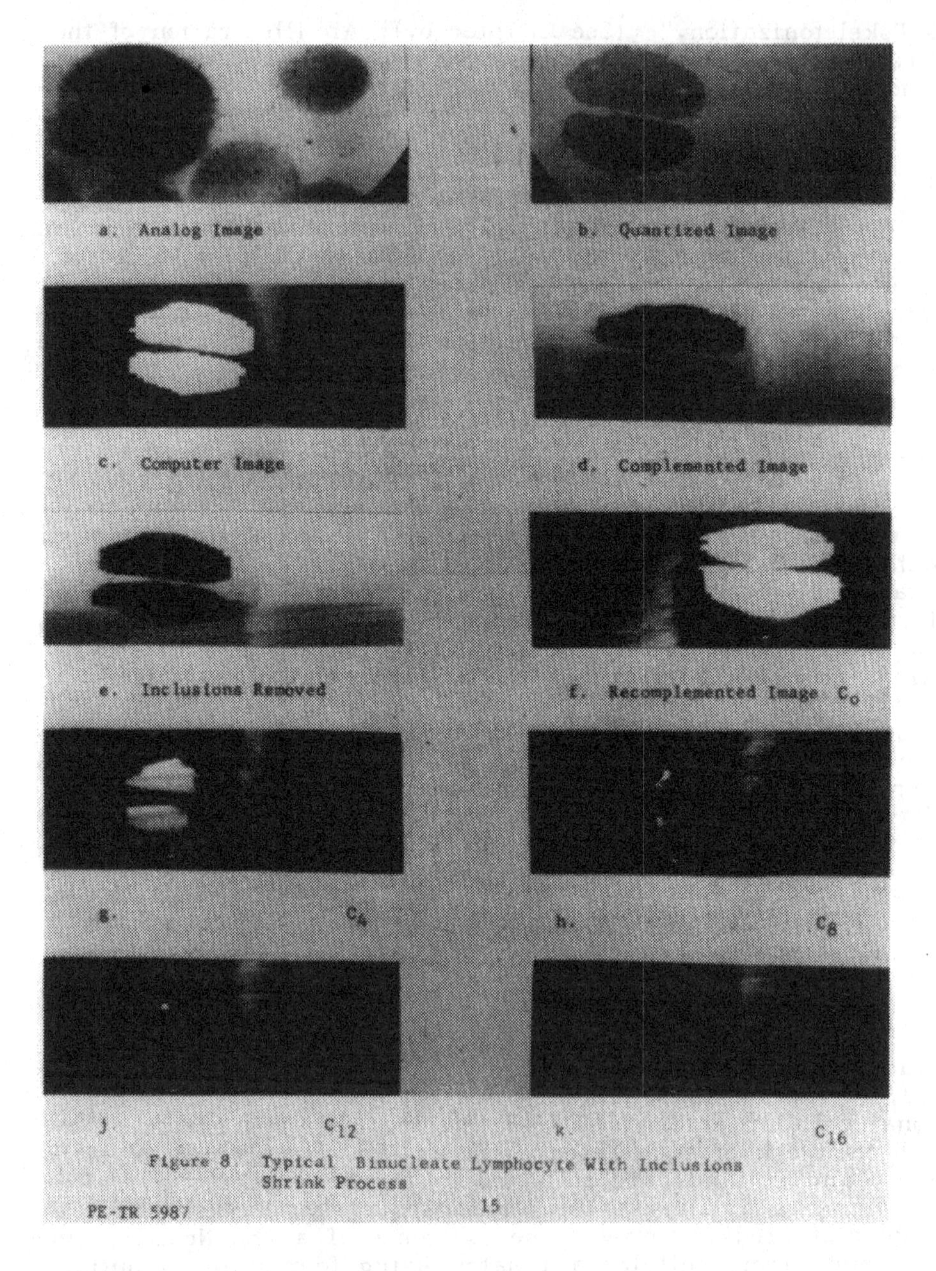

Fig. 1-6 The first known demonstration of connectivity-preserving reduction or *skeletonization* was performed using CELLSCAN (Preston, 1961). (Illustration taken from unpublished Perkin-Elmer Engineering Report number 5987, "Final Report on Atomic Energy Commission Blood Cell Scanner," September 12, 1961).

mal conference on cellular automata was held by the Institute of Electrical and Electronics Engineers in 1975 and its proceedings were published (IEEE 75 CHI052-6C). A second meeting on the subject was sponsored by the Department of Energy at Los Alamos in 1983. Papers from this meeting are being published in the journal PhysicaD (see, for example, Preston, 1984). Surveys of theoretical work on cellular automata have been written by Noguchi and Oizumi (1971), Nishio (1975), and Maruoka (1978). Also of interest are the series of papers by Tojo (1967-1969) as well as that of Minnick (1967).

1.3.2 Cellular Logic Machines

Starting in the 1960s, a series of machines were constructed in various laboratories which carried out sequential cellular logic operations by means of special purpose hardware. These machines were not true cellular automata in that a single processor (or, at most, a few processors) was used to operate sequentially upon an array of numbers as if these numbers resided in an array of computing elements although, in actuality, they were stored in some form of bulk memory.

Nevertheless, the principles developed in these machines and, in some cases, in engineered versions of the machines themselves, have led to important advances in the design and construction of commercially successful image analysis systems; they have let cellular automaton principles be put to good use before technological progress has allowed full scale automata to become practical possibilities. (See Chapter 10.)

1.3.3 Full-Scale Array Automata

Two full-scale cellular automata were proposed in the 1960s. These were the 16×16 Solomon of Westinghouse (Slotnick et al., 1962) and the 32×32 ILLIAC III of the University of Illinois (McCormick, et al., 1968). Construction was commenced on both machines but neither was completed. Finally, a small array automaton (8×8) called the ILLIAC IV, conceived at the University of Illinois, was built by Burroughs (Slotnick, 1971). This machine, fabricated at enormous cost (tens of millions of dollars), was used during the 1970s at the Ames Laboratory of the National Aeronautics and Space Administration and then was replaced by a standard, commercial "supercomputer" in the 1980s. Although small, ILLIAC IV was, in fact, the initial embodiment of von Neumann's original concept. Each computer in ILLIAC IV used about 100 thousand transistors in four thousand dual inline packages (made by Texas Instruments). These furnished 2K of 64-bit memory and permitted both fixed and floating point calculations with a 350ns clock cycle. The main applications of ILLIAC IV were in solving problems in aerodynamics, weather analysis, and remote sensing.

At more or less the same time that ILLIAC IV was being built, one of the authors (Duff) at University College London started the CLIP (Cellular Logic Image Processor) series of cellular automata. Third in this series, CLIP3, was a 12×16 array which became operational in 1973 (Duff, 1975). This was followed by the 96×96 CLIP4 in 1980 (Fountain, 1981) which is comprised of some 30 million transistors integrated at 3000 per chip. This machine as well as the 64×64 DAP (Distributed Array Processor) of International Computers Ltd. (Reddaway, 1979) and the 128×128 MPP (Massively Parallel Processor) of Goodyear Aerospace Corp. (Batcher, 1980) are discussed in Chapters 11 and 14.

1.3.4 Cellular Automata Games

In their studies of cellular automata using two-state computing elements, Schrandt and Ulam (1967) became intrigued by the patterns formed by the groupings of 1-state-elements (1-elements) and 0-state-elements (0-elements) at various stages of the computation. Certain computational rules caused an initial configuration of 1-elements (in a background of 0-elements) to dissipate, i.e., to vanish or be taken over by the 0-elements. Other rules produced the opposite effect. Conway (see Gardner, 1971) discovered a simple rule which had an intermediate effect, i.e., led eventually to either a stable pattern or to patterns exhibiting a repetitive, i.e., oscillatory, stability. Conway's discovery of this rule, based on the square tessellation and on each computing element having connections to eight neighbors, caused an outpouring of studies on what is essentially a game, played, as it were, on an infinite checker board. Although of no practical value, *Conway's Life*, as the game is now called, has had a major impact on cellular automata research. (See Chapter 12.)

1.3.5 Non-Cellular Arrays

It becomes apparent, when the cellular automaton is used in the analysis of image data, that image structure is not always immediately revealed by examining nearest neighborhoods; frequently, neighborhoods of diameters of five or seven computing elements (or even greater) must be inspected in order to detect the presence of certain image features. In extreme cases, such as when a global image property is involved, it may be necessary to include the entire array in a neighborhood. The classical concept of a cellular automaton does not fulfill these further requirements which may, however, be met in practice by providing interconnections to a wider neighborhood. A simpler method, entailing fewer interconnections, is to allow the computing element outputs to be a function of not only the internal states of the element but also its inputs without invoking any minimum time delay restriction. In this way, data can be passed asynchronously across the array before a transformed set of internal states is established in a stable configuration. This *propagation* process is discussed in Chapter 5.

Despite the apparent differences between arrays which allow propagation and the simpler nearest-neighborhood array, it can be shown that any propagation-dependent function can always be represented by a finite sequence of nearest-neighbor operations. Exceptions are non-stable propagation functions which result in oscillatory internal states.

1.4 DEFINITION OF TERMS

Since no two works on cellular automata use exactly the same terminology or notation, this section is provided to guide the reader in using this book. First, all arrays of computing elements discussed so far can be classified as *cellular automata*. It is an essential feature that the *functions* computed shall be functions of the internal states of the computing elements and of the inputs from neighboring elements. At its simplest, the role of a cellular operation is to transform an array of data as it exists at time t, as given by the function s(i,j,t) (i,j are the array coordinates of the elements), into the array of data s(i,j,t+1). At time t+1 each element of the array has a value determined only by its original state along with the original values of its nearest neighbors. It is this inspection of neighboring data which allows cellular automata to be so effective in the analysis of two-dimensional structure or, to use more familiar terms, in pictorial pattern recognition.

The literature on cellular automata uses the terms *computing element* and *processing element* interchangeably. Usually *processing element* (often abbreviated, *PE*) is preferred. The scope of this book is limited to *regular* arrays of processing elements where all processing elements are identical (Figure 1-1). It is also assumed that the array is *tightly coupled* in that each processing element is connected to the immediately adjacent processing elements or *neighbors* rather than to a switching network that would permit connections to an arbitrary set of remote processing elements. This, at first, may appear to be a remarkably severe restriction, but, as is noted above, even a regular, tightly-coupled array is capable of powerful computations of great utility.

The adjacent processing elements to which a particular processing element is connected form the *neighborhood* of that processing element. This book assumes that each neighborhood is *complete*, i.e., the neighborhood consists of all immediately adjacent processing elements. This, of course, does not prohibit the processing elements from being electrically disconnected from any or all of its neighbors during some epoch in the computational process. Also, when operations in *subfields* are executed, certain processing elements may be excluded from taking part in the computation by means of a central control unit or *deus ex machina*. It is further assumed that the operations of the array are *isotropic* in that, when the data stored in the array are shifted to a new location in the array and operated upon, the results are identical to the same operation conducted on the data in the original position.

Operations are assumed to occur in discrete time with each step in time being a *generation*, *iteration*, or *cycle*. It is further assumed that changes in all processing elements occur simultaneously, i.e., the action of all elements in the cellular array is *synchronous*. The particular action which occurs depends on a universal *transition rule* or *transform* that uses as its independent variables the existing state of the particular processing element and the states of its neighbors. Finally, it is assumed that every transform is *deterministic* in that a given *configuration* of the states of the elements of the array has precisely one *successor* configuration for a given transform; although a given configuration may have one or more *predecessors*. Burks (1970) mentions the possibility of *non-deterministic* operations where a given configuration may be transformed into different configurations at different times even when executing the *same* transition rule. Burks also discusses the possibility of *probabilistic* operations. Neither of these possibilities are treated in this book.

This book frequently calls the configuration of the array a *signal* $s(\hat{r})$ where, in some special cases, $\hat{r}$ may be multidimensional. (See Chapter 3.) This implies a non-planar array. Usually $\hat{r}$ is two-dimensional in which case the signal is $s(i,j)$. Using the terminology of *signal processing* a signal is a function which contains information on the physical state of a system under study. *Pattern recognition* is that field of analysis whose purpose is to extract from $s(i,j)$ those measures which are specific to those aspects of the physical state of the system which are important to some decision making process. The signal may be as simple as that generated by scanning an image of coal dust particles where the decision relates to the environmental control required. On the other hand, the signal may be as complex as a set of seismic readings and the decision as important as to whether to commit funds to explore for geological resources. It is on this latter area of the analysis of relatively complex signals that this book focuses.

This book frequently makes the distinction between the *neighborhood* which *excludes* the central processing element) and the *kernel* which includes both the processing element and its neighbors. The symbol used for the kernel is $k(\hat{r})$. Many operations using the cellular automaton or the cellular logic machine may be regarded as convolutions of the signal with its kernel in which case the output states of a two-dimensional array after a single iteration is given by

$$s(i,j,t+1) = s(i,j,t) * k(i,j) \tag{1.1}$$

where * signifies the convolution operator. Since both two-dimensional and multi-dimensional convolutions may be performed with both the cellular automaton and the cellular logic machine, there are obvious applications in both filtering and signal detection in the presence of noise. This makes both the cellular automaton and the cellular logic machine useful in image enhancement, edge and boundary detection, scene segmentation, etc.

The specific structure of the cellular automaton which is shown in Figures 1-1 and 1-2 is frequently used in this book because of its correspondence to the CLIP arrays. Here the array consists of a square array of processing elements, each comprising a processor P, a set of Q internal registers (B1,...,BQ) which can receive data from an M-bit memory M. Each processor has Q inputs from the registers and eight inputs (I1,...,I8) from the neighboring processors in the tessellation. (Note that in an hexagonal array, the quantities I7, I8, P7, and P8 are omitted and the other connections are rerouted to form an hexagonal neighborhood.) The processor generates eight outputs (P1,...,P8) which form the I inputs to the neighboring processors. Output D1 goes to a location outside the array while output D2 can be addressed into a chosen location in the Memory M. M is loaded from the control unit outside the array with an input Di and furnished with an output to the control unit through Do.

When using this structure, it is assumed that an iteration consists of several phases. (Not all of these phases may be necessary for every iteration of the automaton.) The phases are

1. *Input Phase.* The data is placed in the memory M.
2. *Buffer Load Phase.* The data is loaded from the memory M to B1,...,BQ.
3. *Propagation Phase.* P1,...,P8 are established as functions of B1,...,BQ and I1,...,I8 and the flow between the processing elements is allowed to continue until all P and I are constant.
4. *Processing Phase.* D1 and D2 are generated as functions of B1,...,BQ and I1,...,I8.
5. *Memory Load Phase.* D2 is loaded into the memory M.
6. *Output Phase.* Data is output from the memory M.

1.5 SUMMARY

The pattern of development of cellular automata over the last thirty years may be summarized as follows

1. A modeling phase, in which a study of the nervous system of mammals suggested the possible value of similar design principles for automatic computing machinery, i.e., cellular automata, with particular emphasis on self-repair by self-reproduction in a nearest-neighbor interconnected network of simple finite-state automata.
2. A software simulation phase, in which the properties of complete cellular automata were deduced in paper studies and in simulation by conventional computers.
3. A hardware simulation phase, in which machines comprised of a few processing elements were built and operated, providing encouraging evidence for future full arrays and, in the short term, offering good performance on simple tasks.
4. The cellular automata construction phase which did not occur until technological advances made practical the assembly of arrays of several thousand simple processing elements of a non-prohibitive cost. Figure 1-7, which combines figures from Chapters 10 and 11, depicts graphically the advances made during phases 3 and 4.

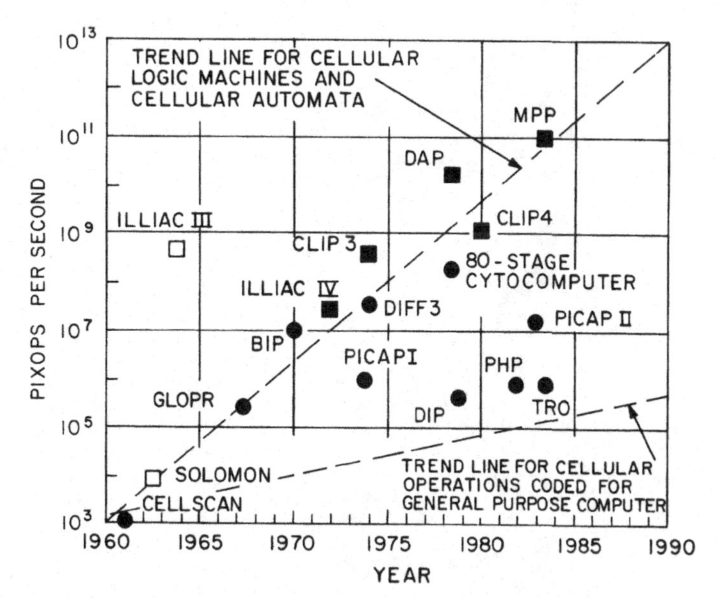

Fig. 1-7 The computing speed of cellular systems has advanced from one thousand picture point operations per second (CELLSCAN) to one hundred billion pixops per second (MPP) in only 22 years. In the above figure closed circles represent cellular logic machines; closed squares, cellular arrays; open squares, arrays proposed but never fully operational (See Chapters 10 and 11 for details.)

Von Neumann's persistent claim that complexity should contain the key to the solution of the problem of obtaining very high performance is still heard today. Very large scale integrated circuit techniques are opening the way to increasingly higher complexities at reasonable cost, size, and power consumption. It has even been suggested that large arrays might actually be *grown* in highly complex chemical structures. Still another significant question has emerged - should cellular automata in the future comprise arrays of automata which, although structurally identical and uniformly interconnected, are not simultaneously performing the same sequence of operations? The problem of controlling such an automaton is undefined but known to be large. The potential advantages are surmised rather than proven. As a compromise, whilst more complete control strategies are being devised, it will be of interest to set up systems in which data in each cell determine the local function to be performed. Thoroughly understanding the behavior resulting from this form of control would be an important step forward.

Fig. 2.7 The computing [illegible] operations [illegible] the number [illegible] above [illegible] required [illegible] for details [illegible]

Von Neumann's [illegible] complexity [illegible] [illegible] the way [illegible] complexity [illegible] [illegible] that large [illegible] [illegible]

2. TWO-DIMENSIONAL LOGICAL TRANSFORMS

2.1 INTRODUCTION

Cellular array transforms were first employed many years ago to perform "noise cleaning" operations on the input images of character recognition machines (Chapter 1). These transforms operated upon the bilevel or "binary" input generated by the character image digitizer and were designed to remove "salt and pepper" noise in the image of the character being read by the machine. By so doing the transforms utilized were in actuality executing two-dimensional, low-pass spatial filtering. Only in recent years, however, have researchers, such as Nakagawa and Rosenfeld (1978), Goetcherian (1980), and Preston (1982), performed the analysis required to fully explain the characteristics of such filters. It is the purpose of this chapter to both summarize and expand the analysis with emphasis on the effects of the array tessellation, the neighborhood configuration, and other important computational parameters.

2.2 THRESHOLDING

Functions which are binary are produced from functions which are real (or integer) by means of thresholding. In this chapter and the chapter which follows such binary functions are called "logical" functions because their element values are either false or true (0 or 1). If $s(x,y)$ is the two-dimensional function to be thresholded, then $s_L(x,y)$ is a logical function which is derived from $s(x,y)$ according to the equation given at the top of the following page.

$$s_L(x,y) = \begin{cases} 1 & s(x,y) \geq T \\ 0 & \text{otherwise} \end{cases} \tag{2.1}$$

where T is the threshold utilized. Frequently a family of logical functions are derived from s(x,y) by using multiple thresholds. There are a variety of methods for selecting such thresholds most of which depend upon an analysis of the probability density function (PDF) of the values of s(x,y). In many cases the range of the PDF is divided into a multiplicity of equal increments and thresholds are selected at each increment. In other more specific cases, when the PDF is multimodal, thresholds may be selected at the modes of the PDF and at the minima which exist between the modes. An example of multilevel thresholding is provided in Figure 2-1. Figure 2-1 displays 8 logical images derived from a variable frequency sinewave by utilizing 8 thresholds equally spaced over the range of the PDF. The frequency of this particular function varies linearly with radial distance from the origin. Its frequency is zero at the origin, increases to 64 cycles across its span at the horizontal and vertical margins, and to $95\frac{1}{2}$ cycles at the corners. This test function is presented in a 512×512 array digitized to 8 bits per element (Figure 2-2).

In the digital computer thresholding is considered a point operation which, when the number of digitizing levels used to represent s(x,y) are few, may be calculated by table lookup. For example, if s(x,y) is initially digitized at eight bits per element, then a table of 256 addresses is constructed. In this table the content of all addresses corresponding to values of s(x,y) equal to or above threshold is 1 and 0 when below threshold. Using this method of thresholding by table lookup it is possible to compute the values of a single logical function within the time required to address the table and assemble the resulting 1-bit results over the entire array. Typically, such an operation takes place in less than 1 second on a modern, general-purpose computer. The total number of logical functions derived from each s(x,y) vary considerably depending upon the task to be performed. Clearly, in the case considered here, 256 logical functions fully represent an eight-bit-per-element digitization of s(x,y). In many cases, however, as few as four to at most 32 logical functions are entirely satisfactory to represent significant features in s(x,y).

2.3 CONVOLUTION

The simplest logical transform $r_L(x,y)$ of the function

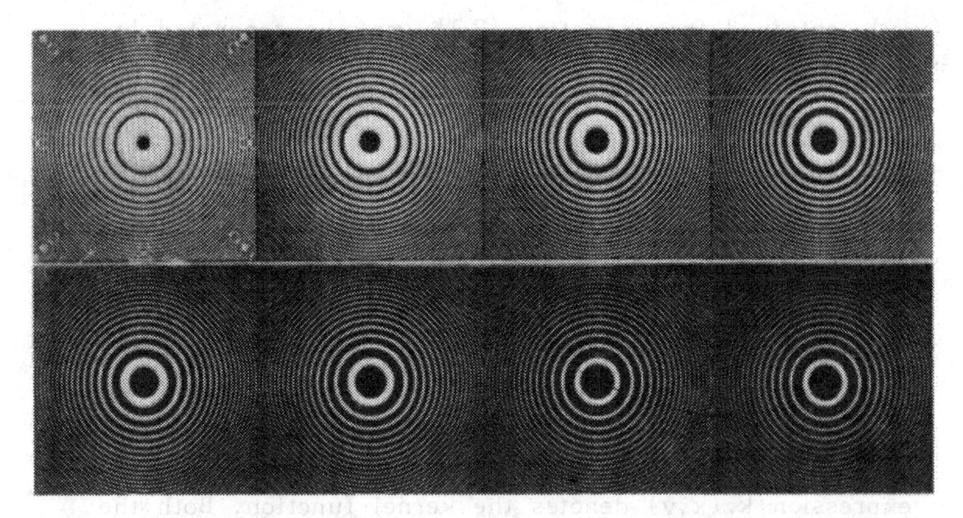

Fig. 2-1 Representation of a sinewave test function using 8 logical functions generated by thresholds equispaced over the range of its probability distribution.

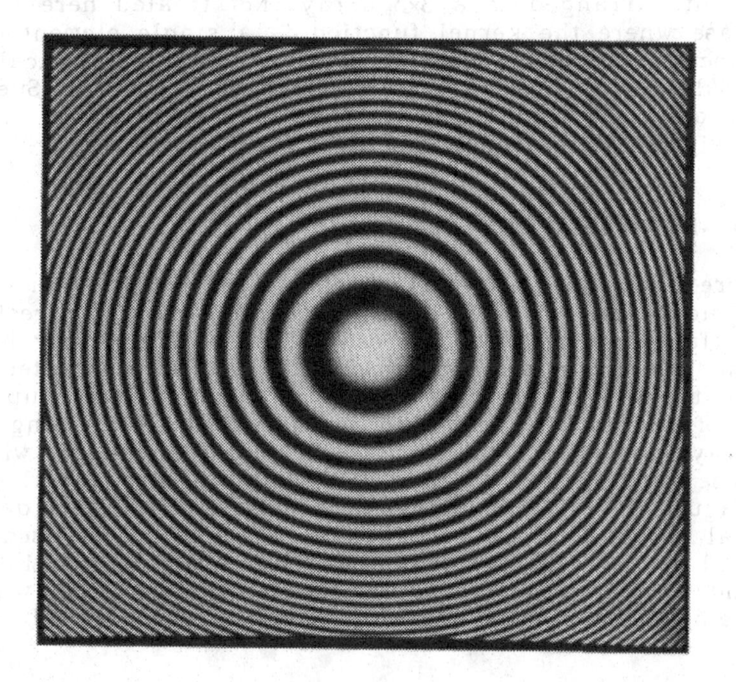

Fig. 2-2 The sinewave test function presented in graylevel format digitized at eight bits per element in a 512×512 array.

$s_L(x,y)$, obtained from equation (2.1), may be generated by means of the convolution function given below.

$$\xi(x,y) = \iint s_L(x',y')k_L(x-x',y-y')dx'dy' \tag{2.2}$$

Thresholding $\xi(x,y)$ yields $r_L(x,y)$ as given by

$$r_L(x,y) = \begin{cases} 1 & \xi(x,y)>\Xi \\ 0 & \text{otherwise} \end{cases} \tag{2.3}$$

The expression $k_L(x,y)$ denotes the kernel function. Both the configuration of the kernel and the threshold Ξ are of fundamental importance in determining the results achieved by this transform. The kernel function defines what is called the "neighborhood" of the logical transform. In this chapter the analysis is restricted to kernel functions which have at most nine elements arranged in a 3×3 array. Not treated here is the trivial case where the kernel function is a single element or "delta function" because, in this case, $r_L(x,y)$ is identical to $s_L(x,y)$. Also not treated are the "extended" kernels of Suenaga Toriwaki, and Fukumura (1974) in that they imply other than nearest neighbor interconnections in the cellular array.

2.3.1 Dual-Point Kernels

There are eight dual-point kernels in the Cartesian coordinate system. Four of these have the second element directly adjacent (face-connected) to the central element and four have the second element in a diagonal position (corner-connected). Two basic types of dual-point kernels are shown in the upper two rows of Figure 2-3. Using an input function consisting of a 3×3 array, the two convolutions ($\xi(x,y)$) which result when using these two types of kernels are shown in the central column of Figure 2-3. The boundaries of these convolutions define the logical result ($r_L(x,y)$) when a value of $\Xi = 0$ is used. When $\Xi = 1$ and $\Xi = 2$, the logical results are as given in the fourth and fifth columns of Figure 2-3. As can be seen, when $\Xi = 2$, both of the results are null, i.e., they consist of the empty set.

Even this simple logical transform using dual-point kernels has interesting and useful properties. Figure 2-4a shows

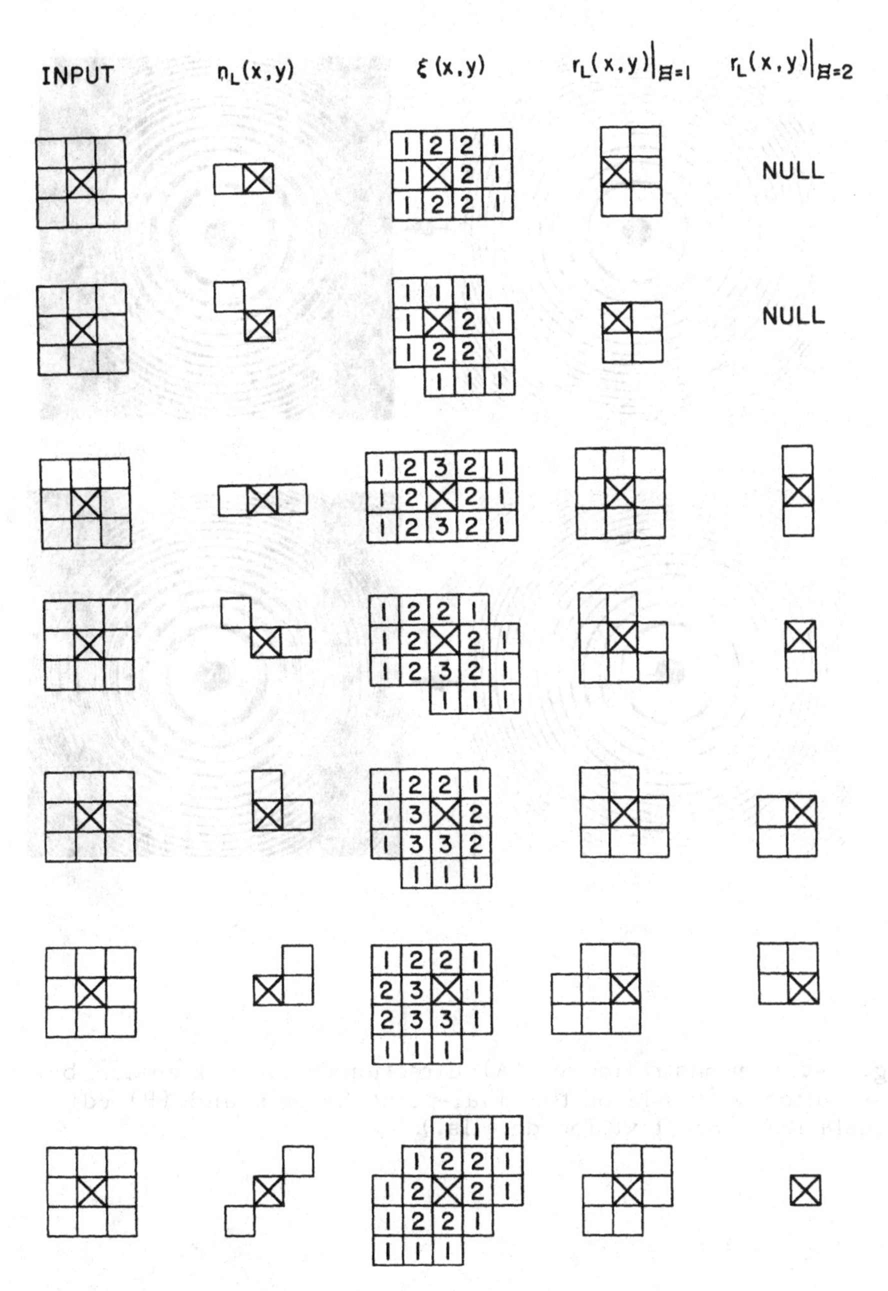

Fig. 2-3 Convolution of a 3×3 input function with both dual-point and triple-point kernels. Results are shown for three values of the threshold Ξ.

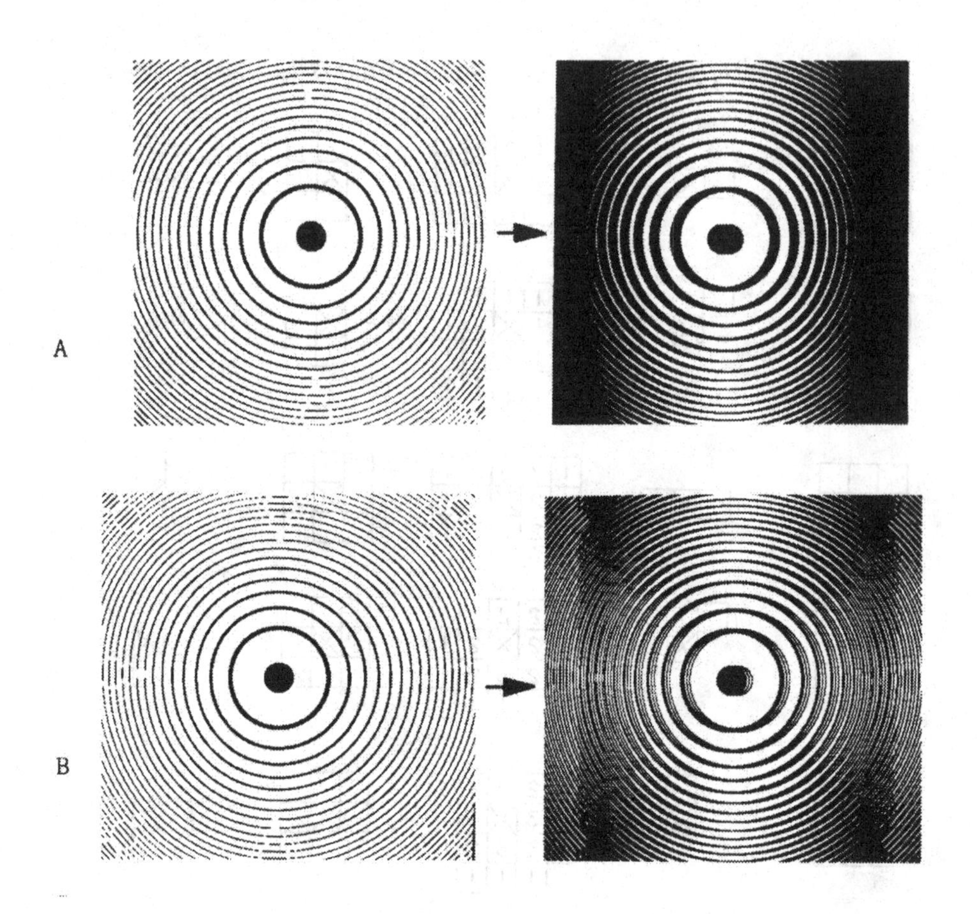

Fig. 2-4 Demonstration of (A) directional edge extension by convolution with one of the dual-point kernels and (B) edge translation. (See text for details.)

what occurs when the face-connected, dual-point kernel (first row of Figure 2-3) is used in applying equation (2.2) to one of the thresholded images in Figure 2-1 with $\Xi = 0$. As can be seen, horizontal extension to the right has been achieved after several iterations of the transform were performed. If, after each transform the result is logically EXORed with the input, then the result is as demonstrated in Figure 2-4b. This figure shows horizontal edge translation to the right. By alternating the dual-point kernel from iteration to iteration translation in any given direction up-down, left-right, etc. may be achieved. Thus the edges in an entire image or in selected portions of an image may be translated in arbitrary directions across the field by use of the dual-point kernels.

One additional and important feature of the dual-point kernels is evident from Figure 2-3. For $\Xi = 0$ the number of 1-elements are always greater than the number 0-elements. Thus $\Xi = 0$ implies augmentation of the total number of 1-elements. For $\Xi = 1$ there is always a reduction in the number of 1-elements so that, after some finite number of iterations, the result will be null. For $\Xi = 2$, the result is always null. There is, therefore, no dual-point kernel which reproduces the signal function in the result function.

2.3.2 Triple-Point Kernels

There are C_2^8 (8 items taken 2 at a time) triple-point kernels in the Cartesian tessellation. The types of triple-point kernels of greatest interest are shown in the last five lines of Figure 2-3 along with their convolution with a 3×3 signal function. In contradistinction to the dual-point kernel the triple-point kernel may be used to reproduce the signal in the result. This occurs for the horizontal (or vertical) triple-point kernel (third line, Figure 2-3) for $\Xi = 1$. Further the diagonal triple-point kernel (last line, Figure 2-3), although it does not generate a result identical to the signal, produces a result which, on further iterations with $\Xi = 1$ does generate a result equal to the original signal. This diagonal triple-point kernel is also of interest because, with the value of $\Xi = 2$, it generates what is called a "residue" (a single element in a background of elements whose values are of opposite polarity).

In general, as with the dual-point kernels, the triple-point kernels show augmentation for $\Xi = 0$, reduction (except for the cases mentioned above) for $\Xi = 1$, and eventual erasure for $\Xi = 2$. In some cases stabilization, i.e., a steady unchanging result occurs after multiple iterations. As mentioned in discussing custering in Section 2.6. The use of triple-point kernels for skeletonization is given in Chapter 6.

Both dual-point kernels and triple-point kernels are of interest because of the extremely simple circuitry by which they may be implemented. This is of importance when large array cellular automata are to be mass-produced. In general, however, the dual-point and triple-point kernels are less powerful computationally than the multi-point kernels described in the following section. In particular, since they are incomplete, i.e., they include only a subset of the immediate neighbors, they cannot be used directly in the detection of connectivity relationships.

The dual-point and triple-point kernels may be alternated in long iterative sequences of cellular logic operations and in this manner may generate augmentation and reduction results much as the multi-point kernels. This is illustrated above in Figure 2-5 for the dual-point kernel. Figure 2-6 shows the result of sequencing (clockwise) using only those triple-point kernels which are horizontal, vertical, and diagonal in a four-iteration sequence. The final result is the generation of a 69-element symmetrical figure which is 9 elements wide and 9 elements high and has the shape of an octagon. Note that the result shown in Figure 2-4 consists of 78 elements (9 elements wide and 10 elements high). One additional iteration using the initial dual-point kernel would lead to a symmetric result. When these two sequences are applied to one of the thresholded versions of the sinewave test function (Figure 2-1), the results are as shown in Figure 2-7. As can be seen, these transform sequences cause the removal of certain high spatial frequencies and a change in the duty cycle of the lower spatial frequencies. Further illustrations of cellular logic spatial filtering operations are contained in Section 2.4.

2.3.3 Multi-Point Kernels

As is well known from computational geometry, only the triangle, the hexagon, and the square can be used to uniformly tile the plane. The nearest-neighbor kernels which result from these tessellations are the three-element triangle, the six-element hexagon, and the 3×3 square. The three-element triangle is one of the triple-point kernels treated in Section 2.3.2 and is, therefore, not discussed in this section. The five-element von Neumann cross (1951) is simply a superposition of two of the triple-point kernels and is also not discussed here. The four-element 2×2 square has no central element, but is treated in Chapters 5 and 6 due to its use in region analysis and skeletonization.

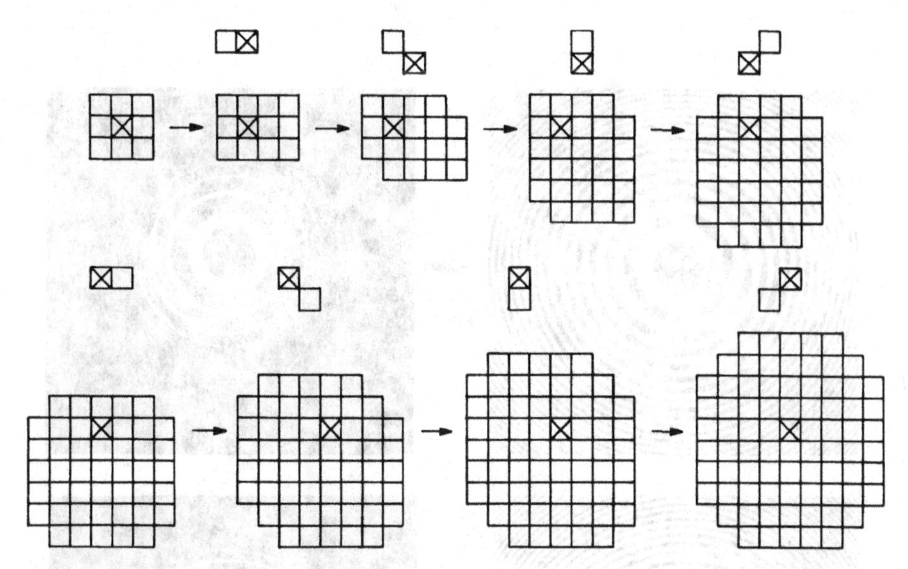

Fig. 2-5 Augmentation by logical convolution by means of an eight-step sequence using dual-point kernels.

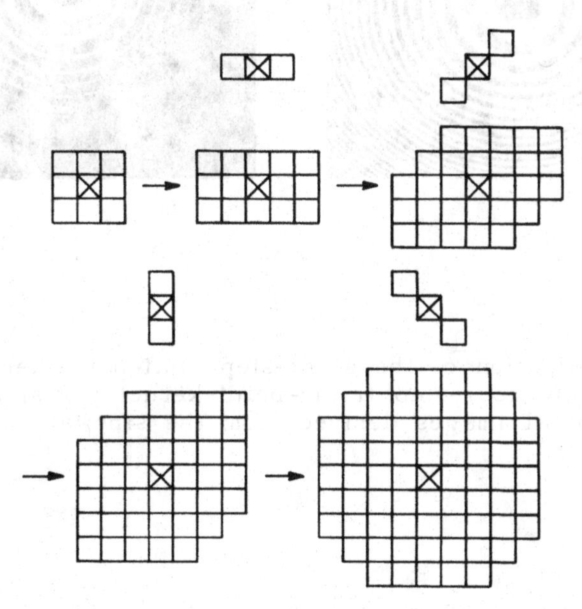

Fig. 2-6 Augmentation by logical convolution using a four-step sequence employing four of the triple-point kernels.

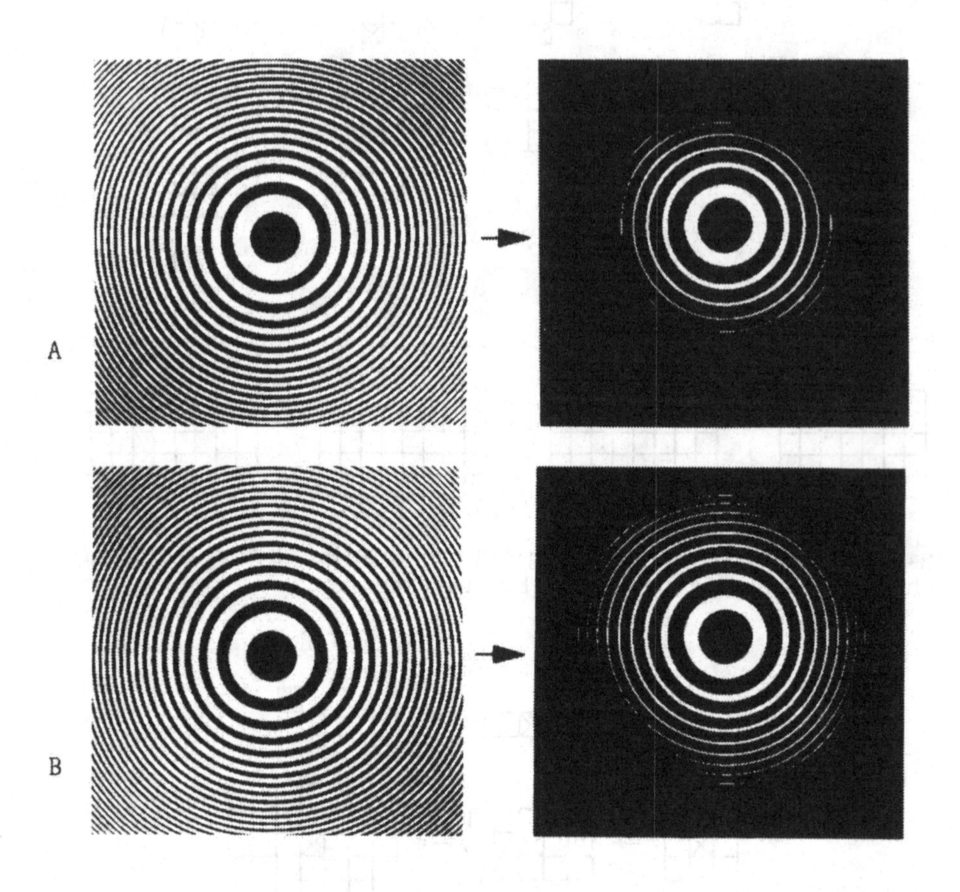

Fig. 2-7 Application of the eight-step dual-point-kernel transform (A) and the four-step triple-point-kernel transform (B) to one of the logical images derived from the sinewave test function (Fig. 2-1).

Augmentation and reduction using the hexagonal and square kernels are illustrated in Figure 2-8. In Figure 2-8 the signal is a 19-element hexagon. When the convolution $\xi(x,y)$ is thresholded with values of $\Xi = 0$ or 1, augmentation results. When $\Xi = 2$, then the result is identical to the signal. For values of $\Xi = 4$ or 5, reduction occurs. In the particular case shown in Figure 2-8a a value of $\Xi = 3$ would yield the same result as $\Xi = 4$. Similarly, $\Xi = 6$ gives an identical result to $\Xi = 5$. Erasure occurs for $\Xi = 7$.

In Figure 2-8 the signal is a 5×5 square which, when convolved with the kernel, yields a 7×7 array of integers. With Ξ equal to 0, 1, or 2 the result is an augmentation of the signal. When $\Xi = 3$, the result is identical to the signal, while for $\Xi = 4$ or greater, reduction takes place with erasure for $\Xi = 9$. A simple calculation shows that in the square tessellation and for $\Xi = 1$ a 57-element octagon is achieved after three iterations and an 81-element octagon after four iterations. These results should be compared with those obtained use the rotating triple-point kernel (Figure 2-6) noting the similarity in the number of iterations and also noting that the expense in circuitry for the nine-element square kernel is three to four times higher than for the three-element triple-point kernel.

Finally, Figure 2-9 shows augmentation from a residue (line A) and from short line segments of length from two through four elements (lines B - D) using the 3×3 kernel and Ξ values of 0, 1, and 2. This figure shows several interesting phenomena: (1) erasure, (2) oscillation, (3) and growth into both the square and the octagon. This leads to the general finding that the action of the cellular logic transform can be described to a large extent by an analysis of boundary propagation as treated below in Section 2.3.4. In the case of the hexagonal tessellation boundaries propagate along the six major axes of the hexagon (quantized at 60 degree intervals). In the case of the square, there are eight major directions every 45 degrees.

2.3.4 Boundary Propagation

As is evident from Figures 2-8 and 2-9, the cellular logic transform often causes modifications in the shape of the boundaries of the signal function. In the situation represented in Figure 2-8b, the values of $\Xi = 0$, 3, or 6 cause the shape of the result to be identical to that of the signal. In other cases the signal is modified so that the result achieves an octagonal form. The reason for this is that the amount of displacement of the boundaries of the signal depends upon the value selected for Ξ and the angle which the boundaries make to the coordi-

A

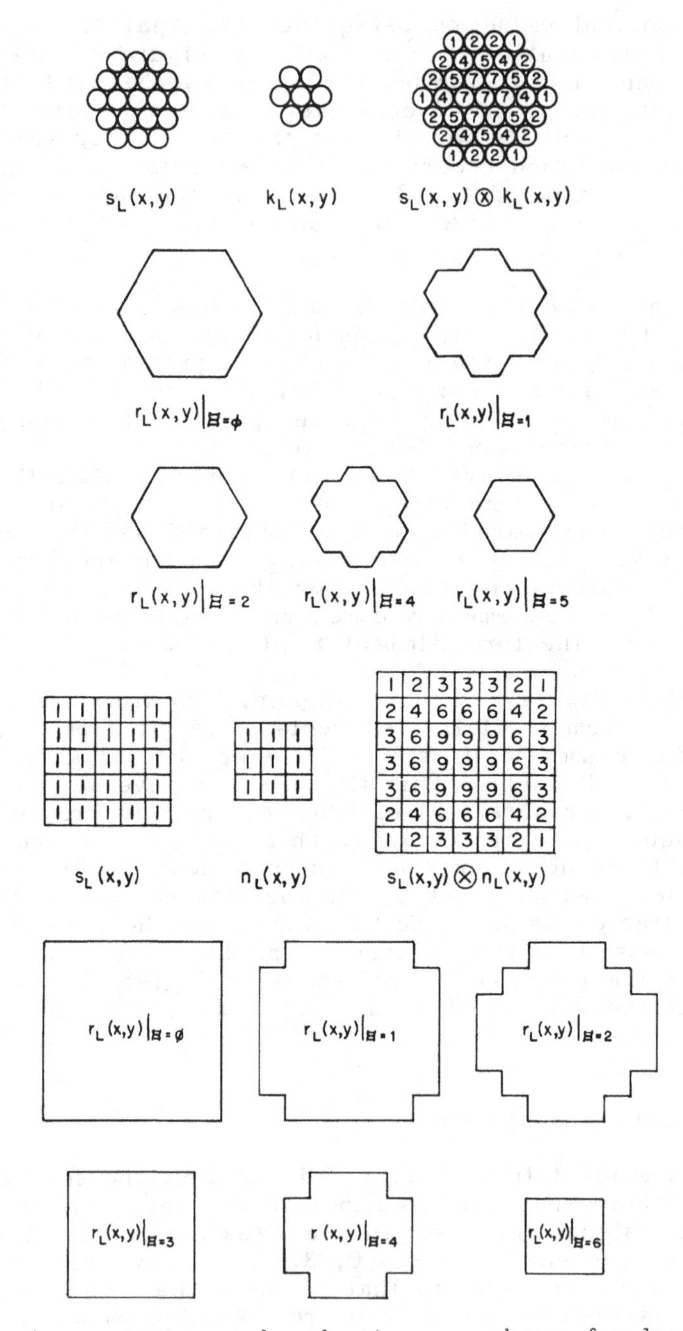

B

Fig. 2-8 Augmentation and reduction are shown for logical convolutions using the hexagonal (A) and square (B) kernels for Ξ values ranging from 0 through 6.

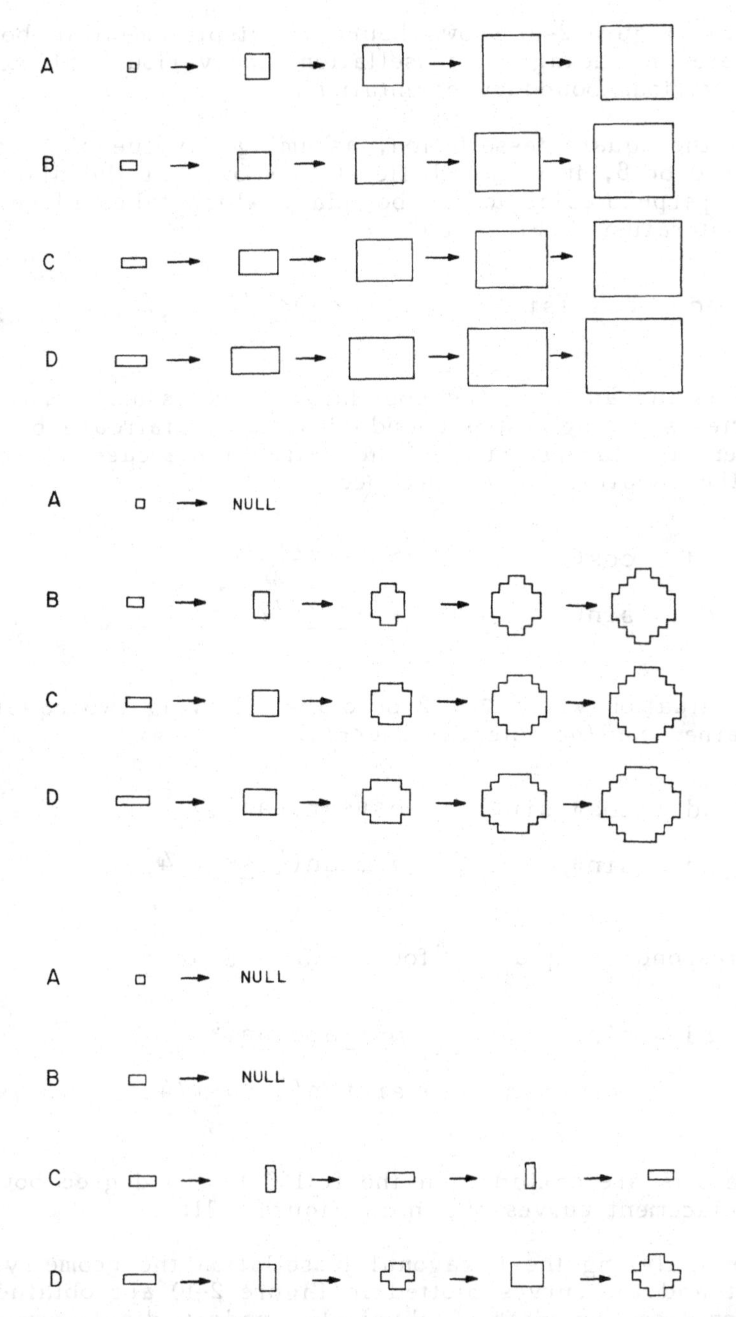

Fig. 2-9 Augmentation of residues and short line segments with the logical convolution employing the square kernel for $\Xi = 0$ (top), $\Xi = 1$ (middle), and $\Xi = 2$ (bottom).

nate axes. Figure 2-10 shows boundary displacement in both the square and hexagonal tessellations for various values of Ξ and for various boundary orientations.

In the square tessellation, assuming a value of Ξ equal to either 0 or 8, it is not difficult to show that the displacement Δd perpendicular to the boundary which takes place during one iteration is given by

$$\Delta d = \cos\theta + \sin\theta \qquad 0 \leq \theta \leq \pi/2 \tag{2.4}$$

where θ is the angle of the boundary. It is assumed that all boundaries are single-step boundaries, i.e., staircase boundaries where the largest rise is one unit. In the case where Ξ = 1 or 7 the equation for Δd becomes

$$\begin{aligned} \Delta d &= \cos\theta & 0 \leq \theta \leq \pi/4 \\ \Delta d &= \sin\theta & \pi/4 \leq \theta \leq \pi/2 \end{aligned} \tag{2.5}$$

For the situation where Ξ = 2 or 6 the following two equations are obtained for the interval $0<\theta<\pi/4$.

$$\begin{aligned} \Delta d &= \cos\theta - \sin\theta & 0 \leq \theta \leq \arctan(\tfrac{1}{2}) \\ \Delta d &= \sin\theta & \arctan(\tfrac{1}{2}) \leq \theta \leq \pi/4 \end{aligned} \tag{2.6}$$

The corresponding equations for Ξ = 3 or 5 are

$$\begin{aligned} \Delta d &= \sin\theta & 0 \leq \theta \leq \arctan(\tfrac{1}{2}) \\ \Delta d &= \cos\theta - \sin\theta & \arctan(\tfrac{1}{2}) \leq \theta \leq \pi/4 \end{aligned} \tag{2.7}$$

These results are combined in the full 0 to 360 degree boundary displacement curves shown in Figure 2-11.

In analyzing the hexagonal tessellation the geometry is different and the curves plotted in Figure 2-10 are obtained by inspection from the plots of physical boundary displacement provided in Figure 2-10. Figure 2-11 shows that there is considerable angular anisotropy for boundary propagation generated by the cellular logic transform in both the hexagonal and square

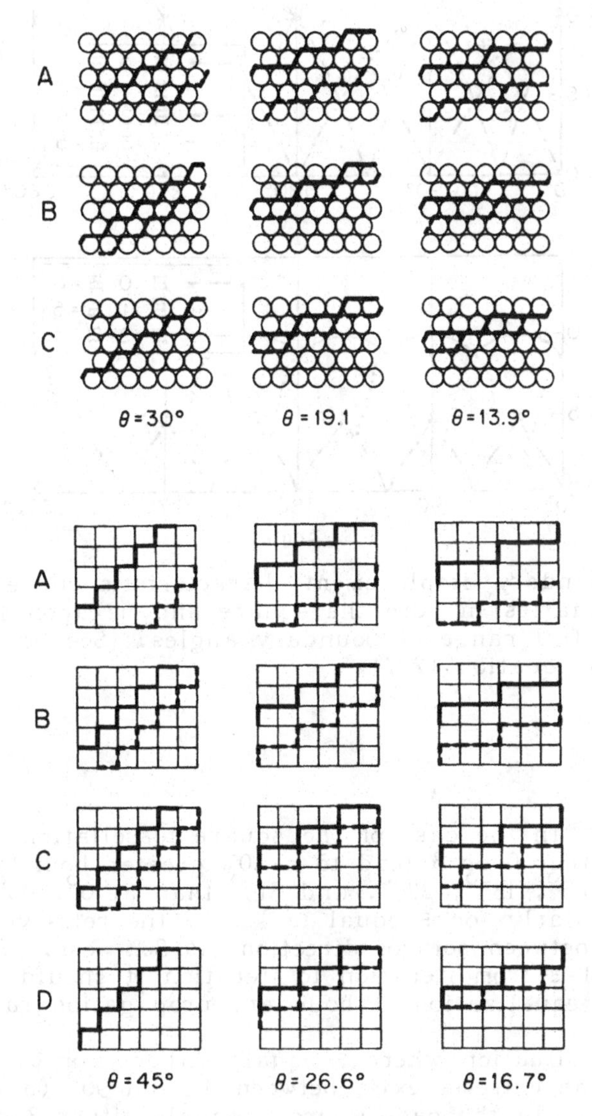

Fig. 2-10 Boundary displacements given two iterations in the hexagonal tessellation (above) and the square tessellation (below) for $\Xi = 0$ (A), $\Xi = 1$ (B), $\Xi = 2$ (C), and $\Xi = 3$ (D).

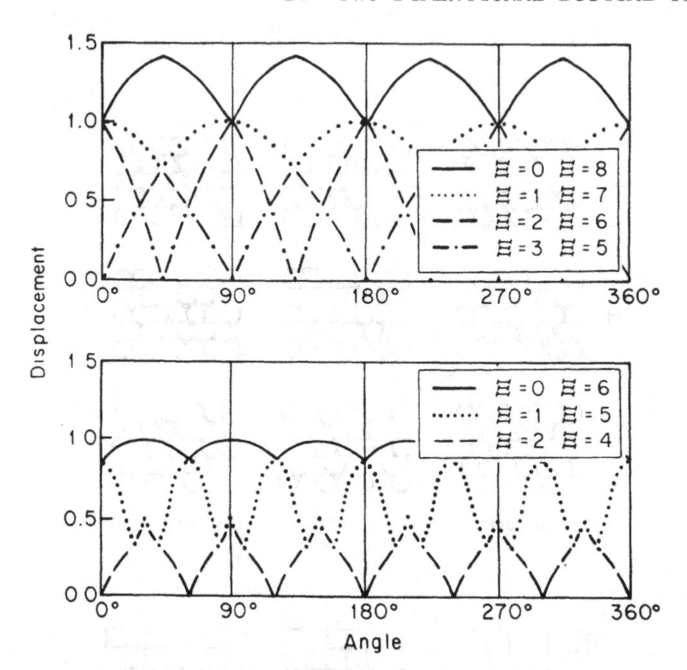

Fig. 2-11 Boundary displacement characteristic curves for single-step boundaries in both the square and hexagonal tessellations for the full range of boundary angles. (See equation (2.4) through equation (2.7).

tessellations. In the case of the square tessellation values of Ξ equal to either 0 or 8 lead to a 40% greater boundary displacement at 45°, 135°, 225°, and 315° than at 0°, 90°, 180°, and 270°. Similarly for Ξ equal to 1 or 7 the relative boundary displacement between sets of directions is 30% less. By alternating values of Ξ from iteration to iteration it should be possible to obtain equalization of boundary propagation rates.

For the situation where Ξ equals either 2 or 6, three rather than one extrema exist between 0° and 90° (and every 90° thereafter). In the case where Ξ equals either 3 or 5, a special case results where there are nulls in the displacement characteristic every 45°. This implies that boundaries in all major directions of the square tessellation do not propagate at all for these values of Ξ. This effect, which is not observed when using other than the multi-point kernels, is a special property which is characteristic of the square and hexagonal kernels. This property is particularly useful in deriving the pseudo-convex hull (Chapter 5).

For the hexagonal tessellation it can be seen that for Ξ equal to either 0 or 6 there is considerable uniformity in the displacement of a boundary irrespective of its direction in the tessellation. There is only an approximately ±10% variation in this case. However, in the case where Ξ equals either 1 or 5, there are the same multiple extrema as exhibited by $\Xi = 2$ or 6 in the square tessellation but with an even greater variation in displacement (almost 3:1). Finally, for the values of Ξ equal to either 2 or 4, the same nulls in the boundary displacement characteristic are found as in the square tessellation for $\Xi = 3$ or 5 except that the increment between nulls is 60° rather than 45° due to the smaller number of principal directions.

Figures 2-12 and 2-13 illustrate boundary propagation anisotropy in the square and hexagonal tessellations. The input signal was chosen as a circle 256 elements in diameter. When this circle is transformed in the square tessellation, using values of $\Xi = 0$, 1, 2, and 3 to produce augmentation, the results are as shown. The circle augments into a square with the value of $\Xi = 0$. This is because propagation along the 0°, 90°, 180°, and 270° directions are slow. Fast propagation in all other directions quickly "catches up" and is then limited by the rate of propagation in the slow directions. In the case of $\Xi = 1$ the converse is true and a diamond results. In the case of $\Xi = 2$ the velocity of propagation is slowest in all eight of the major directions and the result is an octagon. When $\Xi = 3$ an octagon also results but for different reasons. This octagon has its sides along the major directions in the square tessellation where there are nulls in the boundary displacement characteristic for $\Xi = 3$.

For the case of reduction the opposite holds true and boundaries with the most rapid propagation velocity dominate. The circle reduces to a diamond for $\Xi = 8$, to a square for values of $\Xi = 1$ and 2, and to an octagon for value $\Xi = 3$. Similar studies for the hexagonal tessellation are shown in Figure 2-13. Since there are fewer values of Ξ for which augmentation and reduction may be studied for this tessellation, direct augmentation from the hexagonal kernel is also shown. Results are as would be expected from an examination of Figure 2-11.

2.4 FILTERING OPERATIONS

Ordinarily images are digitally filtered either by convolution in the spatial domain or by multiplication in the spatial frequency domain. The latter method employes the Fourier transform. Convolution by a square (N×N) kernel leads to a sinc(x)-sinc(y) transfer function which produces a 180 degree phase shift beyond its cutoff frequency returning to 0 degrees at a frequency twice its cutoff frequency and oscillating thereafter

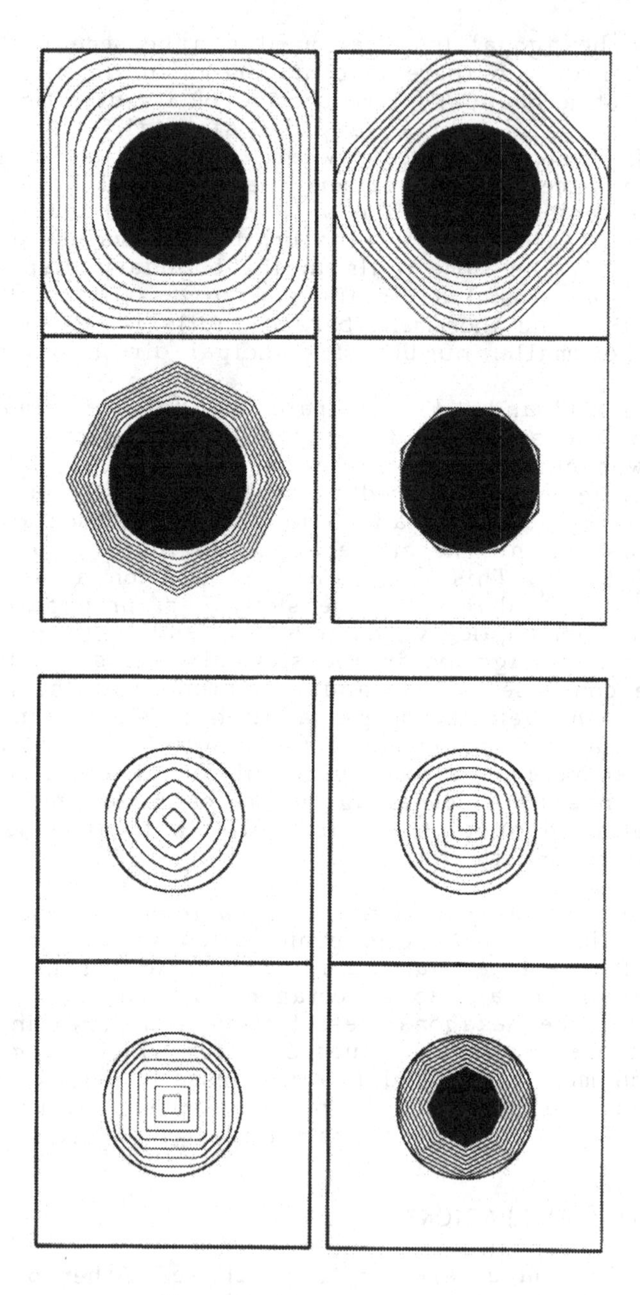

Fig. 2-12 Augmentation and reduction of a 256-element diameter circle in the square tessellation using (left-right, top-bottom) Ξ = 0,1,2,3 (augmentation) and Ξ = 8,7,6,5 (reduction). Contours are shown for each 16 iterations.

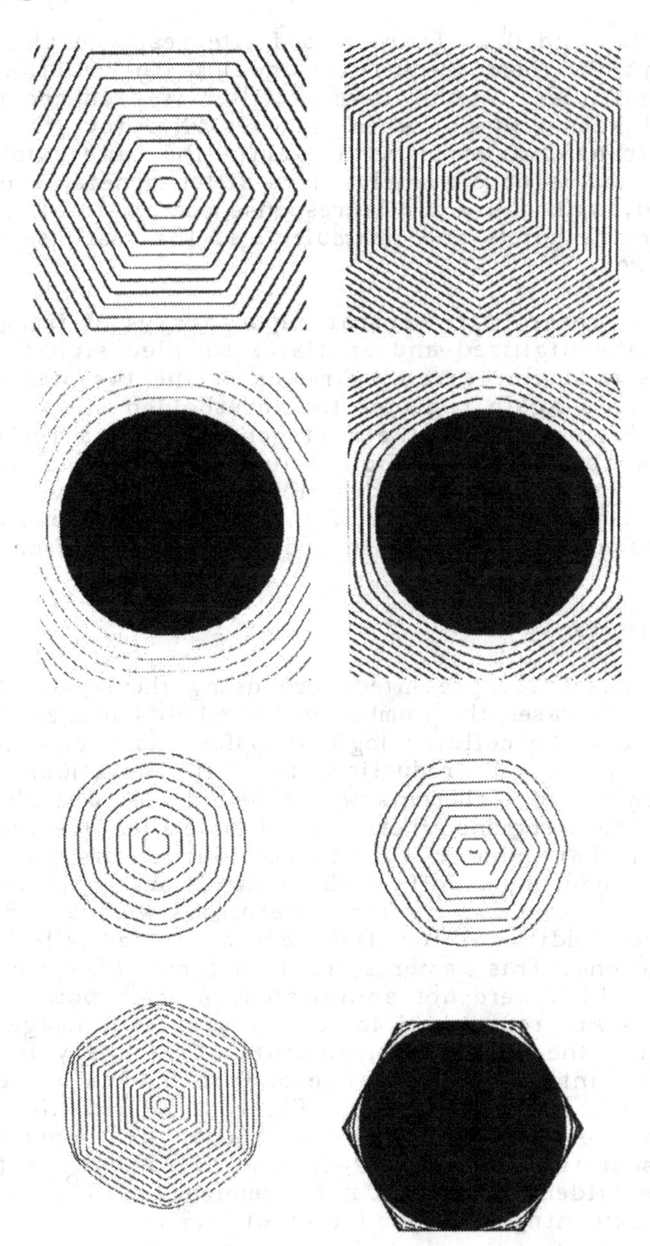

Fig. 2-13 Augmentation of a residue (A-upper) and of a 256-element diameter circle (A-lower) in the hexagonal tessellation using $\Xi = 0$ and $\Xi = 1$ (left-right). Reduction of the circle (B-upper) using $\Xi = 6$ and $\Xi = 5$ (left-right) and $\Xi = 4$ (lower left). Augmentation of the circle using $\Xi = 2$ (lower right).

between 180° and 0°. There is a finite response at all frequencies except for those frequencies equal to the cutoff frequency and its multiples. Another common filter is produced using a pyramidal kernel which, when made $2N\times2N$, provides a $\text{sinc}^2(x)$-$\text{sinc}^2(y)$ transfer function with exactly the same cutoff frequency as the $N\times N$ square kernel. This filter generates no phase shifts and, although a finite response occurs at all frequencies except for the cutoff and its multiples, the response is considerably lower.

All of these filters operate upon arrays of integers which represent the digitized and spatially sampled signal. In order to process a similar array by means of the two-dimensional logical transform again consider the thresholded sinewave function shown in Figure 2-1. In the most general case a different algorithm is applied to the logical function produced at each threshold. In the situation described below, however, the same algorithm is applied to each of the logical functions and then the results are re-combined in order to create a final output.

2.4.1 Examples

Two cases are presented here using the square tessellation. In both cases the number of thresholds utilized is 8. In the first case the cellular logic transform is performed at each threshold with $\Xi = 8$ (reduction) for three iterations. This annihilates regions of 1-elements where boundaries are close together, i.e., those regions where high frequencies are present. Due to the fact that using $\Xi = 8$ produces larger displacements at $\pm45^\circ$, the frequency cutoff in these directions is lower than at 0°, 90°, 180°, and 270°. After 3 iterations with $\Xi = 8$ (reduction), three additional iterations are conducted with $\Xi = 0$ (augmentation). This restores the boundaries of those groups of 1-elements which were not annihilated. At this point the 8 logical images are recombined to form a graylevel image by simply counting the number of 1-elements at each x,y in order to generate an integer in the range of 0 to 8. The binary functions which result are shown in Figure 2-14 with the integer array which equals their direct summation displayed in a graylevel presentation in Figure 2-15. The anisotropic performance is clearly evident with the cutoff frequency at 0°, 90°, 180°, and 270° extending $\sqrt{2}$ beyond that at $\pm45^\circ$.

In order to overcome boundary displacement anisotropy, one may interleave logical transforms using values of both $\Xi = 8$ and $\Xi = 7$ for reduction followed by a similar sequence with both $\Xi = 0$ and $\Xi = 1$ for augmentation. In the example whose result is shown in Figures 2-16 and 2-17 the sequence for Ξ was 7,8,7,0,1,0. This provides almost complete isotropy, yielding an angularly symmetric cutoff frequency.

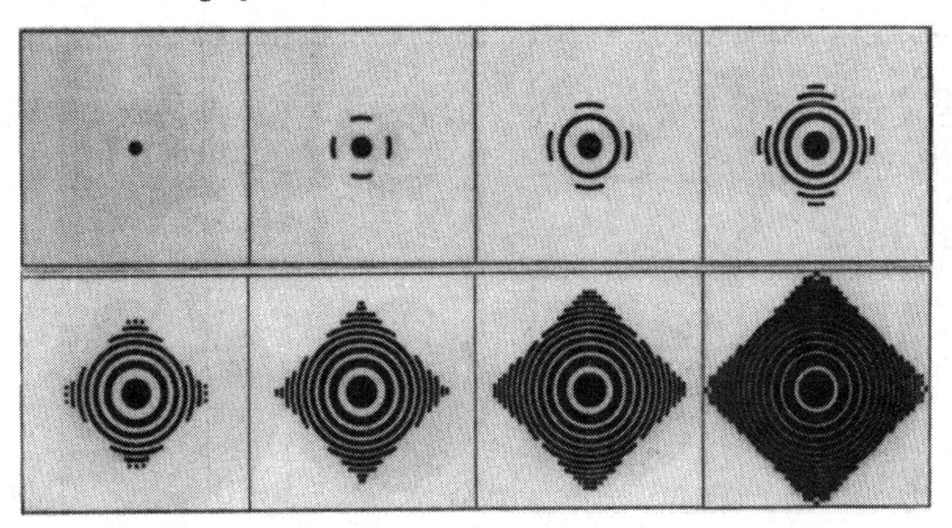

Fig. 2-14 Cellular logic filter applied to the 8 logical functions shown in Fig. 2-1 with three iterations using $\Xi = 8$ followed by three iterations using $\Xi = 0$.

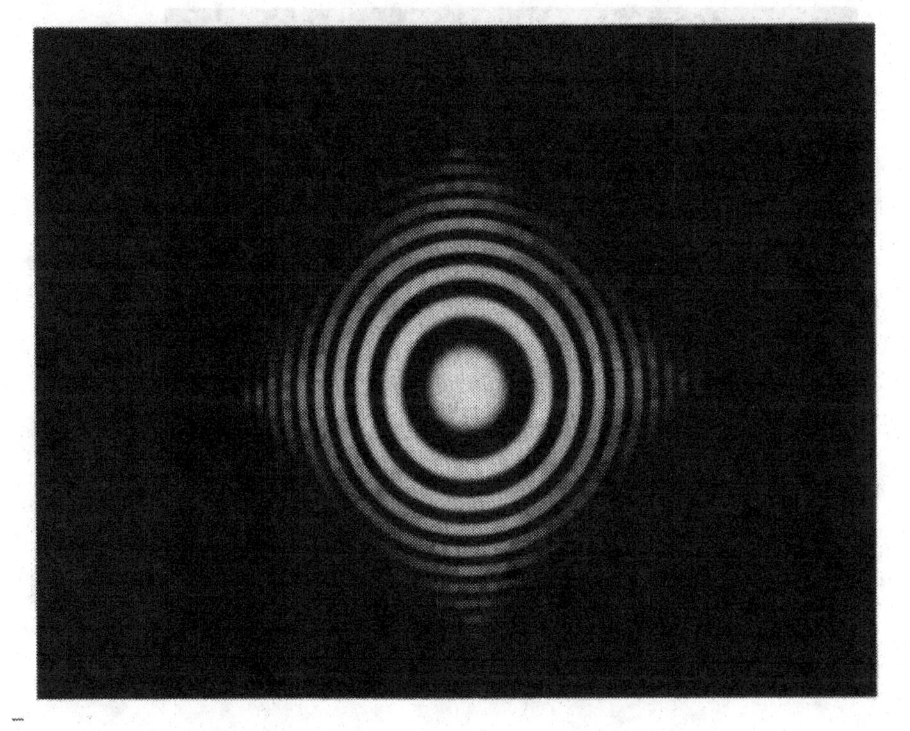

Fig. 2-15 When the 8 logical functions shown in Fig. 2-14 are summed, a graylevel result occurs as given above.

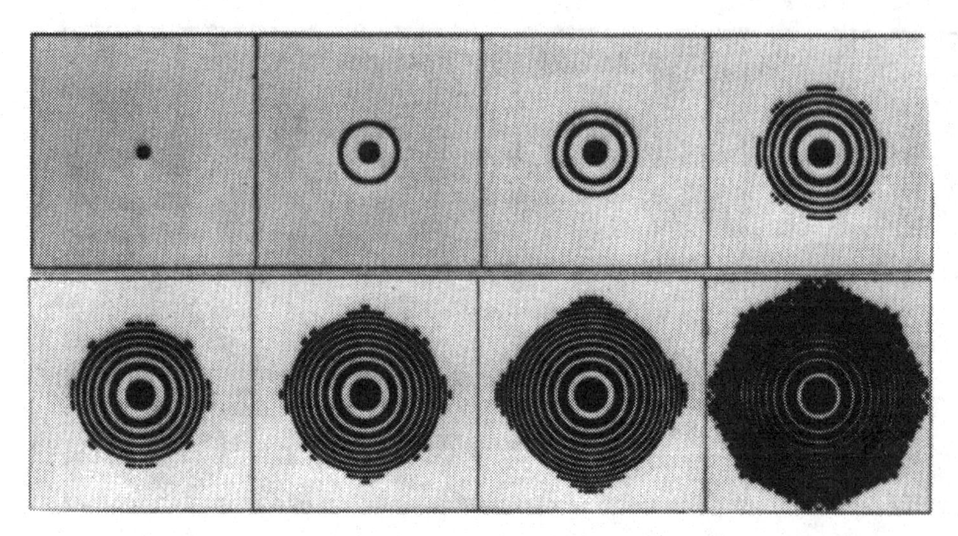

Fig. 2-16 A more isotropic result is achieved by using the sequence Ξ = 7,8,7,0,1,0 in order to balance the boundary displacement at various angles. Compare with Fig. 2-14.

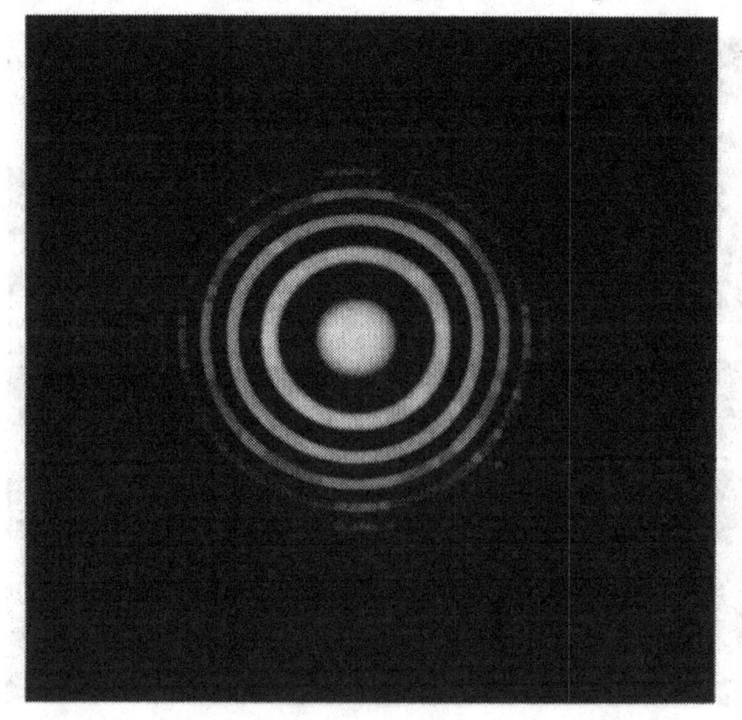

Fig. 2-17 Graylevel result obtained by summing the logical functions shown in Fig. 2-16.

2.4.2 Frequency Response Characteristics

To analyze the frequency response of the cellular logic-filter let n_i be the number of iterations for reduction (followed by n_i iterations of augmentation). Let Δd be the average (isotropic) displacement per iteration. If a sinusoidal component of the signal function has period P and the sampling interval when digitizing the signal is Δx, then this particular sinusoid and all those of higher frequency will be annihilated when

$$n_i = P/2\Delta x \tag{2.8}$$

Thus the cutoff frequency f_{co} of the cellular logic filter is given by

$$f_{co} = 1/2n_i \Delta x \tag{2.9}$$

Lower frequencies, i.e., those which are not annihilated, are clipped at their crests as can be seen in Figures 2-14 through 2-17. It is not difficult to show that their peak-to-peak amplitude A_{p-p} is given by

$$A_{p-p} = \tfrac{1}{2}[1+\cos(\pi f/f_{co})] \tag{2.10}$$

where f is equal to the reciprocal of the period of the general sinusoidal component. Some typical frequency response characteristics are given in Figure 2-18 expressed in terms of the maximum frequency recordable by Nyquist's Theorem for an array of N×N elements, i.e., N/2 total cycles.

The cellular logic filter can be seen to have certain unusual properties in comparison with other digital filters:

1. Absolutely no signal is passed beyond the cutoff frequency.

2. All output frequencies appear with the same phase as the corresponding input frequencies, i.e., the filter is "constant phase."

3. The exact cutoff frequency achieved does not depend on the span of the kernel.

4. The cutoff frequency may be altered by the choice of the number of iterations.

Other examples of the use of the two-dimensional cellular logic filter are provided in Chapters 7 and 9.

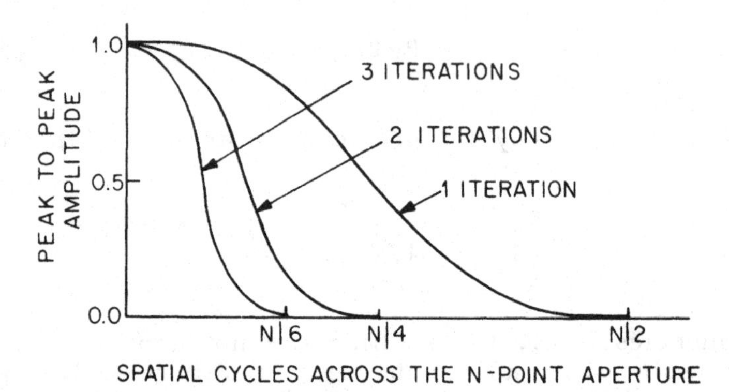

Fig. 2-18 Amplitude versus frequency characteristics for the cellular logic spatial filter.

2.5 CONNECTIVITY RELATIONSHIPS

In addition to use in two-dimensional spatial filtering, the logical transform may be employed for counting and sizing operations and for skeletonization (Chapters 5 and 6). Both of these operations require that a measure of connectivity be employed. In this book the connectivity function is symbolized by $\chi(x,y)$ which is generated by means of the Ω transform expressed by the below equation.

$$\chi(x,y) = \Omega\, s_L(x',y')k_L(x-x',y-y')dx'dy' \tag{2.11}$$

$$\chi_L(x,y) = \begin{cases} 1 & \chi(x,y) \geq X \\ 0 & \text{otherwise} \end{cases} \tag{2.12}$$

In two dimensions the function $\chi(x,y)$ is identical to the crossing number of Deutsch (1969) and, in both the square and hexagonal tessellations, this number is identically equal to the number of 1-element to 0-element transitions which occur in treating the sequence of elements which comprise the neighborhood as a cyclic code. In a logical image the neighborhoods of edge elements of a region comprised of 1-elements yield an element sequence whose crossing number is two. In the case of a graph consisting of branches one element wide connected at nodes, all branch elements have a neighborhood whose element sequence yields a crossing number of four. In the square tessellation a graph may also contain elements at nodes. These elements have neighborhoods which yield a crossing number of either six (at a node of order three) or eight (at a node of order four). In the hexagonal tessellation the highest crossing number is six which indicates a node of order three.

2.5.1 Counting and Sizing

Operations requiring counting and sizing are performed using the logical transform by reducing each connected region in the logical signal $s_L(x,y)$ to a residue and measuring the number of iterations required for the reduction to take place. Once a residue is achieved, it is retained. Periodic counts of the number of residues in the logical result are often used to form a characteristic function of the initial logical signal. (This characteristic function was originally termed the "shrink histogram" by one of the authors (Preston, 1969).) In order to successfully reduce each connected region to a residue, it is necessary to insure that a region is not annihilated before its corresponding residue is achieved. One method for obtaining this result is to carry out the operations expressed in equations (2.11) and (2.12) only for non-adjacent points in the tessellation for any one particular iteration. To do this the tessellation must be divided into several symmetric disjoint subsets whose union comprises all of the points in the tessellation. Such subsets were originally introduced by Golay (1969) and are called "subfields." Figure 2-19 shows not only Golay's original subfields in the hexagonal tessellation but also other subfield configurations, two of which are based on the S1 and S2 neighborhoods of Smith (1969), as developed into subfields by one of the authors (Preston, 1971).

When operating in subfields (Figure 2-20) the sequence in which the subfields are selected, iteration by iteration, is sometimes of importance. If, in the hexagonal tessellation, the subfield sequence is taken as 1,2,3 and continuously repeated, it is found that the resultant residue is displaced somewhat to the right of the centroid of each region of 1-elements. Conversely,

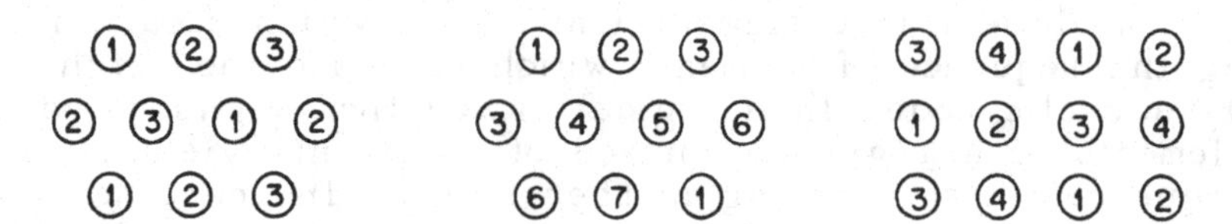

Fig. 2-19 Subfield assignments of Golay for elements in the hexagonal tessellation and those of Preston in Smith's S1 and S2 "templates" for the square tessellation.

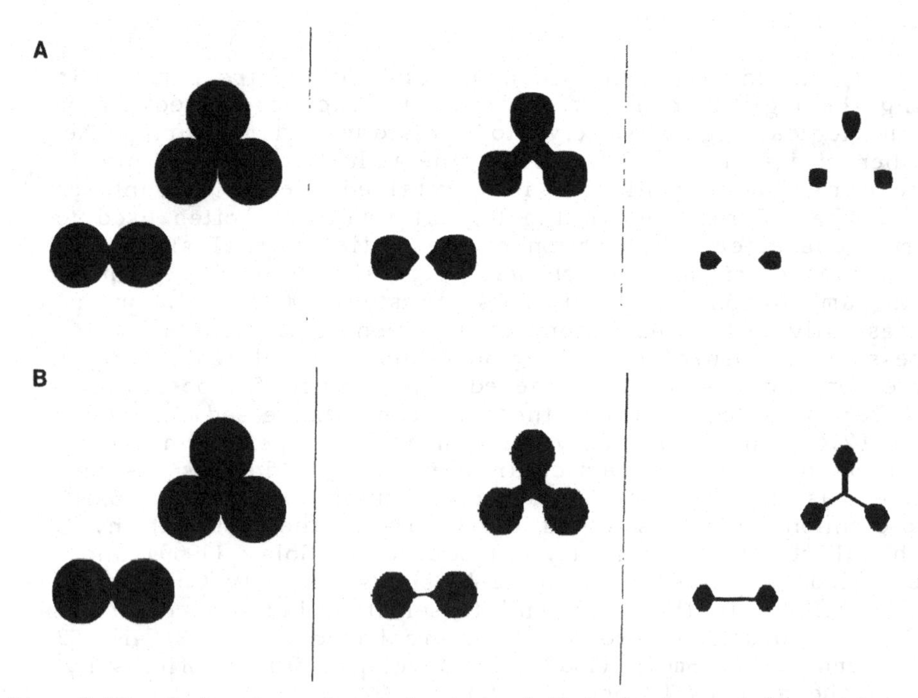

Fig. 2-20 Reduction of synthetic multilobed shape both with (A) and without (B) regard to connectivity. (See text for details.)

taking the subfield sequence 3,2,1, etc., produces a displacement of the residue to the left of the centroid. It is therefore common to alternate these two subfield sequences from major iteration to major iteration. Similarly, in the square tesselation a subfield sequence of 1,3,4,2 is frequently utilized instead of 1,2,3,4 or 4,3,2,1. Some studies of the effects of boundary propagation when using subfields were undertaken in the 1960s by Golay (unpublished), but a boundary displacement theory for operations in subfields has not been developed.

Figure 2-20 furnishes examples of reduction both with and without subfields. The signal consists of two contiguous regions one of which has two and the other three lobes. In Figure 2-20a reduction is performed using a value of $\Xi = 6$ and $X = 9$. Since the highest possible value of X is 8, setting $X = 9$ causes connectivity to be disregarded. The result, as can be seen, is that the three-lobed region splits into three separate regions and the two-lobed region into two. In Figure 2-20b the corresponding values are $\Xi = 6$ and $X = 4$ with the operation conducted in subfields of four with a subfield order of 1,3,4,2. The result is that the connectivity of each region is preserved and, as the reduction process proceeds, first single element wide branches are formed between the lobes and, finally, a portion of the endoskeleton appears. Subsequently the endoskeleton of each connected region would be reduced to a residue. Since equations (2.11) and (2.12) do not provide for the preservation of a residue, a further requirement is that the algorithm contain a provision which makes it impossible for any 1-element whose neighborhood is entirely comprised of 0-elements, i.e., a residue, to be preserved. Further information on skeletonization, including a description of both the endoskeleton and the exoskeleton is given in Chapter 6.

2.6 THE GOLAY TRANSFORM

In most cases involving use of the two-dimensional logical transform it is sufficient to utilize the parameters Ξ and X in addition to the provisions that residues may be retained and that operations in subfields may be carried out where required. Some investigators, however, have gone further and have developed and studied logical transforms which are based on specific neighborhood configurations which, although they have identical values of the parameters Ξ and X, may also have distinguishing characteristics. The best known of these investigations are those of Golay (1969) who initially proposed the hexagonal parallel pattern transformation now known simply as the "Golay transform." In this transform 14 neighborhood patterns called the "Golay surrounds" are employed. These neighborhood patterns comprise all orientation-independent cyclic binary codes of length six. Golay's patterns 0 through 6 all have a crossing number of two with a total count of the number of 1-elements in the neighborhood equal to the number allocated to the pattern itself. Pattern 7 has a crossing number of six, while patterns 8 through 13 have a crossing number of four. Patterns 7 through 9 are each comprised of three 1-elements; 10 and 13, four 1-elements; 11 and 12, two 1-elements. (See Figure 2-21.)

The general form of the Golay transform, as expressed in GLOL (Preston, 1971) is

$$D = M[G(A)B+G'(A)C]N1...N2,N3,N4/N5...N6 \qquad (2.13)$$

wherein the letters A, B, C and D refer to four separate logical arrays. The numerics N1...N2 designate specific Golay neighborhood patterns; N3 is the number of iterations to be performed; N4 the type of subfield; N5...N6 the subfield order. The expression within brackets is in Boolean algebra. In what has been termed the "compound" Golay transform, the arrays A, B, and C are different and distinct. In what has been called the "simple" Golay transform, A = B = C. In analyzing this simple Golay transform there are four basic sets to be considered.

1. The set of all elements in the designated array which have the value one.

2. The set of all elements in the array which have the value zero.

3. The set of all elements in the array which exhibit the specified Golay pattern(s).

4. The set of all elements in the array which do not exhibit the specified pattern(s).

If the array A is the designated array, then these sets are given by A, A', G(A), and G'(A), respectively. The intersection of these sets, as represented by the Boolean algebraic portion of the simple Golay transform are best illustrated by a two-variable Karnaugh (1953) map. See Figure 2-22. There are 16 possible configurations of this map (Figure 2-22a) in what has been called a Karnaugh matrix. The element of this matrix is designated Kij. The diagonal elements of the Karnough matrix represent trivial operations such as erasure (K_{11} and K_{44}) duplication (K_{22}), and complementation (K_{33}). The symmetric elements of the Karnaugh matrix, i.e., Kij and Kji are identical operations if used with complementary Golay patterns, i.e., the two subsets of patterns whose intersection is 0 and whose union

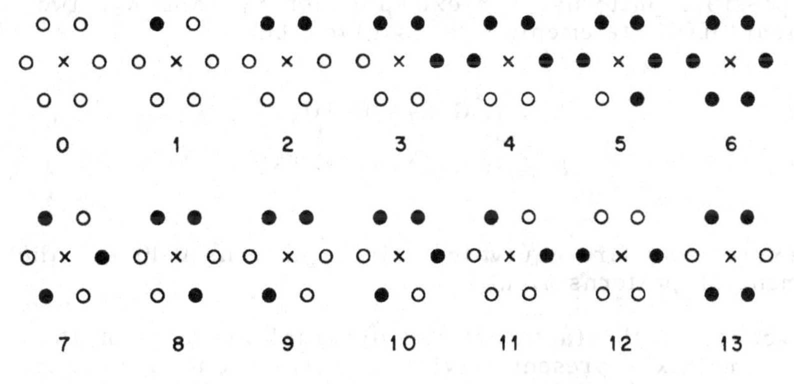

Fig. 2-21 The Golay surrounds.

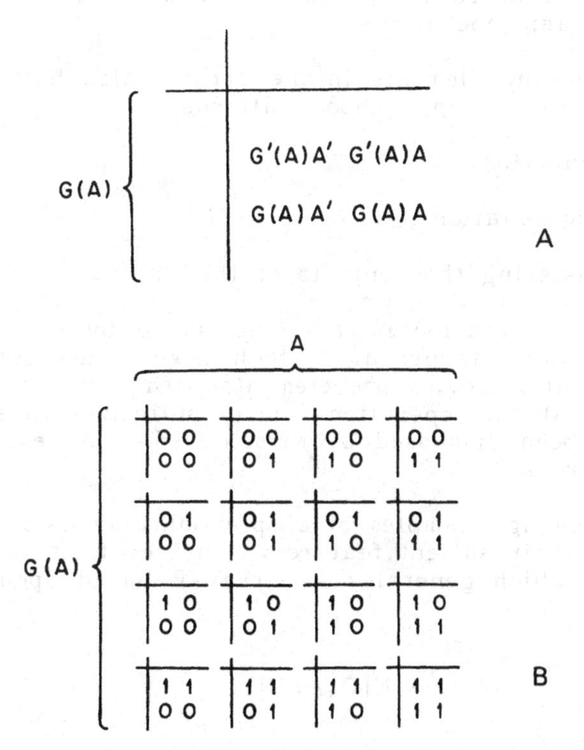

Fig. 2-22 Karnaugh map (A) for the Boolean algebraic portion of the simple Golay transform and the corresponding Karnaugh matrix (B).

is all possible patterns. For example, for K_{12} and K_{21} two equivalent GLOL statements are as given below.

$$A = M[G(A)A]0\text{-}10,,$$

$$A = M[G'(A)A]11\text{-}13,, \tag{2.14}$$

These expressions are equivalent since patterns 0-10 are the complement of patterns 11-13.

Because of the fact that the diagonal elements of the Karnaugh matrix represent trivial operations and symmetrical elements in the matrix represent identical operations when used with complementary patterns, one may describe the properties of all simple Golay transforms in a list having only six entries as given in Table 2-1. Note that K_{12} and K_{13} both represent reduction while K_{24} and K_{34} both represent augmentation. Thus the entries in Table 2-1 may be reduced to a description of only four basic operations

1. Marking elements in the array which have specific neighborhood patterns.

2. Reduction.

3. Augmentation.

4. Custering the contents of the array.

Custering is an operation initially described by Moore (1966) which causes waves to propagate both inward and outward from all edge elements, i.e., those elements with a crossing number of two. The custering operation finally stabilizes in a closed cycle as has been discussed by Kirsch (1957). An example is given in Figure 2-23.

The following examples of simple Golay transforms illustrate some of their salient features. Consider first the following transform which generates an array B based upon the contents of array A

$$B = M[G(A)]N1,, \tag{2.15}$$

In this operation each 1-element in array B corresponds to 1-elements or 0-elements in array A which have the neighborhood pattern N1. (The commas after N1 are required in GLOL in

Table 2-1 Explanation of the Action of Each Element in the Karnaugh Matrix for the Simple Golay Transform

ELEMENT	GOLAY CODE	ACTION
K_{14}	G (A)	MARKS PRIMITIVE
K_{12}	G (A) A	REDUCTION
K_{13}	G (A) A'	REDUCE & INVERT
K_{24}	G (A) + G'(A) A	AUGMENTATION
K_{34}	G (A) + G'(A) A'	AUGMENT & INVERT
K_{23}	G(A) A' + G'(A) A	CUSTER

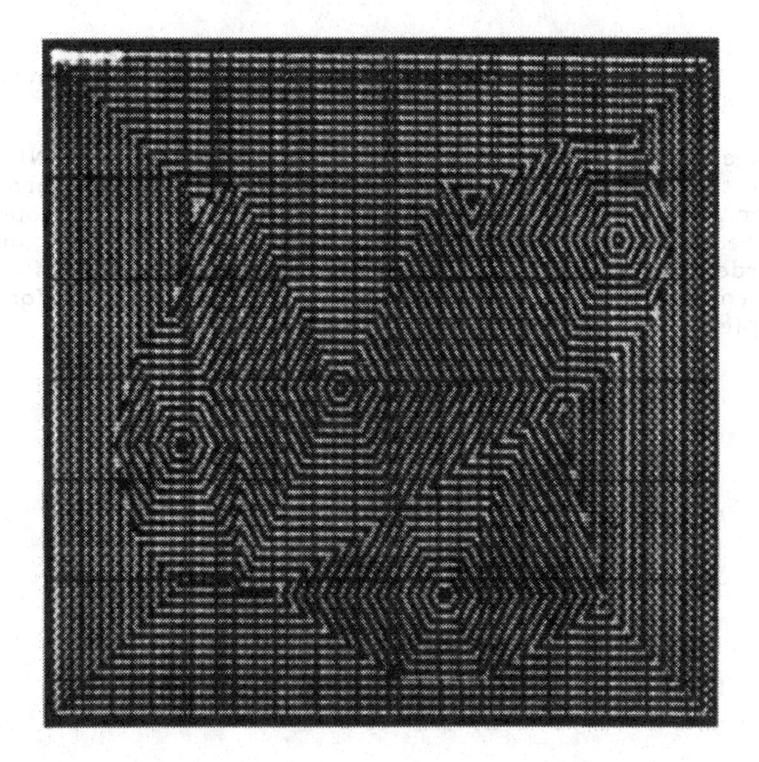

Fig. 2-23 Example of custering using the Golay transform (with permission of the Institute of Electrical and Electronic Engineers, copyright 1971).

order to call the default option where a single iteration is performed and subfields are disregarded.)

The GLOL statement given by

$$A = M[G(A)]N1-N2/N3-N4,, \qquad (2.16)$$

creates in array B a set of 1-elements corresponding to all elements in array A which have any of the patterns N1 through N2 or N3 through N4 in their neighborhoods.

When a GLOL statement is written as a replacement statement, the contents of the specified array is changed. No new array is created. An example is the reduction operation which is given by

$$A = M[G'(A)A]1-4,N1,3/213 \qquad (2.17)$$

where 1-elements are retained in array A after each of N1 iterations if they do not have neighborhoods containing patterns 1, 2, 3, or 4. Further, this GLOL statement indicates that operations are to be performed in subfields of three with the subfield order 2,1,3. Additional examples of the Golay transform are given as GLOL procedures (sequences of Golay transforms) in Chapter 9.

3. CELLULAR LOGIC OPERATIONS IN N-SPACE

3.1 INTRODUCTION

The first studies in cellular logic in spaces having a dimensionality greater than two were undertaken by Ulam (1962) at the Los Alamos Scientific Laboratory in the early 1960's. Much of this work was inspired by von Neumann (1951) who was studying cellular automata for the purpose of determining how computers could be made to reproduce themselves. Ulam and later Schrandt and Ulam (1960) concentrated on developing certain recursive relationships which, when operating upon a starting pattern or residue, would produce interesting patterns of growth (Chapter 12). As stated by Ulam (1962),

> "The objects found in this way seem to be, so to say, intermediate in complexity between inorganic patterns like those of crystals and the more varied intricacies of organic molecules and structures. In fact one of the aims of the present note is to show, by admittedly somewhat artificial examples, an enormous variety of objects which may be obtained by means of rather simple inductive definitions and to throw a sidelight on the question of how much 'information' is necessary to describe the seemingly enormously elaborate structures of living objects."

The only three-dimensional result reported is that given by Schrandt and Ulam (1970) wherein they applied the simple recursion rule which they had developed for planar figures (Chapter 12) to the elements of a cubic space. By carrying out this recursion rule for 30 iterations, a structure, part of which is shown in Figure 3-1, was generated.

Fig. 3-1 View of a three-dimensional object grown from a single cube to the 30th generation. (Excerpted from *Essays on Cellular Automata* with permission of the publisher, copyright 1970, University of Illinois.)

The purpose of this chapter is to report on recent developments in cellular logic in multi-dimensional spaces which permit the user to conduct multi-dimensional data analysis or to process multi-dimensional images. In the following sections the results for two-dimensional logical transforms are extended to spaces of higher dimensionality. Next, some simple examples are given of their application in image processing and data analysis.

3.2 THREE-DIMENSIONAL LOGICAL TRANSFORMS

When the hexagonal kernel of Smith (1969) is extended into three dimensions, the resultant neighborhood is the tetradecahedron (Figure 3-2) in the hexahedral tessellation (Figure 3-3). The tetradecahedron is one of the 13 polyhedra of Archimedes, sometimes called the "cuboctohedron," which was discussed in the 17th century by Kepler (1619). As shown in Figure 3-2, there are 12 elements in the neighborhood of the central element arranged in four interleaved hexagons defined by the sets of vertices 1-3-10-13-11-4, 1-2-8-13-12-6, 2-3-9-12-11-5, and 4-5-8-10-9-6. Each vertex is connected to four other

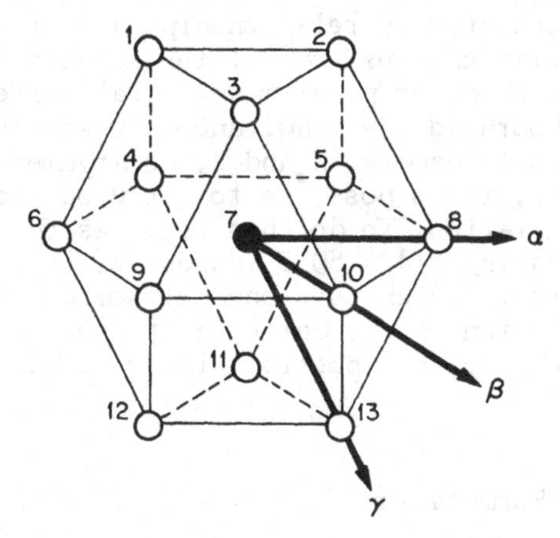

Fig. 3-2 In three dimensions the neighborhood is the tetradecahedron consisting of twelve elements in the hexahedral tessellation.

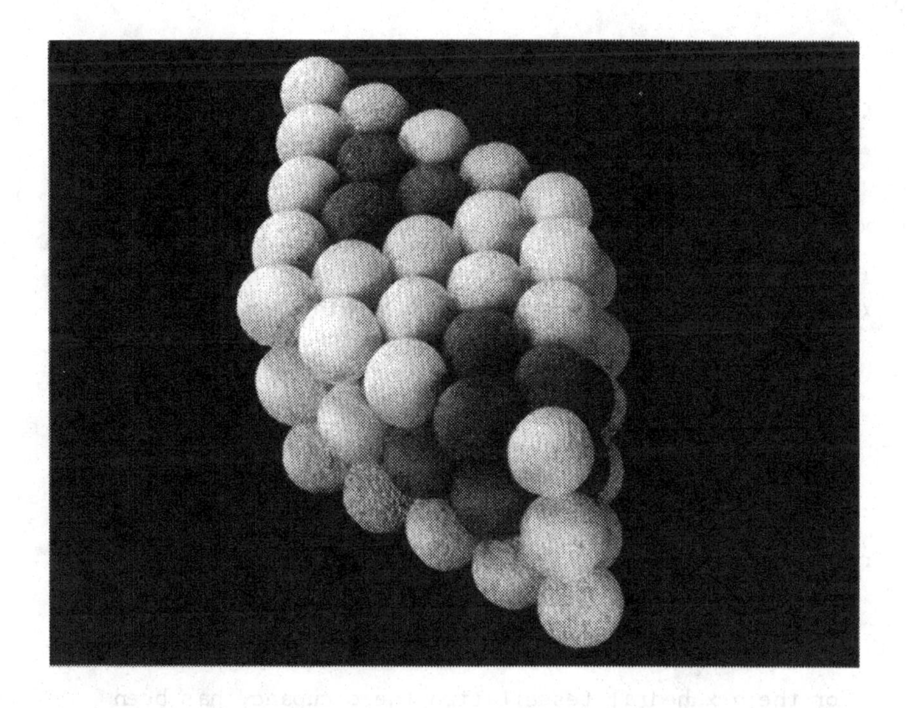

Fig. 3-3 The hexahedral tessellation is identical to the cubic close-packed lattice of inorganic chemistry shown above.

vertices. Their connectivity relationships are given in Table 3-1. Using these relationships, one of the authors (Preston, 1980) has written a computer program which calculates the occupancy for the neighborhood. By simultaneously selecting threshold values for χ and ξ, namely, X and Ξ, as arguments for an augmentation routine, it is possible to carry out logical transforms in this tessellation. To do this requires the construction of a lookup table having 2^{13} = 8912 addresses. This is entirely reasonable in a circuit which uses one LSI semiconductor chip per table. (An extension of the Cartesian tessellation to three dimensions would require an impossibly large lookup table having 2^{27} addresses.)

3.2.1 Mathematical Formulation

In the hexahedral tessellation unit vectors $\hat{\alpha},\hat{\beta},\hat{\gamma}$ directed along the three major axes define the kernel. This furnishes a non-orthogonal set of coordinates corresponding to the cubic close-packed crystal structure of inorganic chemistry (Figure 3-3).

The value of ξ in this tessellation is given by

$$\xi(\alpha_i,\beta_j,\gamma_k) = \sum_{\ell}\sum_{m}\sum_{n} s_L(\alpha_\ell,\beta_m,\gamma_n)k_L(\alpha_\ell-i\Delta\alpha,\beta_m-j\Delta\beta,\gamma_n-k\Delta\gamma) \tag{3.1}$$

The occupancy is calculated from

$$\chi_q(\alpha_i,\beta_j,\gamma_k) = \underset{\ell,m,n}{\Omega_q} s_L(\alpha_\ell,\beta_m,\gamma_n)k_L(\alpha_\ell-i\Delta\alpha,\beta_m-j\Delta\beta,\gamma_n-k\Delta\gamma) \tag{3.2}$$

where q = 0,1 (see below). Finally, the result is computed from

$$r_L(\alpha_i,\beta_j,\gamma_k) = \begin{cases} 0 & \text{if } \xi(\alpha_i,\beta_j,\gamma_k) \leq \Xi \text{ and } \chi_q(\alpha_i,\beta_j,\gamma_k) < X_q \\ 1 & \text{otherwise} \end{cases} \tag{3.3}$$

For the hexahedral tessellation the occupancy has been found by Preston (1980) to vary between 1 and 4, i.e., there can be at most 4 disconnected binary 1's in the 12-element neighborhood. At present no known analytic method for calculating Ω is known. Instead, using the connectivity relationships given in Table 3-1, the computer program conducts an exhaustive search of the 8192 possible configurations of the kernel (a smaller number could be used if symmetry considerations were taken into account) and, in each case, the occupancy is computed. This generates an occupancy table which is utilized in constructing the lookup table for any particular transform.

Unlike the two-dimensional case, where the occupancy for 0's is the same as that for 1's, the occupancies for 0's and 1's are different in three dimensions. Therefore, not only is an occupancy χ_1 for 1's computed, but also the occupancy χ_0 for 0's. These two tables are combined with the table for the value of ξ to obtain the final result.

3.2.2 Results for the Hexahedral Tessellation

A computer program called TRIAKIS was written at Carnegie-Mellon University in 1980 to run on the Perkin-Elmer 3200-series computers for the purpose of conducting the logical transform in three dimensions in the hexahedral tessellation. Figure 3-4 shows the data structure used for computational purposes. The three planes shown (each 5×5) are those given in perspective in Figure 3-3. These planes are drawn both in their true geometry and as the I,J,K array stored in computer memory. As can be seen, the portion of this array in the γ plane contains the Smith S1 kernel and in the $(\gamma-1)$ plane below and the $(\gamma+1)$ plane above there are two mirror-image triangles. The indices of all of the elements of the kernel in the α, β, γ coordinate system are given in the lower half of Figure 3-4. The program TRIAKIS extracts these elements from computer memory, forms a 13-bit word, and uses this word to address the appropriate lookup tables.

A cubic space dimensioned 64×64×64 has been selected for ease of computation and to display the results obtained from the three-dimensional logical transform. The cubic space consists of sixty-four 64×64 planes. These planes may be laid out in 8 rows with each row containing 8 planes as shown in Figure 3-5. In this figure the sides of the cubic space are assumed to contain binary 1's and 1's are colored black. Hence plane 1 (upper left corner) is a black 64×64 square as is plane 64 (lower right corner). The black grid covering the rest of the display is composed of the four other planes bounding the cubic space. These planes are seen edge-on. Figure 3-5 was derived by augmenting a unit kernel placed in plane 28 for 32 total iterations. The process was arrested every fourth iteration and the boundaries of the then-existing three-dimensional figure were computed and recorded. The result is the nested display shown in Figure 3-5. Figure 3-5a shows the

Table 3-1 Connectivity Relationships for the 12-Element Neighborhood of the 13-Element, 3-Dimensional Kernel

Element	Hexagon			
Number	H1	H2	H3	H4
1			3-4	2-6
2		3-5		1-8
3		2-9	1-10	
4	5-6		1-11	
5	4-8	2-11		
6	4-9			1-12
8	5-10			2-13
9	6-10	3-12		
10	8-9		3-13	
11		5-12	4-13	
12		9-11		6-13
13			10-11	8-12

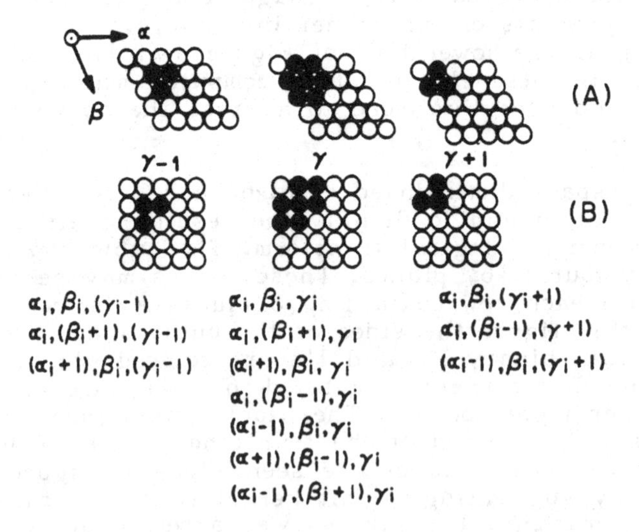

Fig. 3-4 The 5×5×3 array shown in Figure 3-3 may be displayed as three planes (A). These planes may also be arranged as an I,J-array in computer memory (B).

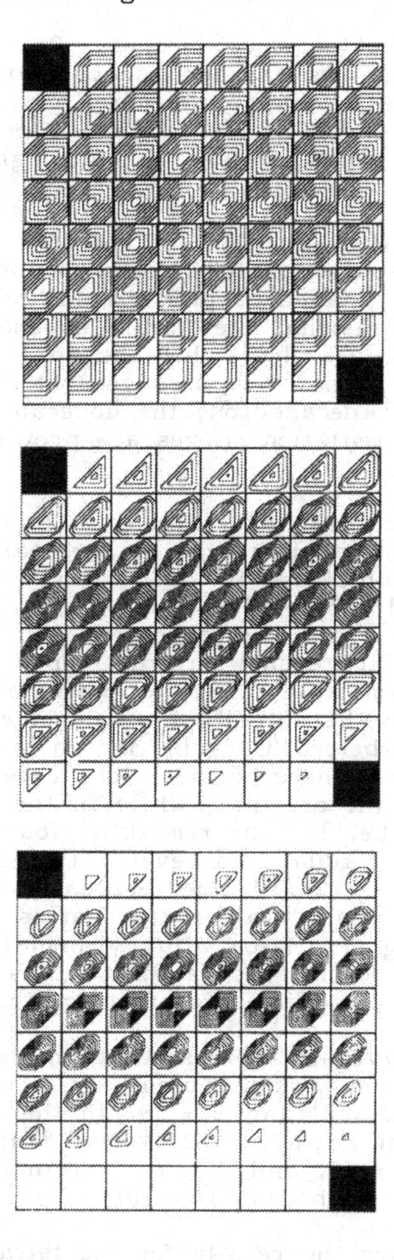

Fig. 3-5 Results of augmentation in a three-dimensional space configured as a 64×64×64 array starting with a unit tetradecahedron located in the center of plane 28. Results are shown for threshold values of zero (A), one (B), and two (C) using a nested presentation, where the boundaries of the surfaces are drawn after every four iterations.

result of augmenting using a value of $\Xi = 0$; Figure 3-5b, $\Xi = 1$; Figure 3-5c, $\Xi = 2$. (As has been demonstrated in Chapter 2, low values of Ξ lead to augmentation.) For values of Ξ equal to 3 and 4, there is neither augmentation nor reduction while, as illustrated below, values of 5 through 12 produce reduction.

Since it is difficult to visualize the figure which is generated in the cubic space from the nested boundary presentation of Figure 3-5, a shaded-graphic display was devised so that a true three-dimensional view could be shown of the structure produced. Figure 3-6 shows this shaded-graphic display for each of the three primary figures generated by augmentation, namely, the tetradecahedron, the dodecahedron, and the tetrahexahedron. Augmentation curves are provided in Figure 3-7.

In order to illustrate three-dimensional reduction, the three primary figures are displayed as the starting figures in the 8 rows of each sub-figure in Figure 3-8. Each starting figure is shown reduced, using a different value of Ξ with the process arrested and a display generated each 10 cycles. The top row of each sub-figure shows results for $\Xi = 12$; the bottom row, for $\Xi = 5$; intermediate rows, for the other values of Ξ between 12 and 5. When the starting figure is the tetradecahedron, erasure occurs before 30 cycles for values of Ξ from 12 through 9. In the first three of these four cases the residual retains the shape of the original, while in the fourth case it is modified to the cube. For the remaining four cases the rate of reduction is more gradual and, even after 30 cycles, the figure is not erased. When $\Xi = 9$, the residual is the previously observed tetrahexahedron while for $\Xi = 7$ it is a different tetrahexahedron consisting of a cube with pyramidal faces. For $\Xi = 6$ and $\Xi = 5$ little reduction occurs with only slight modification of the edges or corners of the starting figure.

Figure 3-8b shows similar results when the starting figure is the dodecahedron. Erasure occurs at 20 cycles for $\Xi = 12$; 30 cycles, when $\Xi = 11$ and $\Xi = 10$. In the other five cases the figure is still not erased at 30 cycles. Shape modification occurs for $\Xi = 11$, $\Xi = 10$, and $\Xi = 7$. Certain new shapes are created, such as the octohedron ($\Xi = 10$).

Figure 3-8c gives the results for the tetrahexahedron where erasure occurs before 30 cycles for values of Ξ of 12 through 10. Shape modification occurs for the values of $\Xi = 10$, $\Xi = 9$, and $\Xi = 7$.

Reduction curves are graphed in Figure 3-9 along with baseline curves for the cube and the sphere calculated on the

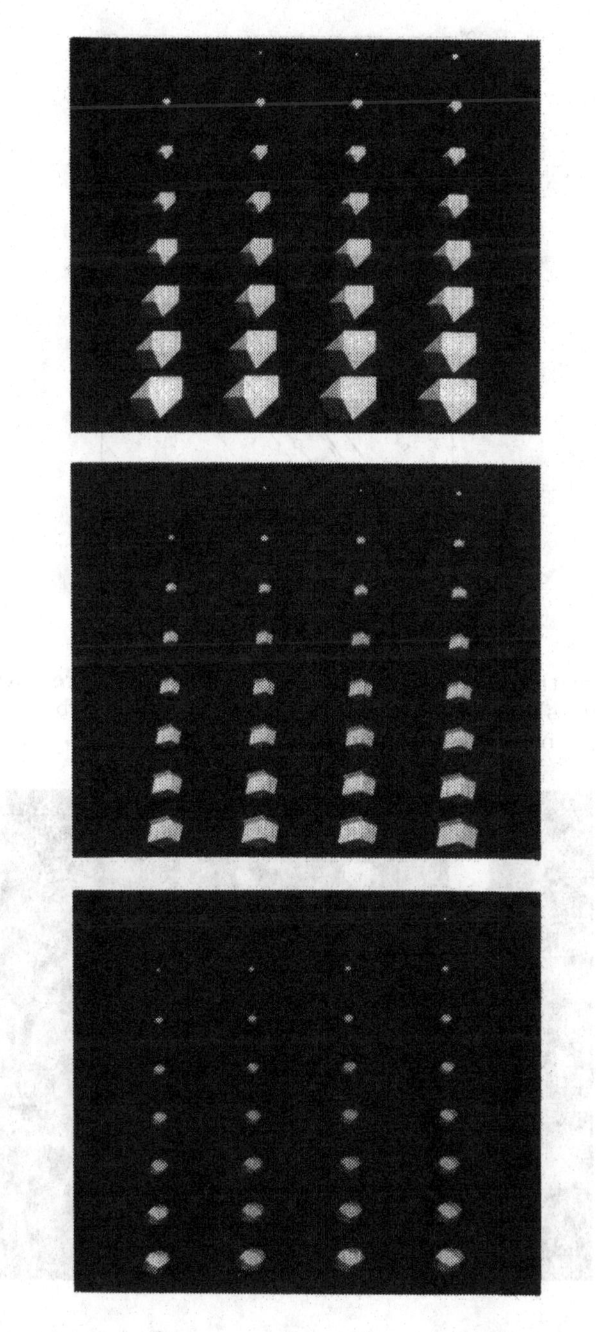

Fig. 3-6 The results shown in Figure 3-5 may be displayed using shaded-graphic techniques so as to permit easy visualization of the augmentation process.

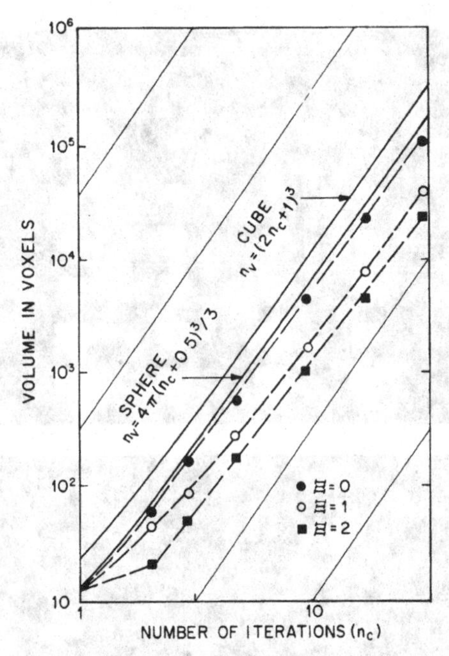

Fig. 3-7 Growth curves corresponding to the three augmentation processes shown in Figure 3-6 are plotted above showing volume as a function of the number of iterations.

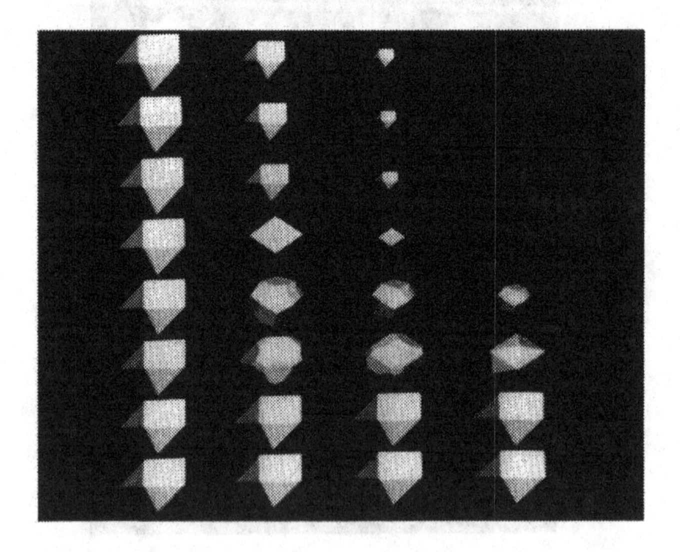

Fig. 3-8a The tetradecahedron shown in Figure 3-6a may be reduced using threshold values from 12 (top line) down to 5 (bottom line) with a display generated every 10 iterations.

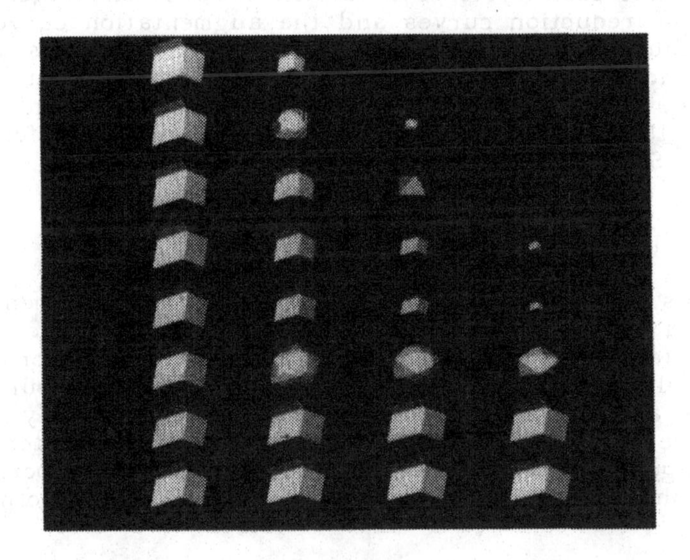

Fig. 3-8b Reduction of the solid shown in Figure 3-6b is shown using the same series of threshold values as were used in the display shown in Figure 3-8a.

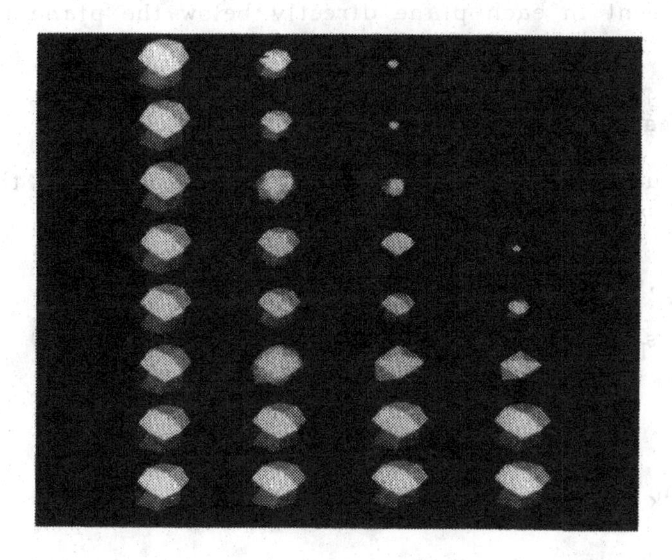

Fig. 3-8c Reduction of the solid shown in Figure 3-6c is shown using the same series of threshold values as were used in the display shown in Figure 3-8a.

assumption that each cycle of reduction removes one layer of voxels. These reduction curves and the augmentation curves given in Figure 3-7 are used in the design of sequences of three-dimensional logical transforms to select the number of iterations required to achieve the desired results. An application of digital filtering in three-dimensions using this transform is provided in Section 3.4.

3.3. FOUR-DIMENSIONAL LOGICAL TRANSFORMS

Just as four interleaved hexagons form the 12-element neighborhood in the three-dimensional tessellation, so do four interleaved tetradecahedrons form the 20-element neighborhood in the four-dimensional octohedral tessellation. These four tetradecahedrons are shown in Figure 3-10. Their connectivity relationships are given in Table 3-2 in terms of the ten interleaved hexagons which can be shown to form the neighborhood. This table shows that each element is connected to six other elements.

The unit vectors of this tessellation are $\hat{\alpha}, \hat{\beta}, \hat{\gamma}, \hat{\delta}$. The δ dimension is the cubic direction in this four-dimensional space. Three adjacent planes in each of three adjacent cubes contain the kernel as illustrated in Figure 3-11. This figure provides the indices of all elements of the neighborhood, giving those for each element in each plane directly below the plane in question.

3.3.1 Mathematical Formulation

The equations which correspond to equations (3.1) through (3.3) in Section 3.2 are given below:

$$\xi(\alpha_i,\beta_j,\gamma_k,\delta_\ell) = \sum_m \sum_n \sum_o \sum_p s_L(\alpha_m,\beta_n,\gamma_o,\delta_p) k_L(\alpha_m - i\Delta\alpha, \beta_n - j\Delta\beta, \gamma_o - k\Delta\gamma, \delta_p - \ell\Delta\delta) \tag{3.4}$$

$$\chi_q(\alpha_i,\beta_j,\gamma_k,\delta_\ell) = \underset{m,n,o,p}{\Omega_q} s_L(\alpha_m,\beta_n,\gamma_o,\delta_p) k_L(\alpha_m - i\Delta\alpha, \beta_n - j\Delta\beta, \gamma_o - k\Delta\gamma, \delta_p - \ell\Delta\delta) \tag{3.5}$$

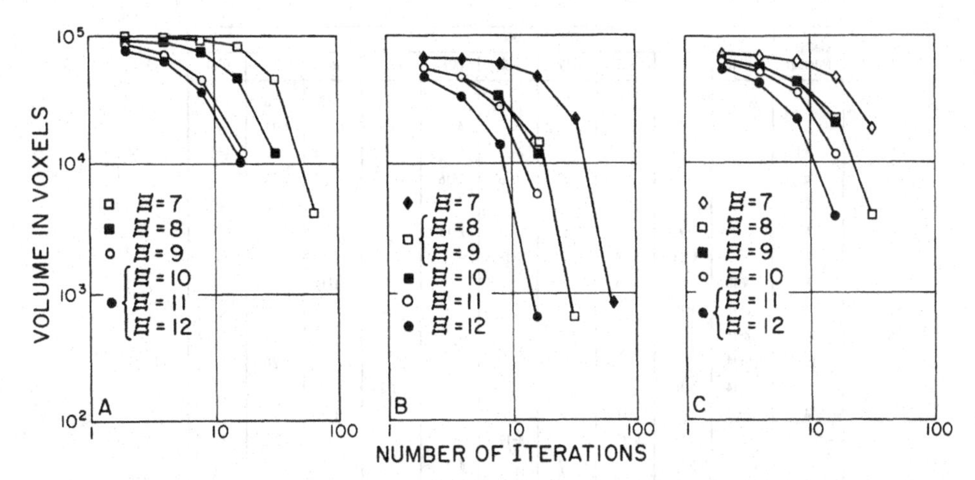

Fig. 3-9 Reduction curves are plotted above corresponding to all of the processes shown in Figure 3-8 for all threshold values which lead to erasure.

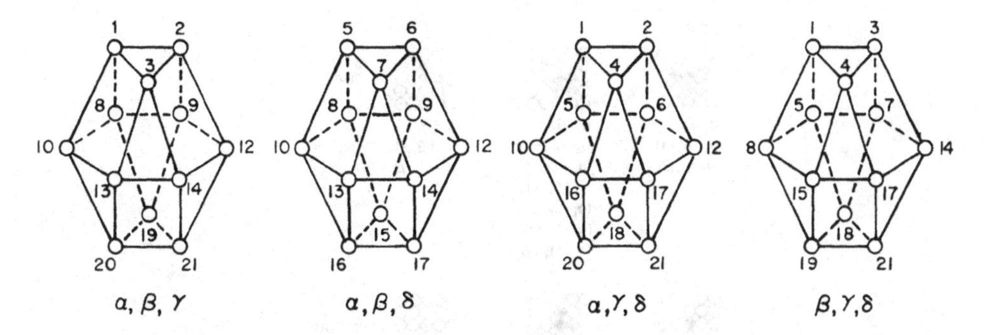

Fig. 3-10 The 20-element neighborhood for use with the logical transform in four dimensions can be considered to be constructed by interleaving four 12-element, 3-dimensional tetradecahedrons.

Table 3-2 Connectivity Relationships for the 20-Element Neighborhood of the 21-Element, 4-Dimensional Kernel

Element Number	Hexagon									
	H01	H02	H03	H04	H05	H06	H07	H08	H09	H 10
1								3-8	2-10	4-5
2						4-6	3-9		1-12	
3					4-7		2-13	1-14		
4					5-13	2-16				1-17
5			7-8	6-10						1-18
6		7-9		5-12		2-18				
7		6-13	5-14		3-18					
8	9-10		5-15					1-19		
9	8-12	6-15					2-19			
10	8-13			5-16					1-20	
12	9-14			6-17					2-21	
13	10-14	7-16					3-20			
14	12-13		7-17					3-21		
15		9-16	8-17		4-19					
16		13-15		10-17		4-20				
17			14-15	12-16						4-21
18					7-19	6-20				5-21
19					15-18		9-20	8-21		
20						16-18	13-19		10-21	
21								14-19	12-20	17-18

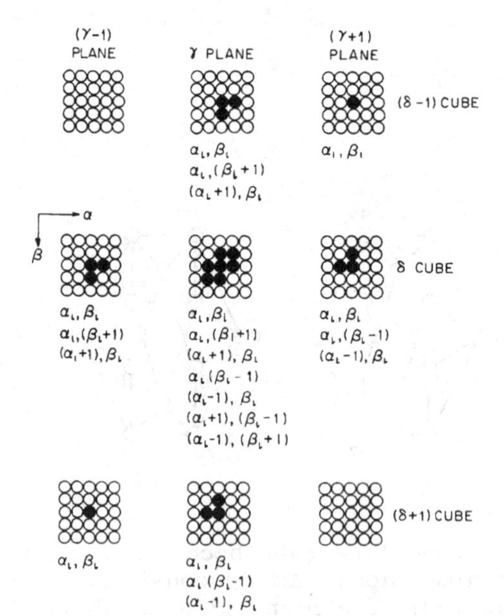

Fig. 3-11 The 4-dimensional kernel (see Figure 3-10) may be displayed in planar form as an I,J,K array. The indices of the ith element and its neighbors are shown.

$$r_L(\alpha_i,\beta_j,\gamma_k,\delta_\ell) = \begin{cases} 0 & \text{if } \xi(\alpha_i,\beta_j,\gamma_k,\Delta_\ell) \leq \Xi \\ & \text{and } \chi_q(\alpha_i,\beta_j,\gamma_k,\delta_\ell) < X_q \\ 1 & \text{otherwise} \end{cases} \tag{3.6}$$

As with the three-dimensional case, there is no known analytic method for evaluating the occupancy of the four-dimensional, 20-element neighborhood. Therefore, a computer program has been written which computes ξ, χ_1, and χ_0 for the 2^{21} values of the neighborhood function. These tables, each containing two megabytes, are stored on disc. For any particular logical transform, whose parameters are the values of Ξ, X_1, and X_0, a lookup table is computed and stored in 256K bytes of core for rapid reference as the transform is executed.

3.3.2 Results for the Octohedral Tessellation

For ease of computation and display, studies of the four-dimensional logical transform have been carried out in a 32×32×32×32 space which can be presented as a tetracube (Figure 3-12). Figure 3-12 shows 12 tetracubes in four rows of three tetracubes each. Each tetracube consists of 32 rows of 32 planes so that each row displays one 32×32×32 three-dimensional cube. The elements of the eight cube faces of each tetracube are recorded as binary 1's, and binary 1's are colored black. Four of these cubic faces produce the four 32×1024 black borders of each tetracube. The black grid is produced by the planes which define the remaining four cubes seen edge-on. The three tetracubes in each row of Figure 3-12 show the result of augmenting by 4, 8, and 12 cycles, respectively. The values of Ξ which correspond to the 4 rows in Figure 3-12 are 0, 1, 2, and 3, respectively. These are the only values of Ξ which produce augmentation. The computer program which produced these results was completed in early 1981 at Carnegie-Mellon University and is called TETRAKIS.

In order to obtain a shaded-graphic display, each of the 32 cubic cross-sections of the four-dimensional space represented by one tetracube can be displayed separately. The figures shown in one final column of Figure 3-12 with augmentation arrested at 12 cycles for Ξ = 0, 1, 2, and 3, respectively, are each displayed in Figure 3-13. One must realize, when studying Figure 3-13, that all three-dimensional objects shown in each presentation are actually part of the same four-dimensional solid.

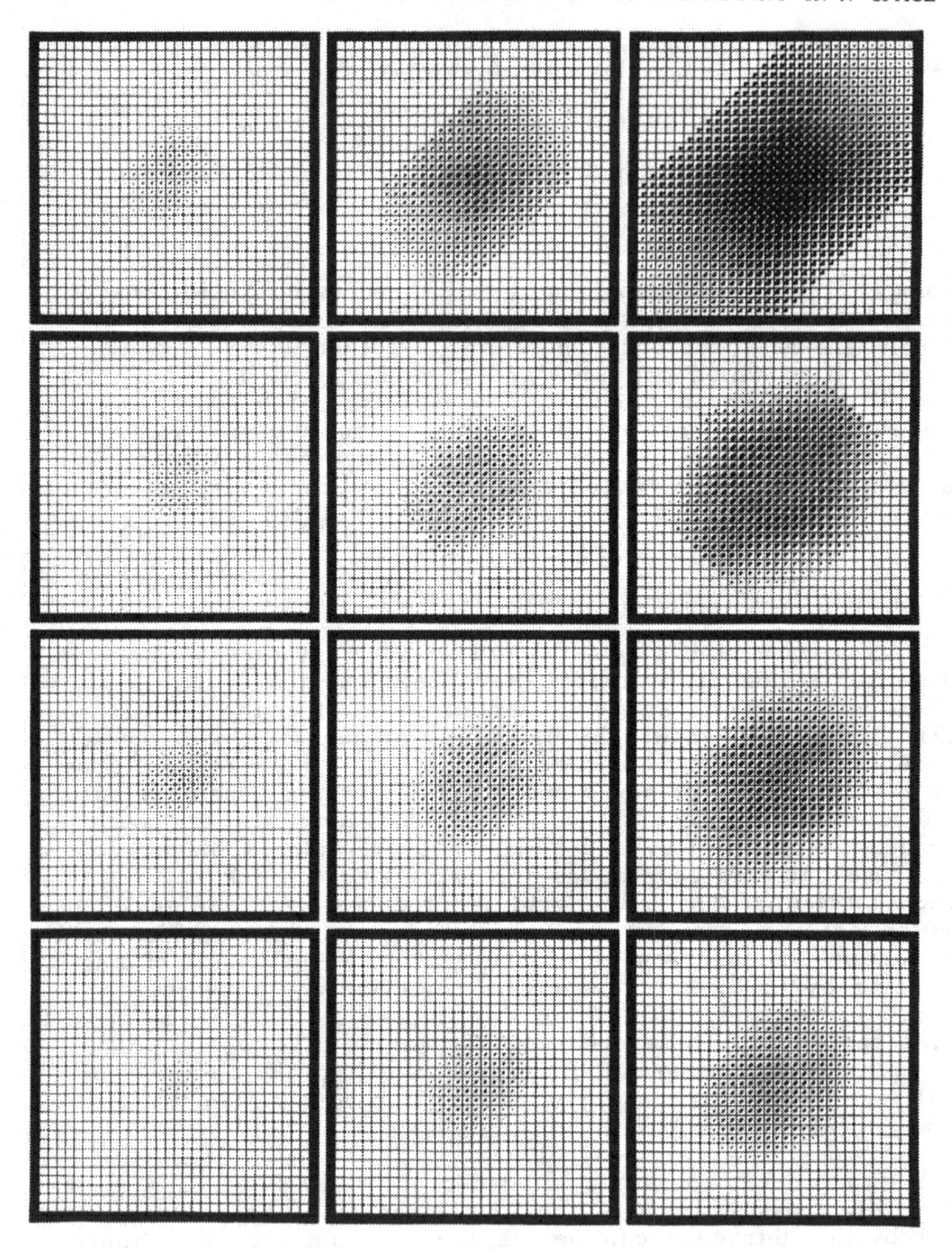

Fig. 3-12 Augmentation in four-dimensions in a 32×32×32×32 tetracube is demonstrated in a planar display as shown above. The four rows of tetracubes shown illustrate augmentation of the kernel arrested every four iterations using threshold values of zero, one, two, and three, respectively (left-right, top-to-bottom).

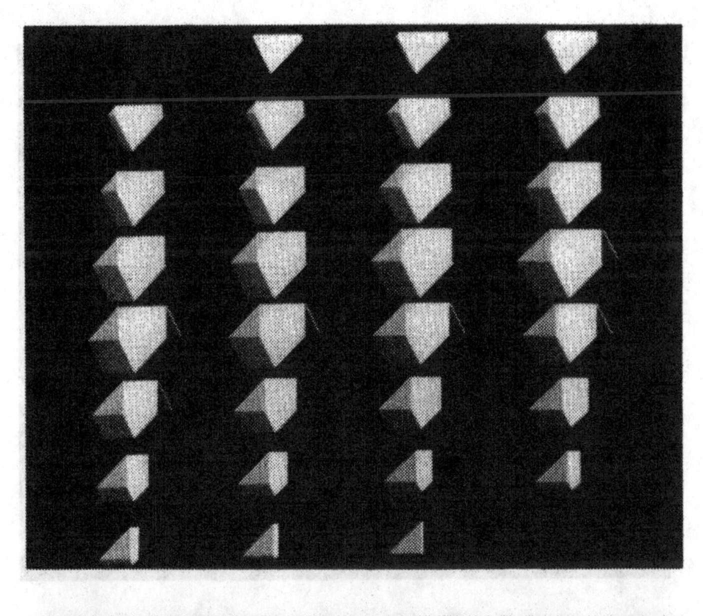

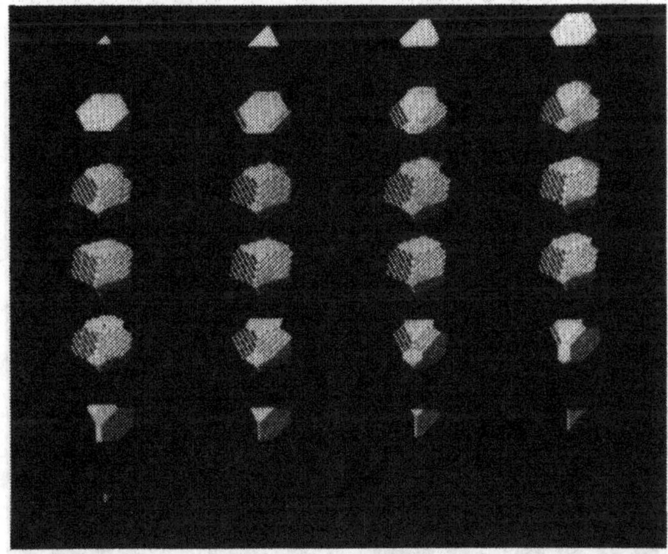

Fig. 3-13 The four-dimensional solids generated by augmentation and shown in Figure 3-12 may also be shown in a shaded-graphics display in which each figure presents the contents of the corresponding cube for each of the 32 cubes of the tetracube.

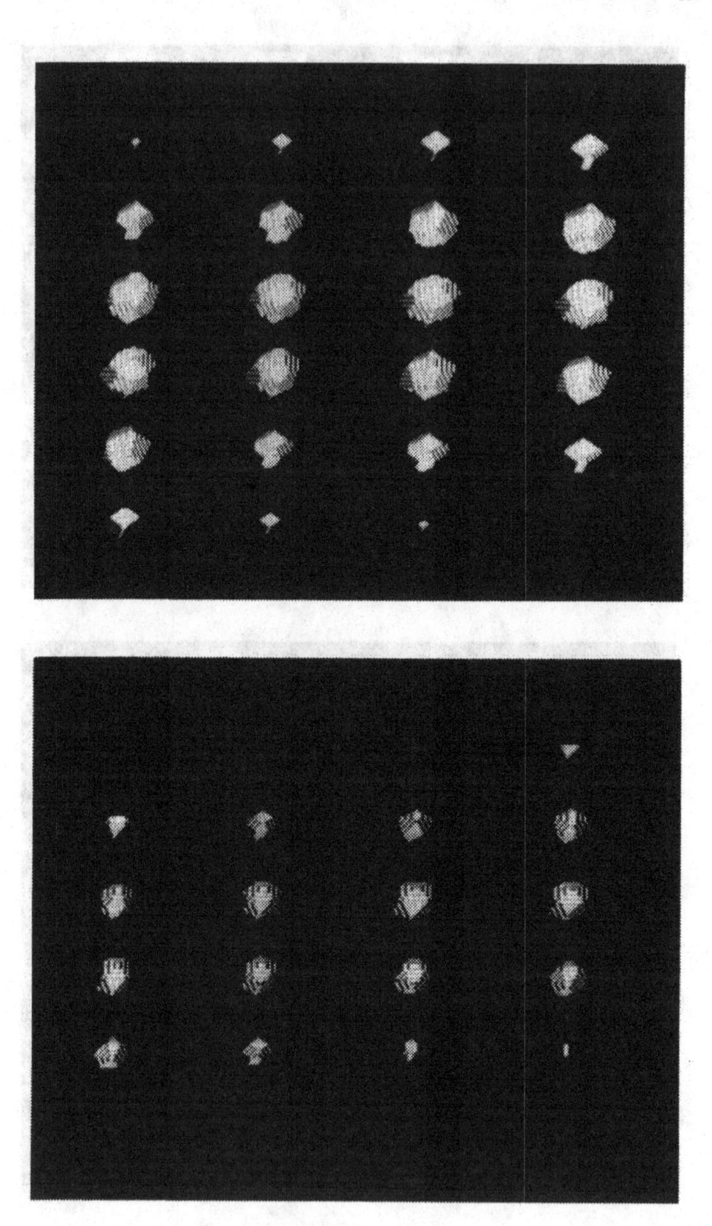

Fig. 3-13 (continued).

Augmentation curves corresponding to the processes illustrated in Figure 3-12 are given in Figure 3-14 compared with baseline curves for the tetracube and the tetrasphere. For the reader unfamiliar with four-dimensional geometry, the formula for calculating the volume of a tetrasphere is given below

$$n_v = \int_{-\pi}^{\pi}\int_{-\pi}^{\pi}\int_{-\pi}^{\pi}\int_{0}^{r} r^3 \cos\gamma \cos\phi \, d\theta \, d\gamma \, d\phi \, dr = 2\pi r^4 \tag{3.7}$$

where n_v is the number of voxels. For the spatially sampled space utilized here, the formula employed for the tetracube and tetrasphere are $n_v = (2n_c+1)^4$ and $n_v = 2\pi(n_c+0.5)^4$, respectively, where n_c is equal to the number of cycles.

Shaded-graphic displays for the reduction processes which occur for values of Ξ between 8 and 20 are too numerous to be illustrated. As would be expected, some reduction processes are shape-preserving, while others are not. A general four-dimensional surface-displacement theory by which one could predict the outcome without empirical investigations has yet to be developed. Reduction curves are provided in Figure 3-15 for certain selected cases, starting with the four space-filling four-dimensional solids whose shapes are those displayed (after 12 cycles) in Figure 3-14.

3.4 APPLICATIONS

This section completes this chapter with two demonstrations of the application of the multi-dimensional logical transform to image processing and to data anlaysis. These illustrations employ the logical convolution function without utilizing the connectivity properties available by employing occupancy. They are intended to illustrate the power and diversity of the multi-dimensional logical transform but are far from representative of all possible applications. Further illustrations may be found in the following chapters.

3.4.1 Image Processing

Three-dimensional cellular logic may be used in spatial filtering in a somewhat similar manner to the two-dimensional filter described in Chapter 2. However, entirely different results are obtained and, at this writing, a theoretical analysis of the transfer function has not been completed. This section furnishes a simple example and presents the empirically

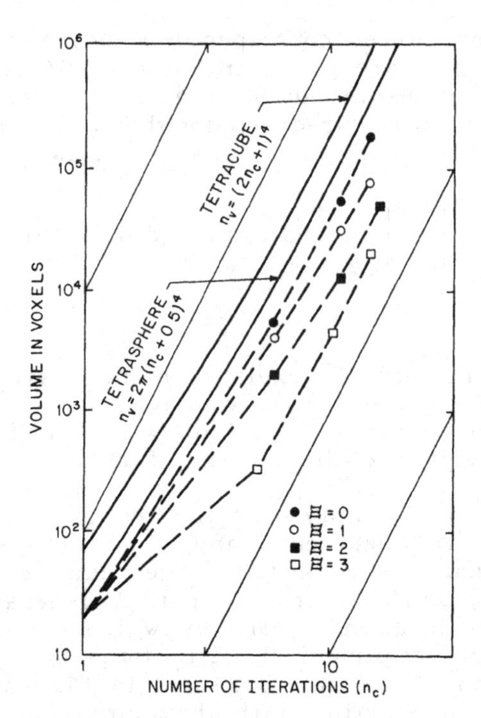

Fig. 3-14 Augmentation curves are shown for the four growth processes illustrated in Figure 3-12 along with theoretical reference curves representing augmentation of the unit tetracube and the unit tetrasphere.

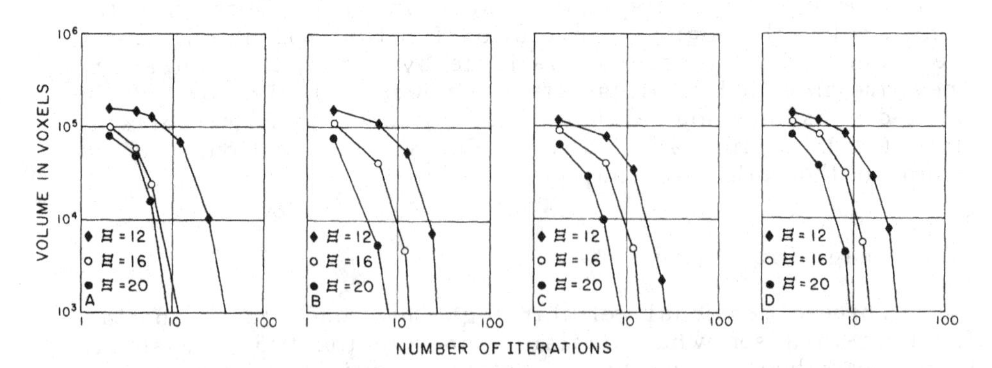

Fig. 3-15 Empirical reduction curves plotted for the four-dimensional solids shown in Figure 3-12 using threshold values of 12, 16, and 20.

derived transfer functions which are obtained. In order to use TRIAKIS to process an N×N image, the image is divided into a multiplicity of 64×64 windows. The integers comprising the 64×64 array in each window are then thresholded using 64 equispaced thresholds. The result is a 64×64×64 binary array which is then entered into the TRIAKIS workspace.

Low-pass filtering was performed on a zoneplate test pattern (Chapter 2) using values of Ξ = 12, 10, 9, and 7, respectively, each for four iterations, followed by Ξ = 0, 2, 3, and 4 for four additional iterations, respectively. Results for the zoneplate test pattern are shown in Figure 3-16. The test pattern is 512×512. It was divided into an 8×8 array of 64×64 windows. The 16 windows which comprise each quadrant of the zoneplate test pattern were operated upon by different filters. For the value of Ξ = 10 (four iterations) followed by Ξ = 2 (four iterations) the most isotropic results were obtained. For other values of Ξ, considerable antisotropy occurred. (See Figure 3-16.)

In order to describe the action of this three-dimensional cellular logic spatial filter, the transfer functions for both the 0° and 45° directions were plotted based on the empirical results obtained. Transfer function curves are shown in Figure 3-17. These curves correspond to similar curves presented in Chapter 2. Several differences are evident. First, the three-dimensional filter does not have a sharp cutoff frequency beyond which no signal is passed. Second, the three-dimensional filter does not clip the sinusoidal components of the zoneplate test pattern. Third, the transfer function curves must be plotted in terms of both the maxima and the minima of the resultant output because, since the filter is nonlinear, the output is asymmetric.

Further applications of the three-dimensional cellular logic filter are given in Chapter 10. It should be noted parenthetically in referring to Figure 3-16 that, due to windowing, there are some edge effects which may be noted as unusual, high values of the output near the borders of the 64×64×64 TRIAKIS window. These edge effects may be completely eliminated by re-dimensioning the TRIAKIS workspace to 64×512×512.

3.4.2 Four-Dimensional Data Analysis

The data gathered by Fisher (1956) on the length and the width of sepal and petal for three species of iris serves to illustrate the use of the logical transform in four-dimensions. Fisher's data has the dimensions sepal length, sepal width, petal length, petal width. Therefore, these data can be dis-

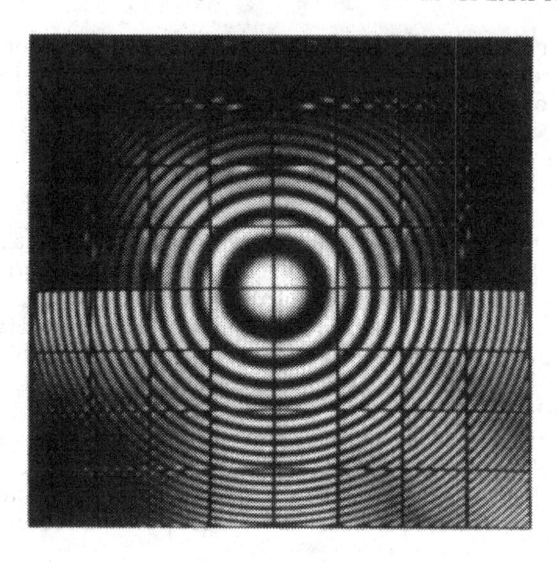

Fig. 3-16 Result of applying a windowed three-dimensional cellular logic spatial filter to a zoneplate test pattern using different values in each quadrant (left-to-right, top-to-bottom Ξ = 12&0; Ξ = 10&2; Ξ = 9&3; Ξ = 8&4).

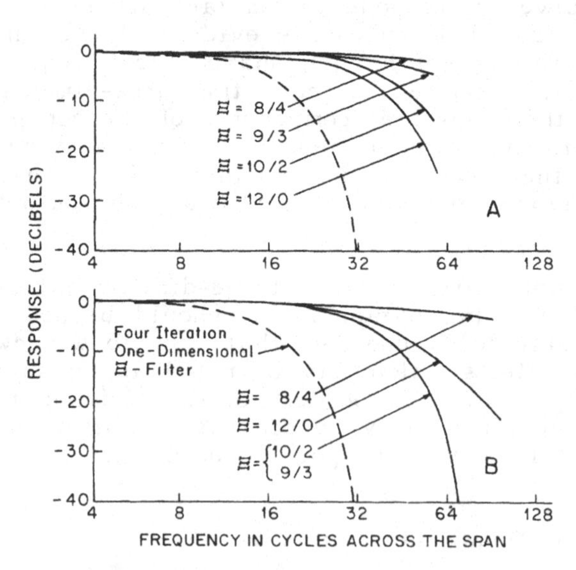

Fig. 3-17 Transfer function curves corrsponding to the four three-dimensional cellular logic spatial filters whose action is illustrated in Figure 3-16. Since the operation of the cellular logic filter is asymmetric, curves are given for both the maxima and minima of the resultant output.

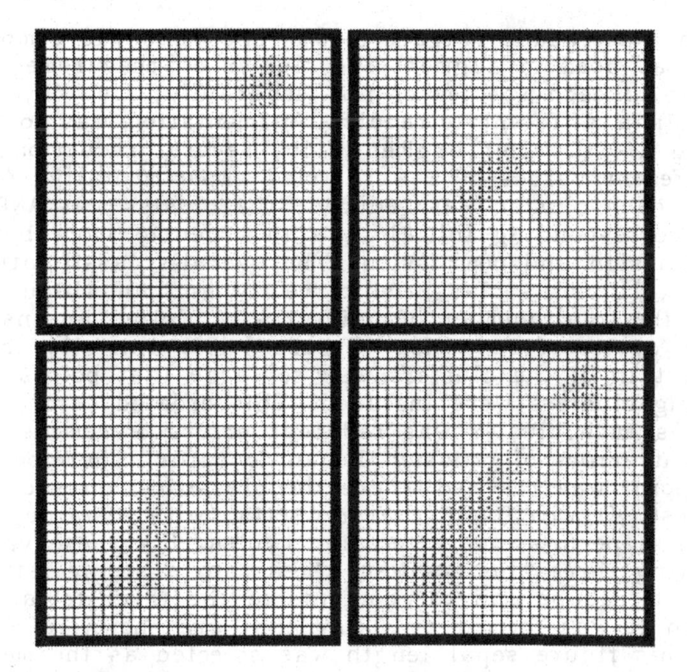

Fig. 3-18 Fisher's Iris data is four-dimensional (sepal width, sepal length, petal width, petal length) and is transformed so as to form 3 clusters in the tetracube. Iris Setosa (upper left), Iris Versicolor (upper right), Iris Virginica (lower left), and a composite are shown.

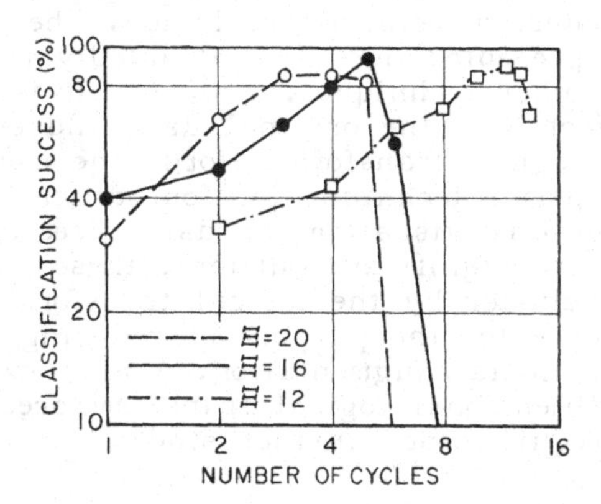

Fig. 3-19 The logical transform may be used to cluster Fisher's Iris data using various threshold values for various numbers of iterations. The success of this clustering may be determined by plotting classification errors as a function of the number of iterations.

played in a tetracube (Figure 3-18). The TETRAKIS computer program was used to augment the cluster of measurements corresponding to each species of iris separately, i.e., by augmenting each in a different tetracube. Augmentation was for 5 cycles using $\Xi = 0$. After augmentation, several reduction cycles using different values of Ξ were tried experimentally. As the volume of each cluster was reduced, the cluster was ANDed with the corresponding initial data in order to score both errors of omission and commission. The overall classifications success rate (all points classified correctly and unambiguously divided by the total number of points) was plotted against the number of cycles of reduction (Figure 3-19). Initially the volume of certain of the clusters was so large that points in neighboring clusters were included. This created errors of commission. As each cluster was reduced, errors of commission decreased but errors of omission began to occur. Therefore, an optimum point was reached where the classification success rate was highest. It was found that reduction with a value of $\Xi = 16$ gave best performance after five cycles. For these values the resulting three clusters were ORed into a single tetracube. It is this tetracube and the contents of the three tetracubes from which it was formed by ORing that are shown in Figure 3-18. In this figure sepal length was selected as the measure for the α-direction; sepal width, the β-direction; petal length, the δ-direction; and petal width, the γ-direction. Petal length and petal width appeared to have the greatest ability to discriminate the two most closely spaced classes and therefore were used as the long dimensions in the tetracube display.

In evaluating the above clustering experiment, it should be noted that clustering methods are legion. The logical transform methodology presented here is undoubtedly not superior to the best of these other techniques. However, it is felt that the tetracube display of both the original data and of the clusters developed by the logical transform affords the user a new and, in some cases, superior technique for four-dimensional data analysis. The user may visualize the data directly and, in the case where there are significant outliers, these will be separated from the main cluster by the logical transform when employing high values of Ξ for the purpose of reducing the clusters obtained after the initial augmentation. A more explicit evaluation of the multidimensional logical transform presented here will, of course, require much further study.

4. NUMERICAL OPERATIONS

4.1 INTRODUCTION

This chapter considers ways in which cellular automata can be used to perform numerical operations. It is undoubtedly true that any array of general purpose processors is able to implement any numerical operation, given sufficient storage and allowing a sufficiently long sequence of instruction cycles. It is also true that the way in which data are represented in the array and the details of the array architecture strongly influence the efficiency of the array as a numerical processor.

4.2 DATA REPRESENTATION

In this discussion it is assumed that each processing element (PE) in an N×N array has assigned to it m memory bits. Let s(x,y,z) represent the array (Figure 4-1). Then s(x,y,k) can be interpreted as the contents of the kth bit plane, specifying just one bit of memory in each of the PE's. Similarly, s(i,y,k) for $1 \leq y \leq N$ specifies the ith column of single bits in the kth bit plane.

A single bit plane can conveniently store N numbers, assigning one column to each number. Two possibilities for representing numbers in a column have proven useful. The first is the conventional two's-complement binary representation in which the least significant digit in the ith number in the kth plane is s(i,1,k), the most significant digit is s(i,N-1,k) and the sign bit is s(i,N,k). Figure 4-2 shows the number +9 being stored in the ith column. The second method represents a number as a

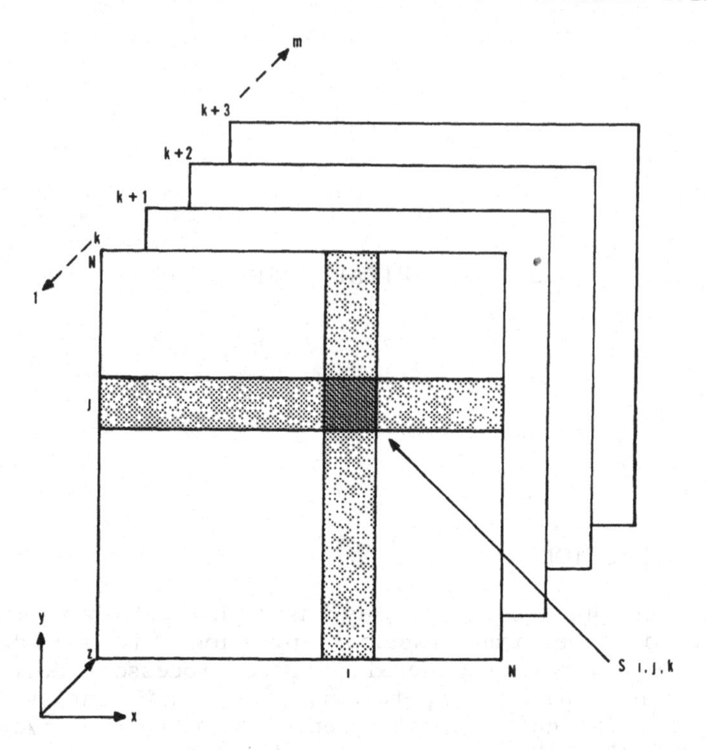

Fig. 4-1 Schematic representation of a single bit in a single processing element in each of m planes.

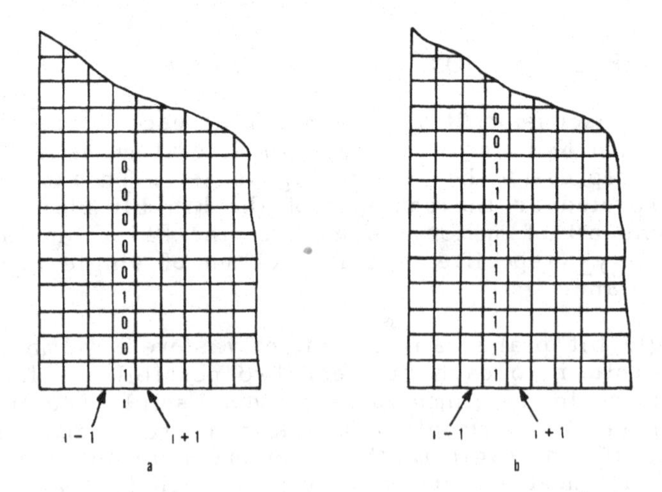

Fig. 4-2 The number +9 represented in the ith column of a bit plane in (a) binary-coded form and (b) by a linear column code.

column of digits whose length is proportional to the number. Figure 4-2b shows +9 stored in this way. This second method is less suitable for representing negative or non-integral numbers and soon fills the array storage unless numbers are small compared with N. For most purposes, the first method is preferred. Nevertheless, the second method has proven useful for use with certain specialized algorithms and as a preliminary to data display. It should be noted, for example, that such a method is ideally suited to histogram construction and display.

The precision with which binary coded numbers can be represented in a single bit plane is N bits. Since in practical automata N will often equal or exceed 32, this precision and range will most usually suffice. Attempts to introduce floating point features into a single bit plane representation have been unsuccessful, mainly due to the requirement in floating point arithmetic that the mantissa and the exponent be treated separately and differently. A way around this is to assign one bit plane to the mantissa and another to the exponent. Thus, for the ith number, s(i,y,k) would be the mantissa and s(i,y,k+1) would be the exponent. With this arrangement, the array would most probably provide range and precision far beyond the needs of most calculations.

A more economical use of the array storage which usually allows better matching between the calculation and the available precision is obtained by using "bit stack" arithmetic in which the array stores N^2 numbers spread across several bit planes. A p-bit binary number stored in location i,j would have s(i,j,1) as its least significant digit, the next more significant digit would be s(i,j,2) and so on. The most significant digit would be s(i,j,p-1) and s(i,j,p) would be the sign bit. Although it is possible to use a linear representation in a bit stack, in analogy with the linear column representation in a bit plane, this has not been found to be of any real value. Conversely, the bit stack lends itself more readily than does the bit plane to floating point arithmetic, primarily because the transition between the processing of the mantissas and the exponents occurs at a certain point within a sequence of instructions rather than at a particular array location.

Figure 4-3 shows the number +21 stored in location in a bit stack using bit planes k to k+4. The main features of this and the various other methods of data representation are summarized in Table 4-1.

4.3 CONVENTIONAL BIT PLANE ARITHMETIC

This section describes methods for performing addition,

Table 4-1 Storage Requirements for Various Methods of Representing Numbers in an Array

Representation		Numbers Stored	Bits/Number
Bit Plane	2's Complement	n x m	n
	Linear Column	n x m	*$\log_2 n$
	Floating Point	1/2 (n x m)	2n
Bit Stack	2's Complement	n^2	m
	Floating Point	n^2	m

*Numbers of bits in equivalent binary numbers

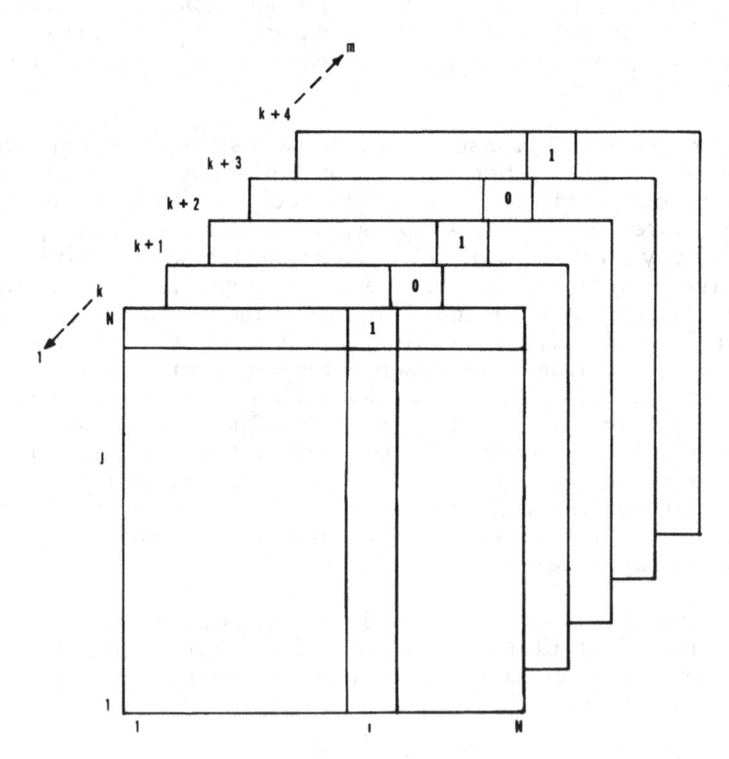

Fig. 4-3 Bit stack representation of the number +21 using a single bit in a single processing element i,j in each of five planes.

subtraction, multiplication, and division using the bit plane representation of numerical data.

4.3.1 Addition

Using the representation illustrated in Figure 4-2a, functions have been developed to perform arithmetic operations between binary numbers stored in corresponding columns of two bit planes. The first task is to implement a one-bit full adder in each processor. Since no communication is required between columns within the processing array, it can be assumed that only vertically upwards interconnection inputs and outputs, I_6 and P_2 (Figure 4-4) are enabled. It should be realized that the P_2 outputs form the "carry" signals between adjacent processors. The numbers to be added can be held in the buffers B_1 and B_2, with all other buffers being cleared. With these conditions, a simplified PE can be used, as shown in Figure 4-4.

The processors form sums of the corresponding columns in say, the bit planes k and k+1, with the sums being stored in bit plane k+2. The buffers are loaded so that

$$B_1(i,j) = s(i,j,k)$$

$$B_2(i,j) = s(i,j,k+1) \tag{4.1}$$

The sums appear at the D_2 outputs and are stored so that

$$s(i,j,k+2) = D_2(i,j) \tag{4.2}$$

For a full adder, the truth table given in Table 4-2 must be implemented. The Boolean functions producing this truth table are

$$D_2 = (B_1 \ominus B_2) \ominus I_6$$

$$P_2 = (B_1 \cdot B_2) + I_6(B_1 + B_2)$$

$$E = 0 \tag{4.3}$$

where $\ominus$ symbolizes the Boolean exclusive OR.

The input I_6 to the least significant bit processor is

Table 4-2 Truth Table for the Full Adder

Inputs			Outputs	
B_1	B_2	I_6	D_2	P_2
0	0	0	0	0
0	0	1	1	0
0	1	0	1	0
0	1	1	0	1
1	0	0	1	0
1	0	1	0	1
1	1	0	0	1
1	1	1	1	1

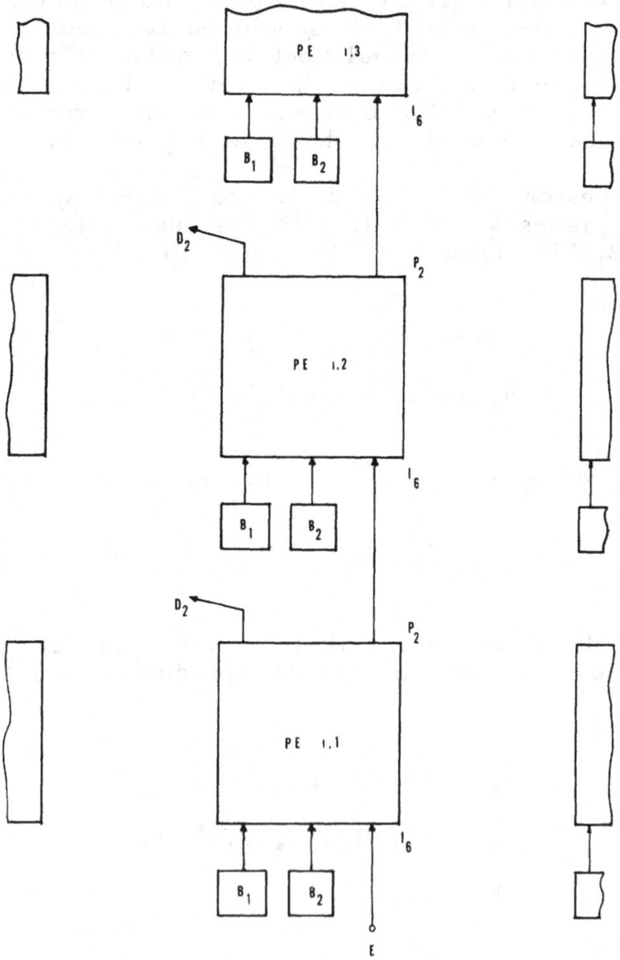

Fig. 4-4 Processing element array showing data storage and information flow for performing addition.

shown as coming from the edge value E. This is the value of the bus surrounding the array and is connected to all the "overhanging" interconnections. During the addition process, E is held at zero since no carry is required from the non-existent processor outside the array.

4.3.2 Subtraction

Subtraction is achieved using the identical array structure given above but with slightly modified functions. To understand the process, note first that in a simple binary subtraction using the process of one's-complementing, if X is an n-bit binary number, then 2^n-X is the two's-complement of X, and 2^n-1-X is the one's-complement of X. Consider the case of n = 5 where

$$2^n = 1\ 0\ 0\ 0\ 0\ 0$$

$$2^n-1 = 0\ 1\ 1\ 1\ 1\ 1 \qquad (4.4)$$

The second number above has the useful property that the result of subtracting any five-bit binary number from it produces a difference which is merely the same number with individual bits inverted. Thus

$$(1\ 1\ 1\ 1\ 1) - (1\ 1\ 0\ 1\ 0) = (0\ 0\ 1\ 0\ 1) \qquad (4.5)$$

so that the one's-complement of a binary number is formed by inverting its digits. Finally, subtraction can be expressed in the form

$$Y - X = Y + \text{2's-complement}(X) - 2^n$$

$$= Y + \text{1's-complement}(X) - 2^n+1 \qquad (4.6)$$

In terms of an array structure, a subtraction can be performed by the following steps

1. Invert X to obtain X'.

2. Load X' into B_2.

3. Load Y into B_1.

4. Add B_1 to B_2.

5. Add 1 to the result.

However, these five steps can be reduced to three by employing slight modifications to the PE functions. Rather than inverting X before loading it into the buffers, an inverting function is selected for the B_2 inputs, yielding

$$
\begin{aligned}
D_2 &= (B_1 \theta B_2') \theta I_6 \\
P_2 &= (B_1 \cdot B_2') + I_6 (B_1 + B_2') \\
E &= 1
\end{aligned}
\tag{4.7}
$$

The edge input E is set at 1 so as to provide an automatic increment in the final sum, thereby achieving a correct difference in two's-complement representation.

4.3.3 Multiplication

An important component of many arithmetic operations is shifting. In binary arithmetic, a one-bit shift corresponds to either multiplication or division by 2, depending on the direction. Using the conventions developed earlier in this chapter, a shift vertically upwards implies multiplication, whereas division requires a shift downwards.

The process of multiplication is more complicated. It will be recalled that binary multiplication is achieved by adding the multiplicand into an accumulator if the least significant digit of the multiplier is 1, then right shifting the multiplier and left shifting the multiplicand before repeating the operation. The process is summarized in the flow diagram given in Figure 4-5.

The two shifts and the addition can easily be performed with the array structure already described, but the conditional loop depending on the value of the least significant digits of the multiplier requires careful consideration. Since the bit plane will contain N numbers, one in each of its columns, the least significant digits will not usually be the same in different columns. Thus the conditional loop must be traversed in different directions for different columns, which is clearly impossible. A way round this difficulty is to introduce an additional step in which a mask is formed by propagating the whole row of least significant digits vertically upwards in the array.

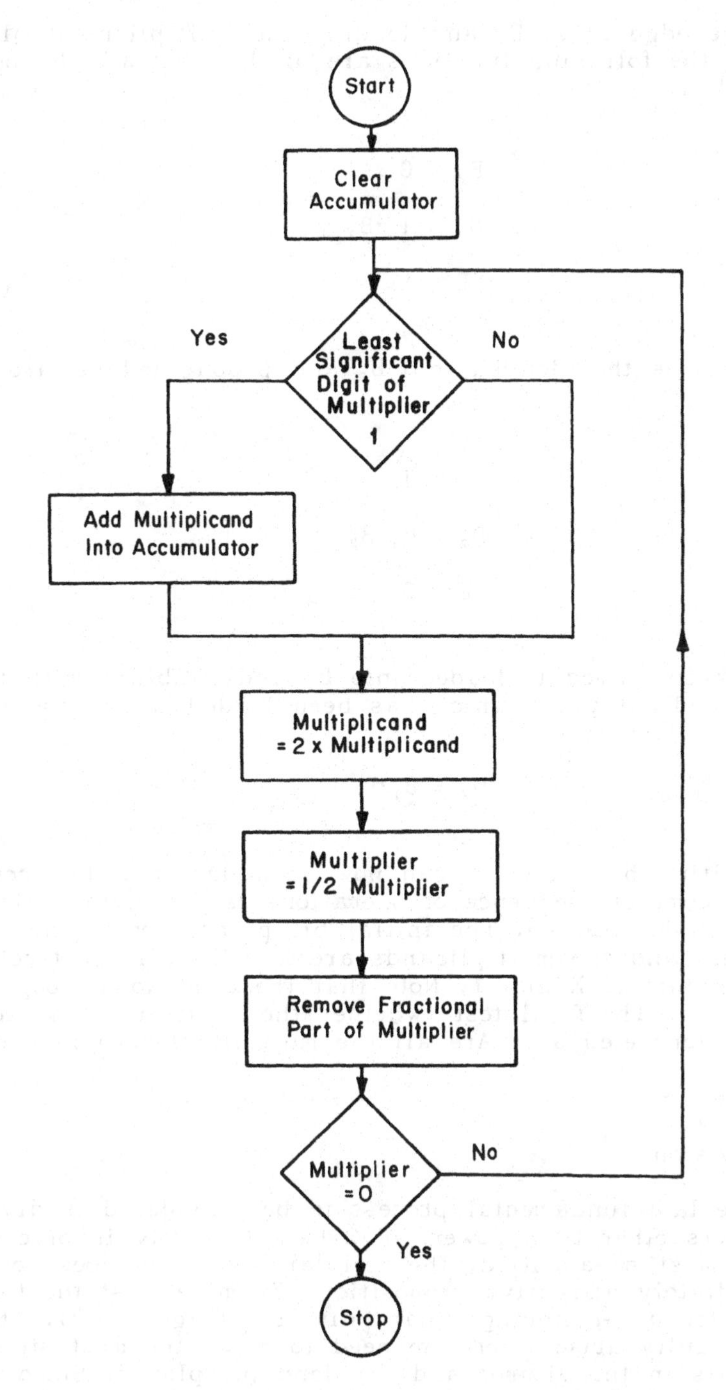

Fig. 4-5 Flow chart describing bit-plane multiplication.

Using the edge input E, and loading the multiplier bit plane into B_2, the following functions are used to extract the bottom row of digits

$$P_2 = 0$$

$$D_2 = I_6 \cdot B_2$$

$$E = 1 \qquad (4.8)$$

The output is then loaded into B_2 and propagated upwards using

$$P_2 = I_6$$

$$D_2 = P_2 + B_2$$

$$E = 0 \qquad (4.9)$$

The mask so formed is loaded into B_2 and combined with the multiplicand bit plane which has been loaded into B_1 so that

$$D_2 = B_1 \cdot B_2 \qquad (4.10)$$

The resulting bit plane D_2 can now be added into the accumulator. The complete sequence of operations is shown in a flow diagram in Figure 4-6. The initial bit planes containing the multipliers and the multiplicands are X and Y, respectively; Z is the product of X and Y. Note that the conditional loop has disappeared. The final test (whether the multiplier X is zero) can be interpreted as, "Are all the multipliers zero in every column?"

4.3.4 Division

The last fundamental process to be considered is division by numbers other than powers of 2 (which merely involve shifts). Here, it must be admitted, the cellular automaton does not offer an immediately attractive proposition. Examine first the flow diagram for non-restoring binary division (Figure 4-7). The first difficulty arises from the need to align the most significant digits in the divisor and dividend bit planes. Since this must be done for each column separately, the process is some-

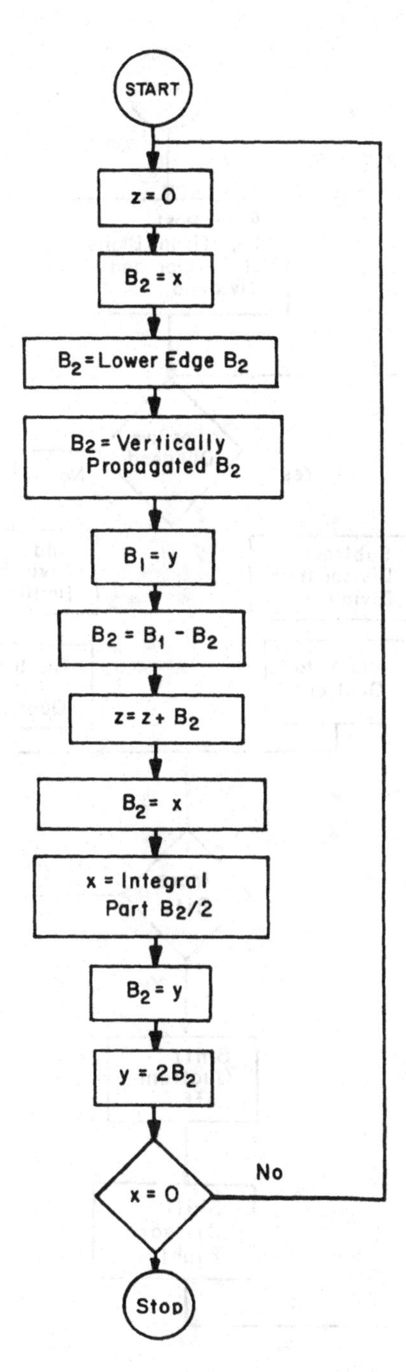

Fig. 4-6 Operational sequence for bit-plane multiplication.

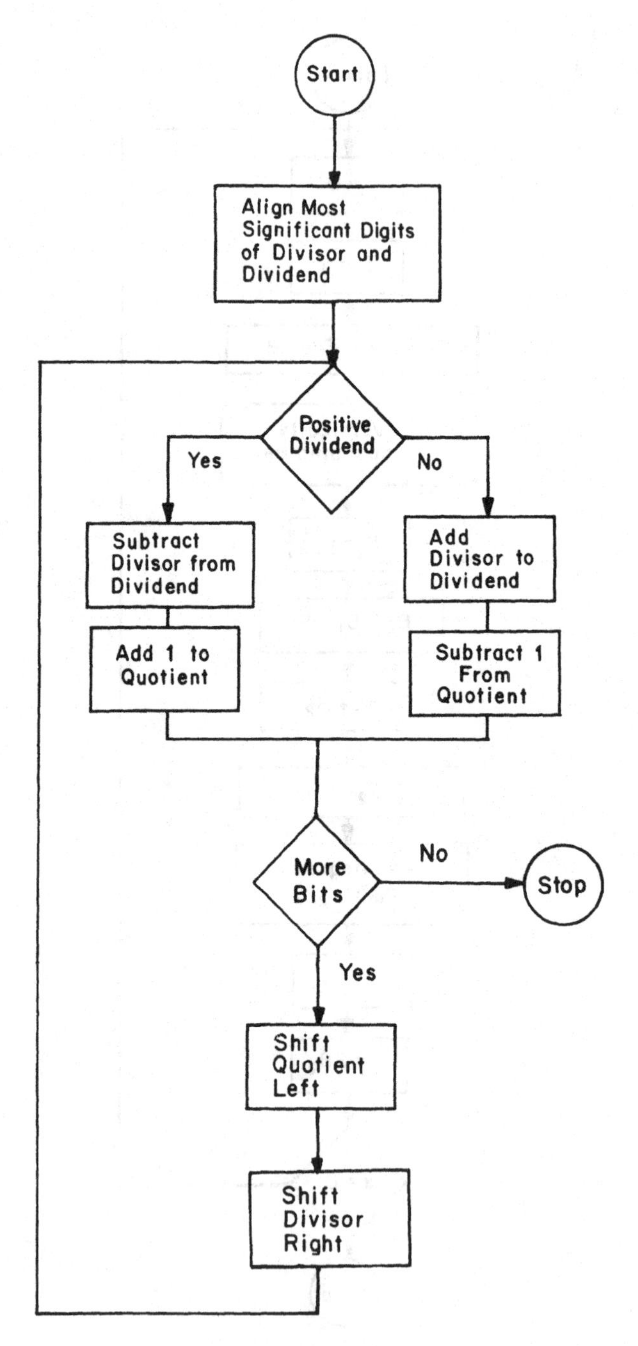

Fig. 4-7 Flow chart for non-restoring binary division.

what cumbersome. The steps to be executed in each bit plane are given below.

1. Extract the top row of the plane X.

2. Form a mask by propagating the extracted top row downwards.

3. AND this mask with the bit plane X.

4. OR the result into an initially empty plane Y.

5. NAND the mask with X and return the result to X.

6. If X is empty, jump to 9; otherwise, continue.

7. Shift the contents of X one row upwards.

8. Return to 1.

9. Stop, the resulting aligned columns being all contained in Y.

This process must be carried out on both the divisor and dividend planes as is shown in Figure 4-8.

The second difficulty stems from the need to implement the conditional branch on the sign of the dividend. One possible procedure, again cumbersome, is as follows

1. Extract the sign row in the dividend plane.

2. Propagate the row downwards to form a mask.

3. AND the mask with the divisor plane and put the result into X.

4. NAND the mask with the divisor plane and put the result into Y.

5. Add X to the dividend plane.

6. Subtract Y from the dividend plane.

7. Extract the bottom row of the mask and put into X.

8. Extract the bottom row of the inverted mask and put into Y.

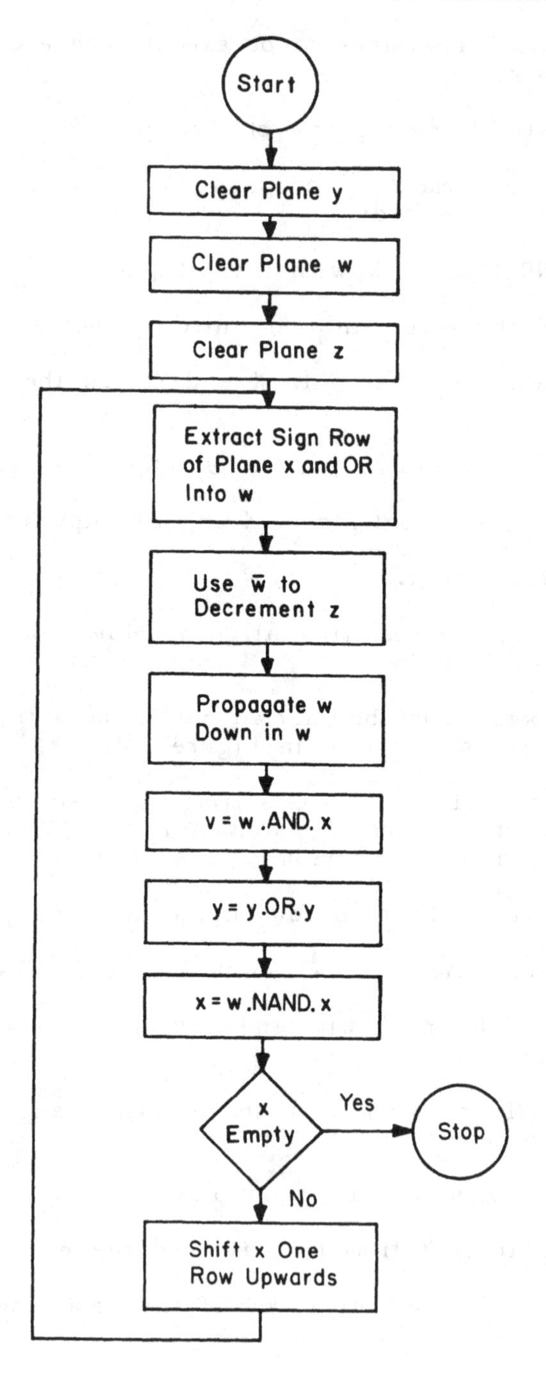

Fig. 4-8 Initial operational sequence for bit-plane division.

9. Use X to decrement an initially empty quotient plane.

10. Use Y to increment the quotient plane.

11. Shift the divisor plane one row down.

12. Shift the quotient plane one row up.

13. Return to 1 unless satisfactory precision has been achieved.

A further factor which must be taken into account concerns the position of the binary points. During the initial alignment procedures, each time a column is shifted upwards in the dividend, a binary one must be taken from the corresponding column in a plane which will contain exponents. Similarly, a single shift upwards in the divisor array is matched by adding one to the exponent. This can be effected by a procedure equivalent to that used in incrementing or decrementing the quotient array, described above.

A small saving can be effected by using the extracted top rows, both in the calculation of the exponents and in the calculation of the quotients, to increment arrays which are upside down, thus avoiding the steps of extracting the bottom row of the masks. The quotient and exponent arrays would then have most significant (or sign) digits in the bottom rows of their arrays (y = 1) and least significant digits at the top (y = N).

Another saving can be made by examining steps 5 and 6 more carefully. These steps are both slow, involving propagation along the length of each column (i.e., N steps). By inverting Y and ORing it into X, the addition and subtraction can be simultaneously carried out. However, ones must be added into each of the inverted columns (to correct one's-complement to two's-complement representations); this will be found possible in certain PE structures by performing an additional fast "load" operation into selected locations.

The complete process is shown in the flow diagrams in Figures 4-8 and 4-9. The first process is repeated for both dividend and divisor alignment; the second process is executed only once. To estimate the length of the process assume, as a first approximation, that each simple operation takes T units of time, but that operations involving propagation through n_e elements involve an extra $n_e PT$ of time where P is the propagation coefficient (usually less than one). Making the simplifying assumption that all propagations involve N elements, we can take the time for a propagating instruction to be (T+NPT) units.

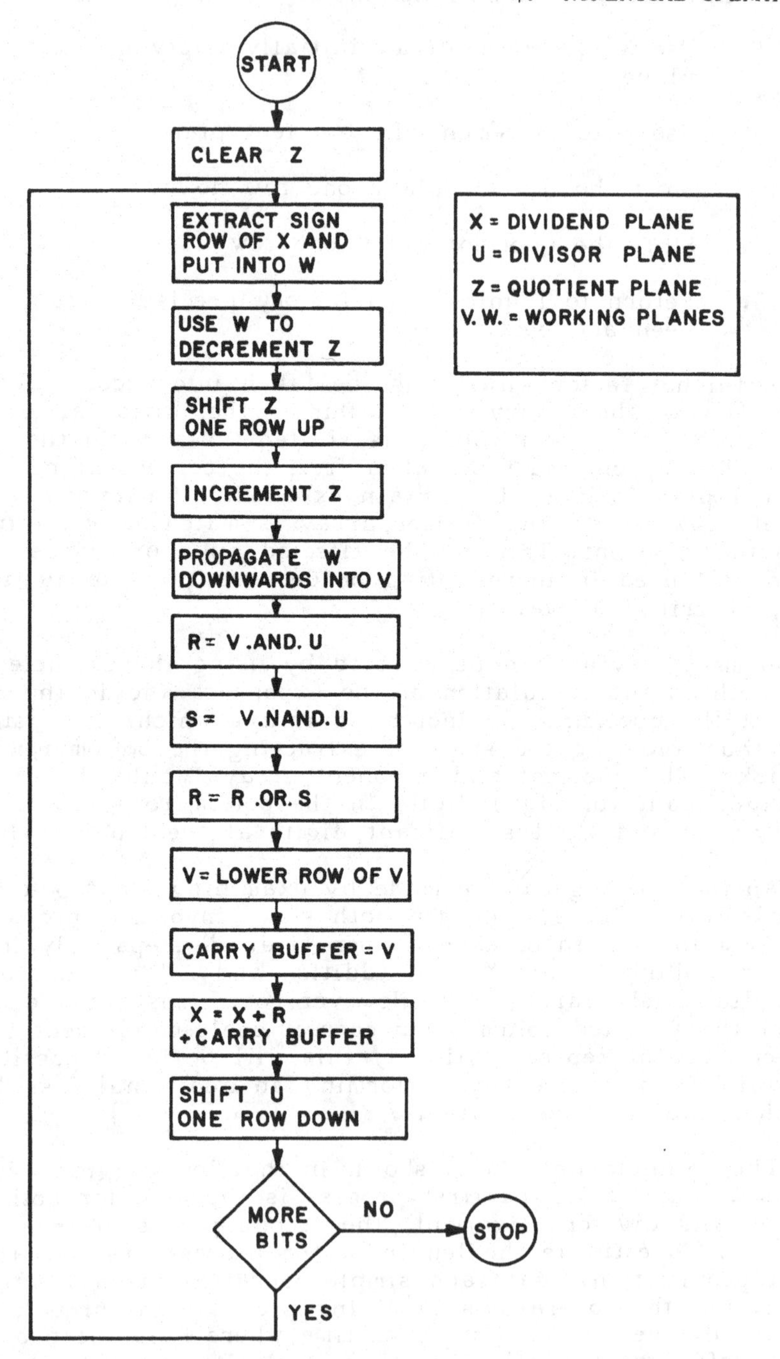

Fig. 4-9 Final operational sequence for bit-plane multiplication.

The alignment operation requires 3T+L(9T+2NPT) units where L is the largest number of steps needed to align the most misaligned column. As its limits, L would have the values N-1 or zero, so N/2 would seem to represent a reasonable compromise. Since this process is used twice, this part of the program requires 2(3T+(N/2)(9T+2NPT)) units of time.

The second part of the program requires T+(14T+4NPT)B units of time where B is the number of bits in the answer columns. It would be difficult to use other than full columns in the array so one is forced to equate B with N and the time required is T+N(14T+4NPT). The total time for all parts of the program may be obtained by taking the sum of the separate parts yielding $T(7+23N+4PN^2)$. The dependence of total time on N^2 is seen to be very unsatisfactory. However, calculations based on the CLIP4 system (for which N = 96, T = 25μs and P = 0.125) yield a division time of only 2ms/column. However, restricting N to 32, in order to work to a more typical precision, yields an overall division time of 30ms or about 0.3ms/column. Finally, a TTL version of CLIP4 would have T ≈ 100ns so that division times in the region 1μs/column would be readily obtainable.

In summary, it can be concluded that bit plane division is one of the less satisfactory operations to perform on a cellular automaton, involving both slow execution speeds and complex programming. As a footnote, it is interesting to observe that division can be carried out extremely rapidly when the divisor is a power of two and the same for every column. Under these special circumstances, division reduces to shifting, involving execution times of only NT. Even when the common divisor is not a power of two, the flow diagram can be substantially reduced.

4.4 LINEAR COLUMN METHODS

Turning our attention to the method of representing numbers illustrated in Figure 4-2b, simple techniques exist for incrementing and decrementing numbers. However, as mentioned above, the precision of addition, multiplication, and so on is limited because the maximum size for any number which can be represented is N.

4.4.1 Linear Column Incrementation

If we assume that the mth column is to be incremented, then much the same PE structure can be used as was employed for binary addition in the previous section. Paths I_6 and P_2 are enabled and the array containing the columns to be incremented is loaded into B_1. A second mask array is required having elements in its bottom row equal to one for each column selected for incrementation. This mask is loaded into B_2. The object of the exercise is to start a propagation in the appropriate B_1 columns so that only the elements having the value 1 transmit the propagation. At the finish of the propagation, all elements receiving the propagation are output into D_2 as ones. These ones correspond to ones in the original columns plus an additional one for each column, which is stored in the element immediately above the topmost element containing a one in each column. The required functions are

$$P_2 = B_2 + (I_6 \cdot B_1)$$

$$D_2 = P_2$$

$$E = 0 \qquad (4.11)$$

4.4.2 Linear Column Decrementation

Decrementation is most easily performed by a similar process, starting with the decrementing bits in the top row of the B_2 array ($y = N$). In this case the propagation is downwards through the 0-elements, effectively adding another 0-element to the selected columns (i.e., deleting the topmost 1-element). The corresponding functions are

$$P_6 = B_2 + (I_6 \cdot B_1)$$

$$D_2 = P_2'$$

$$E = 0 \qquad (4.12)$$

4.4.3 Linear Column Addition

Probably the most efficient way of adding two bit planes is repeatedly to extract the lowest row of the augend plane, using it to increment the addend plane, and then shifting down the entire augend plane. This process is continued until the elements of the augend plane no longer contain ones. However, although it may be the most efficient method which can be devised, it can be easily seen that it will involve times comparable with those for multiplication in the normal binary representation method. It therefore hardly seems worth while to consider this process in any further detail.

4.5 CONVENTIONAL BIT STACK ARITHMETIC

Since N×N cellular logic arrays will most usually be employed to perform operations on N×N arrays of data, bit stack arithmetic will provide the natural way of performing the required operations. In its simplest forms, the operations will be confined to each array address, i.e., will make no use of connections between PE's in the plane. Using the notation introduced at the beginning of this chapter, assume that there are two bit stacks, each with p+1 bit planes.

$$\text{Stack A} = s_a(x,y,a),\ldots,s_a(x,y,a+p)$$
$$\text{Stack B} = s_b(x,y,b),\ldots,s_b(x,y,b+p) \quad (4.13)$$

where s(x,y,a) and s(x,y,b) are the planes for the least significant bits and s(x,y,a+p) and s(x,y,b+p) for the most significant bits in the two stacks.

4.5.1 Addition

For addition, the requirement is that, for the elements in the array address i,j, the contents of the result stack R is given by

$$\text{Stack R} = s_r(i,j,r),\ldots,s_r(i,j,r+p)$$

where

$$s_r(i,j,r+\ell) = (s_a(i,j,a+\ell)\oplus s_b(i,j,b+\ell))\oplus s_c(i,j,c)$$
$$s_{c'}(i,j,c') = s_a(i,j,a+\ell)\cdot s_b(i,j,b+\ell)+s_c(i,j,c)$$
$$\cdot(s_a(i,j,a+\ell)+s_b(i,j,b+\ell)) \quad (4.14)$$

where $s_c(i,j,c)$ is the carry resulting from the previous bit plane addition and where $s_{c'}(i,j,c')$ is the new carry to be carried forward. Implementation of this algorithm can be achieved using the structure shown in Figure 4-10 in which

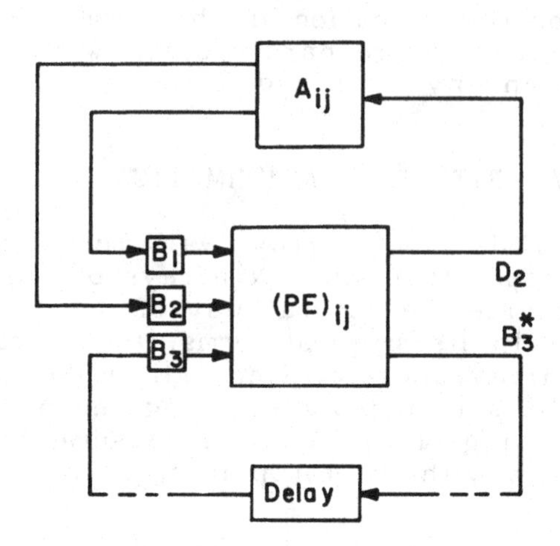

Fig. 4-10 Processing element configured for bit-stack addition.

$$D_2 = (B_1 \ominus B_2) \ominus B_3$$

$$B_3^* = (B_1 \cdot B_2) + (B_3 \cdot (B_1 + B_2)) \qquad (4.15)$$

and $B_3(t+\Delta t) = B_3^*(t)$ where Δt is one operation cycle. The initial value of B_3 is zero.

This arrangement suggests the use of a buffer for the B_3^* output which can be loaded into B_3 after a one operation cycle delay. Depending on the detailed structure of the PE, this may be obtained by using the D_1 output or, more probably, by diverting a propagation output (say P_n) back to its "own" processor at B_3. If neither structure is permissible, then the addition process must be implemented in two cycles, using the D_2 output on both occasions, the first cycle producing the sum and the second cycle producing the new carry which will then be returned to the store A(i,j).

4.5.2 Subtraction

Subtraction is performed using the same general structure but modifying the functions, much as discussed in the previous section dealing with bit plane addition. If we assume the subtrahend is loaded into B_2 and the minuend into B_1, then

$$D_2 = (B_1 \ominus B_2')\ominus B_3$$

$$B_3^* = (B_1 \cdot B_2')+(B_3 \cdot (B_1 + B_2')) \qquad (4.16)$$

Also, the initial value of the carry input (B_3) is set to one to provide the incrementation necessary to correct the one's-complement result to two's-complement.

4.5.3 Multiplication

In contrast to bit plane arithmetic, bit stack arithmetic does not usually require repeated shifting of data during arithmetic operations. The same effect is achieved by address modification. Thus, if all the values in a stack are to be multiplied by the factor 2, it is only necessary to subtract one from the addresses of each plane in the stack. The least significant bit plane will then be an empty plane located beneath the original stack. More specifically, if the contents of a stack are given by

$$\text{INPUT STACK} = s(x,y,k),\ldots,s(x,y,k+p) \qquad (4.17)$$

then the stack of elements with values each twice those of the corresponding elements in the input stack is to be found at

$$\text{OUTPUT STACK} = s(x,y,k-1),\ldots,s(x,y,k+p) \qquad (4.18)$$

where $s(x,y,k)$ and $s(x,y,k-1)$ are the least significant bit planes in the input and output stacks respectively and $s(x.y, k-1)$ is an empty plane.

On the other hand, shifting individual elements in the stack is a complex process. Consideration of this process is provided in Section 4.5.4 dealing with division.

The fundamental process of multiplication described in Section 4.3.3 is again used for bit stack arithmetic. As before it is assumed that the product of the contents of stacks A and B, the result stack, is R. The contents of R will be substantially larger than A or B. The multiplication process can be interpreted as a procedure in which the following sub-processes are performed

1. R is set equal to zero throughout.

2. The (b+t)th bit plane of B (initially t = 0 so this plane will start as the least significant bit plane) is ANDed with every plane in A and the result added into R.

3. The index is incremented and the process repeated until t = q, adding A into R at one plane higher in the R stack on each occasion.

This process is shown in detail in the flow diagram given in Figure 4-11. It will be noticed that the two conditional branches relate to the current bit-plane addresses, not to the bit-plane contents. One disadvantage of the method illustrated is that storage is required for a temporary stack s(x,y,n),.., s(x,y,n+p). This can be largely eliminated by interleaving the ANDing operation with the addition process, thereby reducing the extra storage requirements to one bit plane.

4.5.4 Division

The major problem in performing division between bit stacks is in the initial alignment of divisor and dividend. The requirement is that all elements in both stacks be shifted up in the stacks until all non-zero elements have one's in the most significant digit planes, local counts being maintained of all the shifts executed.

The simplest way to implement this process is repeatedly to add the stack to itself in all elements not yet aligned. In more detail, the process proceeds as follows

1. Clear an "exponent" stack.

2. Extract the most significant bit plane of the divisor stack and invert it.

3. Use this plane to increment the exponent stack.

4. Also form a temporary stack by ANDing the inverted plane with the divisor stack.

5. Add this new stack to the divisor stack.

6. Repeat steps 2-5 until the inverted most significant bit plane is empty.

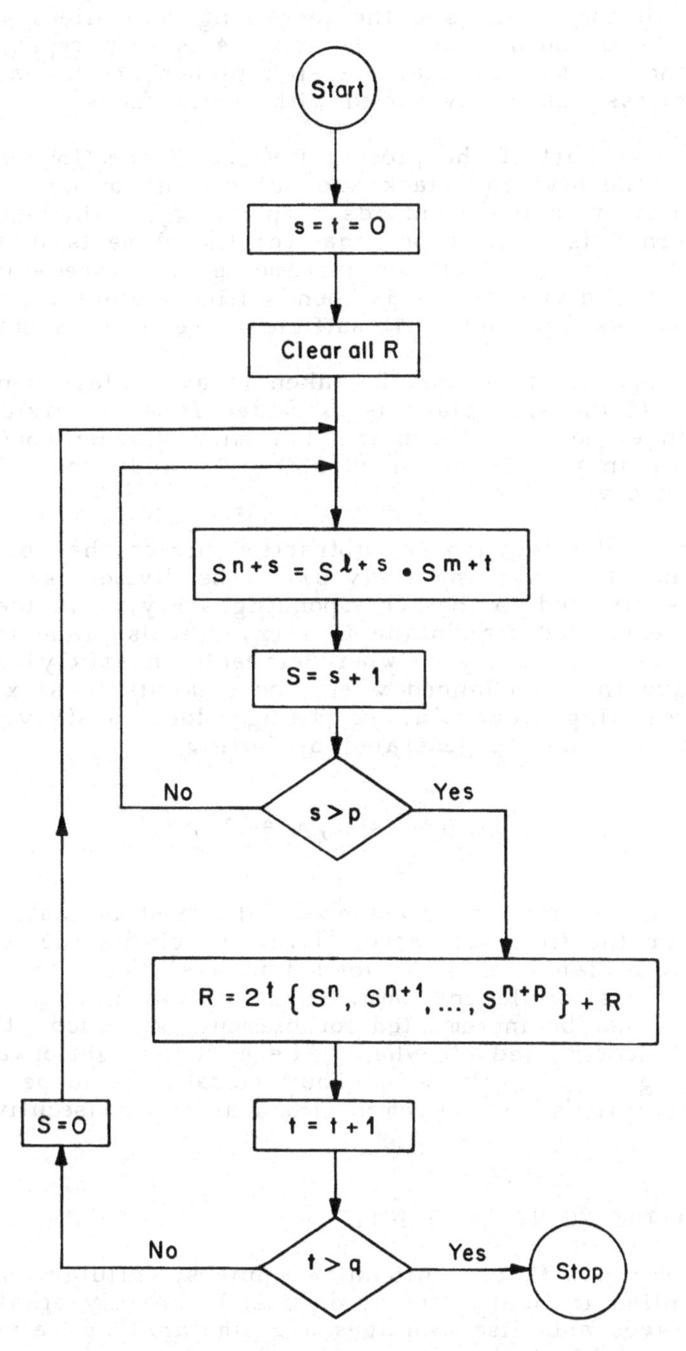

Fig. 4-11 Flow diagram for bit-stack multiplication.

At the end of this process, all elements are realigned in the top of the divisor stack and the correcting multipliers are stored in the exponent stack. The process is then repeated for the dividend stack. Note that the sign planes are not aligned in this process but merely moved with their stacks.

The next part of the process involves inspection of all elements in the dividend stack and subtracting or adding the divisors from or to the dividends, depending on whether the element concerned is positive or negative; the elements in the initially empty quotient stack are incremented or decremented accordingly. The divisor stack is then shifted down (i.e., halved) and the process repeated until sufficient precision is obtained.

Once again, steps must be taken to avoid data dependent branching. If the sign plane is extracted from the dividend stack, then elements in the plane determine whether corresponding elements in the divisor stack are to be added or subtracted from the dividend plane.

Suppose the addition or subtraction process has proceeded to the point where a plane $s(x,y,u)$ in the divisor is to be added or subtracted to the corresponding $s(x,y,v)$ in the dividend. The extracted sign plane is $s\pm(x,y)$; this plane is used to invert the plane $s(x,y,u)$ where elements in $s\pm(x,y)$ are zero and to leave them unchanged where the elements in $s\pm(x,y)$ are one, the resulting plane $s(u,y,w)$ being added to $s(x,y,v)$. The required values can be generated by letting

$$s(x,y,w) \leftarrow (s(x,y,u) \ominus s\pm(x,y))' \tag{4.19}$$

Two other factors must be considered. The first is that the carry plane for the first subtraction (i.e., involving the least significant digit planes) must be loaded with $s'\pm(x,y)$ to correct results to two's-complement form. The second is that the quotient stack must be incremented for elements in which $s\pm(x,y)$ is one and decremented elsewhere. The most straightforward way of doing this (involving minimum storage) is to perform the incrementations and decrementations as two consecutive processes.

4.6 FLOATING POINT ARITHMETIC

In common with conventional computers, cellular automata can be applied to floating point data and, broadly speaking, the advantages and disadvantages are similar. The main advantage lies in the fact that floating numbers use less memory

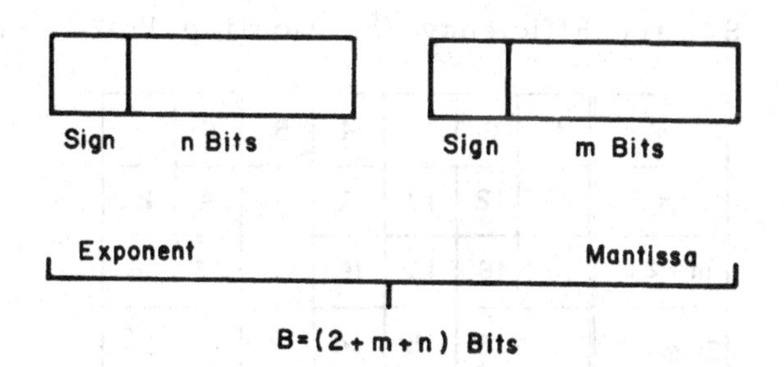

Fig. 4-12 Exponent and mantissa arranged for floating point arithmetic.

(smaller stacks) than fixed point numbers to store the same range of data with the same precision. Referring to Figure 4-12, (2+m+n) bits of memory, in which n bits are assigned to the exponent and m bits to the mantissa, plus two more bits for the sign of each, can be used to represent numbers in the range $((2^{n-1})$ to $+(2^{-n})) \times ((2^{m-1})$ to $(2^{-m}))$.

To accommodate this range in fixed point format, (1+m+(n-1)+n) = (m+2n) bits would be required, interpreting the extremes of the exponents as right and left shifts of the maximum precision (m-bit) mantissa. Assuming that it is required to represent numbers with m-bit precision, then the relative percent efficiency E of the two methods can be expressed in terms of the storage requirements in the form

$$E = 100((m+2n)-(2+m+n))/(m+2n)$$

$$= 100(n-2)/(m+2n) \qquad (4.20)$$

and, if m+2n=16 is a typical word length, then E is given by 6.25(n-2)%.

The quantity E is a measure of the storage which can be saved by changing to floating point, as a percentage of the fixed point storage. Table 4-3 lists E as a function of n for an original 16 bit word. It can easily be seen that the highest economies are obtained where n becomes large, but this entails the penalty of low precision. A uniform precision over a large range has been traded against a high precision over a smaller range. It is therefore advantageous if the relative values of m and n can be adjusted to suit particular problems. This feature can readily be met in the cellular automaton.

Table 4-3 Storage Efficiency for Floating Point Arithmetic

n	1	2	3	4	5	6	7
m	14	12	10	8	6	4	2
m + 2n	16	16	16	16	16	16	16
2+m+n	17	16	15	14	13	12	11
E %	$-6\frac{1}{4}$	0	$+6\frac{1}{4}$	$+12\frac{1}{2}$	$+18\frac{3}{4}$	+25	$+31\frac{1}{4}$

A detailed study of the floating point algorithm is not presented here. In general terms, since there will always be two components to be processed for each number, the algorithmic structure will be more complex than for fixed point representations. This is particularly true in view of the fact that the processes performed on the two components are not the same. Thus, in multiplication, the mantissas are multiplied whereas the exponents are added. On the other hand, the number of cycles to be executed in an algorithm will depend linearly on the number of bits in the numbers. It has been shown above that this number is smaller in floating point than fixed point (for $n>2$, $m<12$ in 16 bit numbers).

4.7 OTHER ARITHMETIC OPERATIONS

In principle, there is no reason why the arithmetic structure of a cellular automaton should be restricted to bit-serial, single-bit operations, as described in the previous sections, except in so far as this offers maximum flexibility at, it must be said, a sometimes high penalty when it comes to writing the programs for the arithmetic processes. In this section, a few extensions of the capabilities of the simple PE will be considered, with particular reference to the enhancement of arithmetic capability.

4.7.1 Multiple-Bit Processors

If the single-bit arithmetic unit in a PE is replaced by a 4- or 8-bit processor with automatic carry, obvious savings in processing time will result (at the cost of increased complexity of the PE hardware). Similarly, multiple parallel communication channels between processors can result in substantial increases in efficiency in implementing, for example, convolutions between 3×3 masks and an image.

Ultimately, every PE could include a sophisticated, floating point, double precision arithmetic unit, but the cost of a useful array of such PE's is presently prohibitive. In practice, arrays constructed so far have all had single-bit PE's, the cost limitation being immediate and dominant.

4.7.2 Bit Stack Shift Registers

Examination of the algorithms described in Section 4.5 reveals a serious weakness in the bit-serial PE's so far described: there is no "vertical" communication between the bit planes. It would be very convenient if it were possible to shift data up and down the stack, i.e., in the z direction, from plane to plane. Ideally, this will be achieved by parallel loading of a shift register from a set of successive planes in the array memory, followed by execution of a set of shift instructions. To complete the structure, an "activity bit" would be stored in the PE to enable the shift operation. It should be possible to test the state of the most significant end of the shift register (possibly by a parallel transfer into memory), partly with a view to setting the activity bit. A structure such as this would greatly speed both multiplication and division in the array.

4.7.3 Threshold Gates

A large class of local neighborhood operations on images involves counting the number of binary one's in the immediate 3×3 neighborhood and then comparing the count with a programmed threshold. Since the interconnection structure already exists in simple PE arrays the addition of a linear or digital threshold gate might not represent too costly an exercise.

4.7.4 Maximum-Minimum Gates

If we assume that a pixel value is communicated between neighboring PE's as a serial data stream, then it is very easy to compare this new data with that stored in a shift register in the receiving PE so as to replace the stored value with the greater of the incoming and stored values. A possible circuit is shown in Figure 4-13 in which the square symbols indicate elements of shift registers. The process starts with ΔA2 and ΔB2 containing one's and the stored value is in the long shift

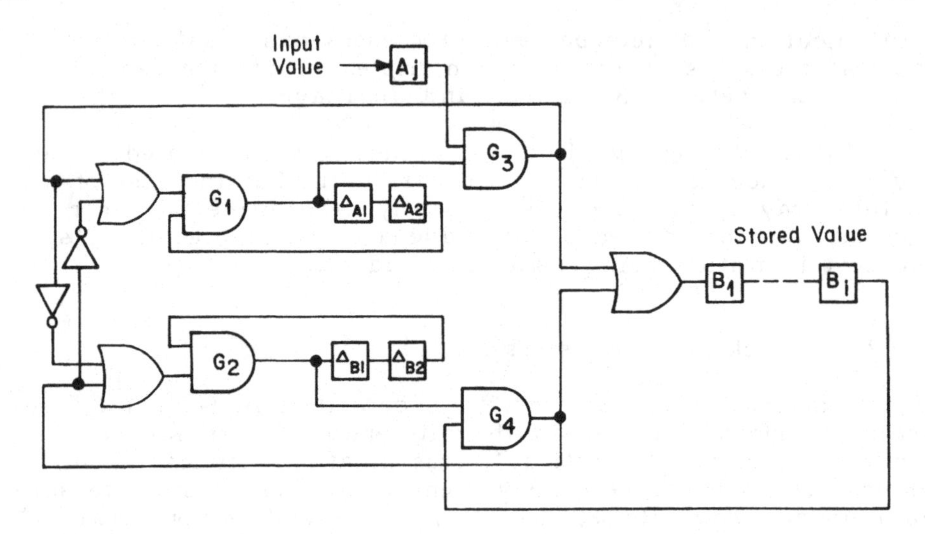

Fig. 4-13 A processing element structure for MAX-MIN computations.

register. The circuit is arranged so that at the beginning both Aj and Bj are one and will continue to be one until the two streams of digits (A_1, A_2, A_3, ...) and (B_1, B_2, B_3, ...) diverge. At this point Aj will be one and Bj will be zero indicating A to be the larger of A and B (or vice versa). Thereafter, the values of Bj will always be zero and Aj will always be one so that the new stored value will be A. By switching the connections (G_1-G_3 and G_2-G_4 to G_1-G_4 and G_2-G_3), the circuit can be made to store minimum rather than maximum values.

This circuit can obviously be extended simultaneously to compare and store the maximum or minimum of any number of serially presented binary numbers. Circuits such as this may not be thought of strictly as arithmetic circuits. However, their effect is to replace conventional arithmetic circuits where MIN and MAX functions are to be computed so, at least to this extent, they can claim to be special purpose arithmetic processors.

4.7.5 Counting Bits in a Plane

One problem which has proven particularly frustrating to designers of cellular automata is that of counting the total number of bits set to one in an array. At root, the problem does not readily lend itself to parallel implementation as counting is, by its very nature, a serial process, or so it would seem. Nevertheless, it is worth considering some human analogies in order to suggest possible parallel approaches. Firstly, voting papers are usually counted by assigning arbitrary bundles to several individuals who then, having counted their own bundles, present partial totals to a supervisor for grand totaling. Secondly, coins can be counted by stacking and, in effect, measuring the total stack height. Thirdly, uniform objects can be counted by weighing them, and similar integration techniques can be applied to other variables such as electric charge and light intensity. Finally, a frequently used counting technique involves running the objects down a chute past a mechanical or optical counting head.

Analogs of all of these techniques have been employed by array designers, with varying degrees of success. Ultimately, the speed of any particular architecture and algorithm will depend strongly on the efficiency with which data can be shifted between various parts of the array. The fastest technique, which also proves to be inexpensive in component cost, is to connect the array to an external "tree" of parallel adders. One such system, implemented on CLIP4, uses six 16-bit encoders to convert the contents of the 96 long right hand column of the array into six 5-bit binary numbers. Three stages of parallel addition produce a 7-bit sum which is accumulated, column by column, as the bit plane is shifted out of the array edge. Obviously, it would be possible to subdivide the array so as to examine several columns simultaneously and, in the limit, an adding tree could be attached to each column or sub-array. Thus a pyramid structure with $\log_2 N$ layers of counters could be envisaged as a means for bit counting. In practice, economic considerations will place a limit on the number of counters that can be afforded and, hence, on the rate at which counting can be achieved. It remains only to say that the "chute" method is often regarded as fast enough for most real applications and that column addition (as in CLIP4) has proven to be considerably faster than any "stacking" algorithms which have so far been devised.

5. SEGMENTATION

5.1 INTRODUCTION

This chapter concerns the problem of how to partition an array of numbers $s(\hat{r})$ into a group of subarrays, each of which has some common property, e.g., a regional *likeness* or *homogeneity*. Specifically, the chapter is concerned with the case where $\hat{r}$ is two-dimensional so that $s(\hat{r})$ becomes $s(x,y)$. Also assume spatial sampling of $s(x,y)$ over an i,j array yielding $s(i,j)$.

If $s(i,j)$ represents a digitized image, then our concern is with *scene segmentation* or *scene analysis*. Clearly a single element taken from $s(i,j)$ has infinite homogeneity. It is only when the value of a single number in the array is compared with the values of other numbers in the array that non-homogeneity can be measured. When the comparison taken is between numbers from adjacent points in the array, then the computation of homogeneity is readily performed by the cellular automaton. Comparison *at a distance* can also be performed by the cellular automaton using *propagation* (see below). More complex techniques for making these comparisons, e.g., those using the *extended neighborhoods* of Toriwaki (1979), are not treated in this book.

5.1.1 Noise Removal

It is unusual to find that the image to be analyzed is already neatly and naturally divided up into directly analyzable parts. Most images are degraded by non-uniform illumination, shadowing, vignetting, etc. All of these effects seem to be inevitable consequences of obtaining images from the *real world*. Furthermore, the borders between regions of interest and the background may be obscurred by structure in the background itself. Thus the process of abstracting an object from the background requires noise removal as a precursor to the segmentation procedure or, perhaps, as part of the segmentation procedure. Once segmentation has been successfully performed, then, of course, detailed analysis of the segmented components of the image is executed.

5.1.2 Knowledge Considerations

It should be remembered that, in actual practice, segmentation is a little understood process which leans heavily on the knowledge of the nature of the regions (objects) in the image and of other critical features. A wide variety of segmentation techniques has been proposed (see, for example, the survey by Fu and Mui, 1981). The technique eventually found to be successful will often be discovered intuitively. Often this involves extensive human action. Therefore, an important feature of successful interactive segmentation is to control human interaction in an organized manner. In this vein, Reynolds (1983) has demonstrated a pilot system in which interaction is reduced to a minimum and the user is employed to assess each attempt made by the program to generate a segmentation scheme in a progressive and ordered manner. Even in this pilot system, the human supervisor is an important element in the process. (For information on unsupervised techniques for exploring the features of s(i,j) the reader is referred to Chapter 7.)

5.1.3 Other Considerations

It is impossible in a single chapter to cover a multitude of segmentation methods and techniques. Therefore, in line with the general purpose of this book, this chapter concentrates on those processes which are readily implemented with the cellular automaton.

Since the existence of likeness or homogeneity within a region implies a change at region boundaries, *edge detection* is a frequently used operation for discerning such boundaries. As pointed out by Kittler et al. (1983) there are both local and global methods for conducting this operation. The cellular automaton is, naturally, the technology of choice for executing local edge detection algorithms. Also, since the cellular automaton may be used for filtering (Chapter 2 and Chapter 7), global methods for edge detection may also be implemented by iterative cellular logic transforms. A major section of this chapter concerns the detection of edges. This section is followed by sections concerning the implementation of *propagation* and a brief account of *post-processing*.

5.2 BOUNDARY DETECTION

As described in Chapter 2, a sampled graylevel image s(i,j) may be converted to a bilevel image by means of thresholding. After that, cellular logic may be used to detect the boundaries between regions of 1-elements and the background of 0-elements. The numerical methods described in Chapter 4 may also be employed directly on the graylevel values of s(i,j) without thresholding. Both of these approaches are reviewed below.

5.2.1 Histogram Analysis for Thresholding

The selection of thresholds is frequently aided by the appearance of modes in the probability density function. An example is furnished in Chapter 9 wherein the graylevel image of a single human white blood cell is segmented readily into nucleus, cytoplasm, and background by thresholding at the minima which appear between the nucleus, cytoplasm, and background modes. Such images will be referred to as "inherently bilevel" or "inherently multilevel."

5.2.2 Segmentation of Bilevel/Multilevel Images

Even when the original image is bilevel, the boundaries of the regions (or objects) themselves may still be corrupted by noise due to, for example, random fluctuations in the image acquisition hardware. Also, this noise might be actual image detail from dust or irrelevant small objects in the original scene being imaged. Furthermore, edges of some objects of interest may be partially occluded or bridged to other objects.

As is discussed in Section 5.3, the segmentation of binary objects may be reduced to a labelling process by means of propagation. However, noise will sometimes join nearby regions so that they acquire the same label. It is, therefore, a sensible precaution to apply some form of noise filtering to the image before performing the propagation labelling. Unfortunately, since such filters are characteristically of a lowpass nature, they tend to eliminate small objects. A knowledge-based choice of the cutoff frequency is, therefore, implied. Thus, it is apparent that there is no image-independent way of eliminating noise reliably without the risk of corrupting the objects retained.

5.2.3 Segmentation of Gray Images

A region in a graylevel image generally will be characterized by having either a mean graylevel different from that of its surroundings or by texture, i.e., statistical properties, which differ from the texture of the surroundings. Segmentation techniques for defining regions, therefore, divide into two broad cases. In the first case, it is assumed that the region will be defined by a boundary or edge in which the graylevel gradient is locally maximal. In the second case, the boundary will again be definable in terms of the slope of a statistical image parameter, but, in this case, the variable will be a measure dependent upon the texture type being evaluated.

5.2.3.1 *Gray Edges*

In case 1 above, the edge of a region is defined by the local graylevel gradient. Segmentation methods in this case are all approximations to local two-dimensional differentiations of the numerical values of the graylevel distribution. Note, however, that the discrete nature of a digitized image prohibits the direct application of continuous differentiation. For a continuous image with intensity given by

$$I = s(x,y) \tag{5.1}$$

the vector whose magnitude is

$$\{(ds/dx)^2 + (ds/dy)^2\}^{\frac{1}{2}} \tag{5.2}$$

and whose direction is

$$\arctan \{(ds/dx)/(dy/ds)\} \tag{5.3}$$

is called the gradient of s(x,y) and is a measure of the magnitude and direction of the steepest rate of change of image intensity at the point x,y.

If both equations (9.2) and (9.3) are applied, the result will be in the form of two new images. The first will have graylevels which are proportional to the magnitude of the local gradient. The second will have graylevel values representing the direction of the maximum local gradient in the original image. For most purposes, results need not be very precise since the exact location and magnitude of the edges is usually not of interest to the overall image segmentation task. Thus, approximations which save computation time are well worth consideration.

Most evaluation of gradients in digital images involve the calculation of weighted summations of the graylevels in local neighborhoods. The weights can be listed as a numerical array in a form corresponding to the neighborhood. Some simple examples follow.

The process of gradient evaluation in digital images which is equivalent to differentiation in continuous images is differencing. For adjacent image elements s(i,j) and s(i-1,j), the difference in the x direction is merely

$$s(i,j) - s(i-1,j) \tag{5.4}$$

and the corresponding expression in the y direction is

$$s(i,j) - s(i,j-1) \tag{5.5}$$

The magnitude of the digital gradient is, therefore

$$\{[s(i,j) - s(i-1,j)]^2 + [s(i,j) - s(i,j-1)]^2\}^{\frac{1}{2}} \tag{5.6}$$

This function decomposes into the following sequence: (1) shift, (2) difference, (3) square and add to accumulator, (4) shift, (5) difference, (6) square and add to accumulator, and (7) square root. This is obviously computationally expensive and it is usually satisfactory to use, instead, either

$$(|s(i,j) - s(i-1,j)| + |s(i,j) - s(i,j-1)|)$$

or $$\max(|s(i,j) - s(i-1,j)|,|s(i,j) - s(i,j-1)|) \tag{5.7}$$

Each of these expressions degenerates into a one-dimensional difference when the maximum gradient direction is parallel to either x direction or y direction. At other angles, the expressions are not exactly equivalent. At worst they may vary by as much as a factor $\sqrt{2}$.

The Laplacian

$$[(d^2s/dx^2) + (d^2s/dy^2)] \quad (5.8)$$

is also used to detect edges. It performs better with points, line ends, and lines than it does with edges. The digital form of the Laplacian derives from the sum of the second differences

$$\begin{aligned}&[(s(i,j+1) - s(i,j)] - [(s(i,j) - s(i,j-1)]\\&+[(s(i+1,j) - s(i,j)] - [(s(i,j) - s(i-1,j)]\\&= s(i,j+1) + s(i,j-1) + s(i+1,j) + s(i-1,j) - 4s(i,j) \quad (5.9)\end{aligned}$$

Note that this is equal to five times the difference between a "blurred image" and the original image, when the "blurred image" is the image formed by averaging each pixel with its four neighbors. (When used in photographic processing, this process is called "unsharp masking.")

Both the digital equivalent of the Laplacian and that of the gradient can be extended to include differences along the diagonals of the neighborhood with the option of reducing the diagonal weights to compensate for the increased distance between the diagonal elements and the center element. This produces a measure less noise sensitive than the direct gradient. This method can be further extended by averaging over larger areas to eliminate the effects of more extensive noise.

The functions discussed above are summarized in Table 5-1 and treated in detail by Rosenfeld and Kak (1976). The results of applying these operators to the same image are shown in Figure 5-1. Note that the intensities in Figure 5-1 have not been square-rooted in the first and final examples.

5.2.3.2 *Multithresholding*

Chapter 2 gives examples of cellular logic filtering using multithresholding. Multithresholding may also be used in edge detection. Binary images are formed by thresholding at several graylevels. The edge images are formed and then ORed together. A contour map results where contours merge where the graylevel gradients are steep. The image is then skeletonized so that the merged edges are thinned to single-line edges. Closed contours are then used to define regions. Sometimes segmentation will be fragmented if slowly varying graylevels are present at a region boundary and some merging of regions may be necessary for object abstraction.

Table 5-1 Summary of Local Digital Edge-Detection Methods

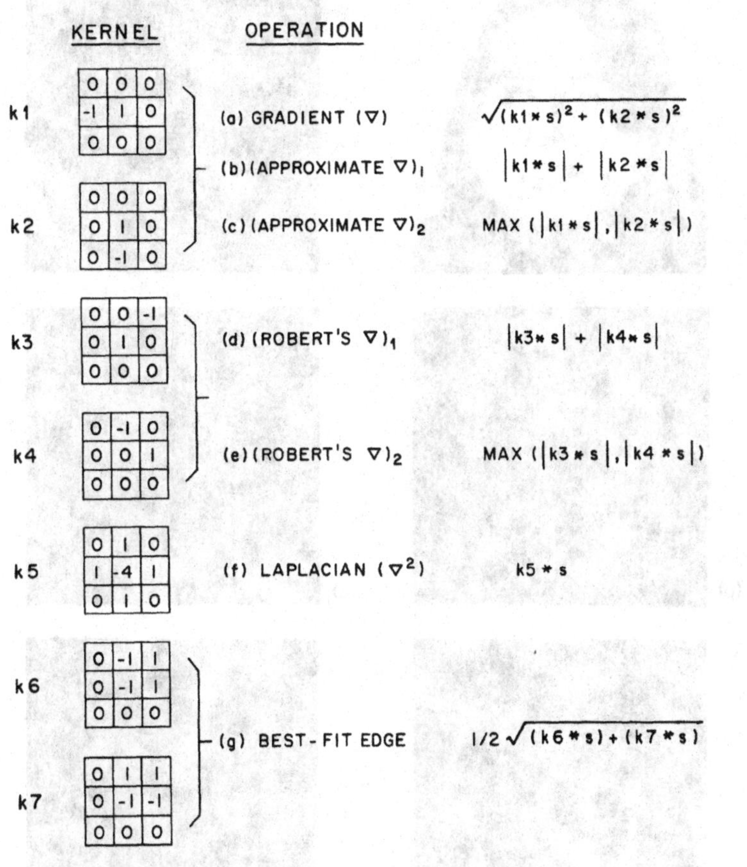

	KERNEL	OPERATION	
k1	0 0 0 / -1 1 0 / 0 0 0	(a) GRADIENT (∇)	$\sqrt{(k1 * s)^2 + (k2 * s)^2}$
		(b) (APPROXIMATE $\nabla)_1$	$\lvert k1 * s \rvert + \lvert k2 * s \rvert$
k2	0 0 0 / 0 1 0 / 0 -1 0	(c) (APPROXIMATE $\nabla)_2$	MAX $(\lvert k1 * s \rvert, \lvert k2 * s \rvert)$
k3	0 0 -1 / 0 1 0 / 0 0 0	(d) (ROBERT'S $\nabla)_1$	$\lvert k3 * s \rvert + \lvert k4 * s \rvert$
k4	0 -1 0 / 0 0 1 / 0 0 0	(e) (ROBERT'S $\nabla)_2$	MAX $(\lvert k3 * s \rvert, \lvert k4 * s \rvert)$
k5	0 1 0 / 1 -4 1 / 0 1 0	(f) LAPLACIAN (∇^2)	$k5 * s$
k6	0 -1 1 / 0 -1 1 / 0 0 0	(g) BEST-FIT EDGE	$1/2 \sqrt{(k6 * s) + (k7 * s)}$
k7	0 1 1 / 0 -1 -1 / 0 0 0		

There is a somewhat analogous process which can be applied directly to the graylevel image by replacing the OR function in the binary multithresholding operation by a maximum function in order to give a gray expansion as explained in Goetcherian (1980). The edge image is then given by

$$I \leftarrow MAX(I) - I \tag{5.10}$$

This causes the central pixel value in a 3×3 neighborhood lying just to one side of a bright edge to be replaced by the highest value of pixels lying in the edge. Subtraction of the original image will then difference the central and edge values.

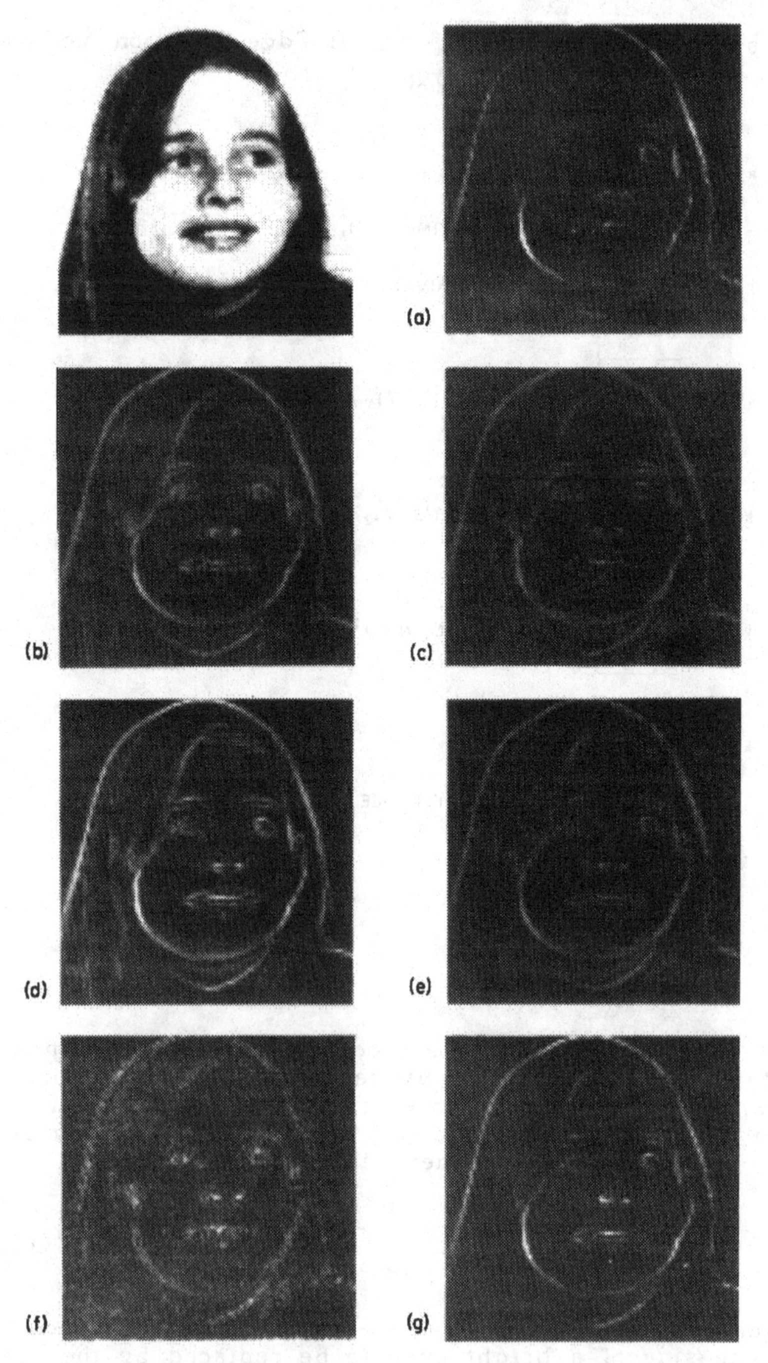

Fig. 5-1 Examples of using the edge detection methods listed in Table 5-1.

Since the maximum function selects values from all the immediate neighbors, the operation will be orientation independent. This process is, therefore, similar to the more familiar selection of the maximum difference between the central pixel and each of its neighbors.

5.2.3.3 *Texture Edges*

Gray edge detection schemes are generally based on averaging processes over regions of a prescribed radius followed by a differencing between the average intensities in adjacent regions. Texture edges are found by applying processes which quantify texture over regions of a prescribed radius and then differencing as before. However, whilst the averaging is only necessary to eliminate noise, and can sometimes be omitted, the quantifying of texture is fundamental to the whole process of texture edge detection. Coarse textures require large regions for their quantification and will, therefore, result in edges which are not well defined.

A short cut to the finding of texture edges can be taken when the textures of either side of the edge reduce to different average graylevels when blurred or spatially filtered. Thus, although the mean graylevel of two textures may be identical, the application of graylevel filtering operations on the two textures will usually produce images of differing mean graylevels. This is particularly true when directional reduction and augmentation operators are used. A general treatment of texture analysis is beyond the scope of this chapter. It has been treated fully by several authors in the literature, particularly Haralick (1973).

5.3 PROPAGATION

At its simplest, segmentation can be defined as the process of dividing up an image into regions. An object in an image may then be represented by one region or by more than one adjacent regions. The next task will often be to label the objects so formed. This task is carried out in the cellular automaton by a process involving *propagation*.

5.3.1 Connectedness

The concept of propagation appears to be fundamental in cellular automata and is strongly related to connectedness in image analysis. Consider a binary image A in which there are groups of 1-elements in a background of 0-elements. Two elements in A, a_1 and a_2, are said to be connected if there exists a path between a_1 and a_2 which is composed entirely of 1-elements in A. A path is a sequence of 1-elements such that all 1-elements in the sequence are local neighbors of 1-elements immediately before and after them in the sequence.

Extending the idea of connectedness, 1-elements in A are said to be connected if, for any pair of 1-elements a_1 and a_2 there exists a path between a_1 and a_2 consisting entirely of 1-elements that themselves belong to A. A region in A is simply connected when there exists a path between all pairs of 0-elements which are adjacent to the region. In other words, background elements adjacent to the region must be connected, implying that the region has no holes in it.

5.3.2 Binary Propagation

In a cellular automaton, a processing element can be programmed to send a propagation signal to all its neighbors if it satisfies a particular combination of the following conditions

1. It receives a propagation signal from one of its own neighbors.
2. It does not receive a propagation signal from one of its own neighbors.
3. It is a 1-element.
4. It is a 0-element.

A further refinement is that directional sensitivity can be added to either or both output or input.

By adopting the conditions (1) and (3) above, a cellular automaton can be used to label all elements in a connected object. Propagation may be initiated at any element in the object. Alternatively, conditions (2) and (4) can be used to flood the background with the propagation signal, thereby differentiating between the background 0-elements and 0-elements which are holes. In the same way, the outer edges of objects can be identified since they are 1-elements receiving, but not transmitting, a propagation signal.

5.3.3 Gray Propagation

Goetcherian (1980) extended the concept of binary propagation to encompass gray propagation, using the maximum and minimum ranking transforms. If s(i,j) is the array in which propagation is to take place and, if a second array p(i,j) originally contains an isolated element from which propagation is to be initiated, then the propagation process can be expressed by the relationship

$$p_n+1 = \min_p(s, \max_{dn}(p_n)) \tag{5.11}$$

where p_k is the graylevel of an element in p at step k of the algorithm, $\max_{dn}(p_n)$ is the deleted neighborhood maximum of p_n (the highest graylevel in the neighborhood of p_n, but not including p_n)

and $\min_p(a,b)$, where a and b are two arrays, is the image whose elements are a or b, whichever is the smaller.

This process causes a propagation from the initial element copying the image s(i,j) so long as the image continues to increase locally in value and provided that the value of the initial element is at least as bright as the corresponding point in s(i,j). The process stops wherever pixels in s(i,j) are encountered which are of a lower graylevel (darker) than the local value of p_n.

Like its binary equivalent, gray propagation is a convenient way to label connected objects, although gray connectedness needs careful redefinition in order to relate to the propagation described.

5.4 EXAMPLE

An example of a comparatively simple segmentation operation is illustrated in Figure 5-2. This sequence shows how the image of the central cell in a tissue section can be abstracted for subsequent analysis and measurement.

The original 64-graylevel image is shown in (a) and the result of thresholding at the trough of the bimodal pixel density histogram is represented by the binary image in (b). Frame (c) shows the result of removing edge-connected objects (sets of 0-elements from which there are paths through other 0-elements to the edges of the picture) using propagation from the edges. In (d), the holes in the remaining objects have been filled by filtering. Frame (e) shows the result of several cycles of reduc-

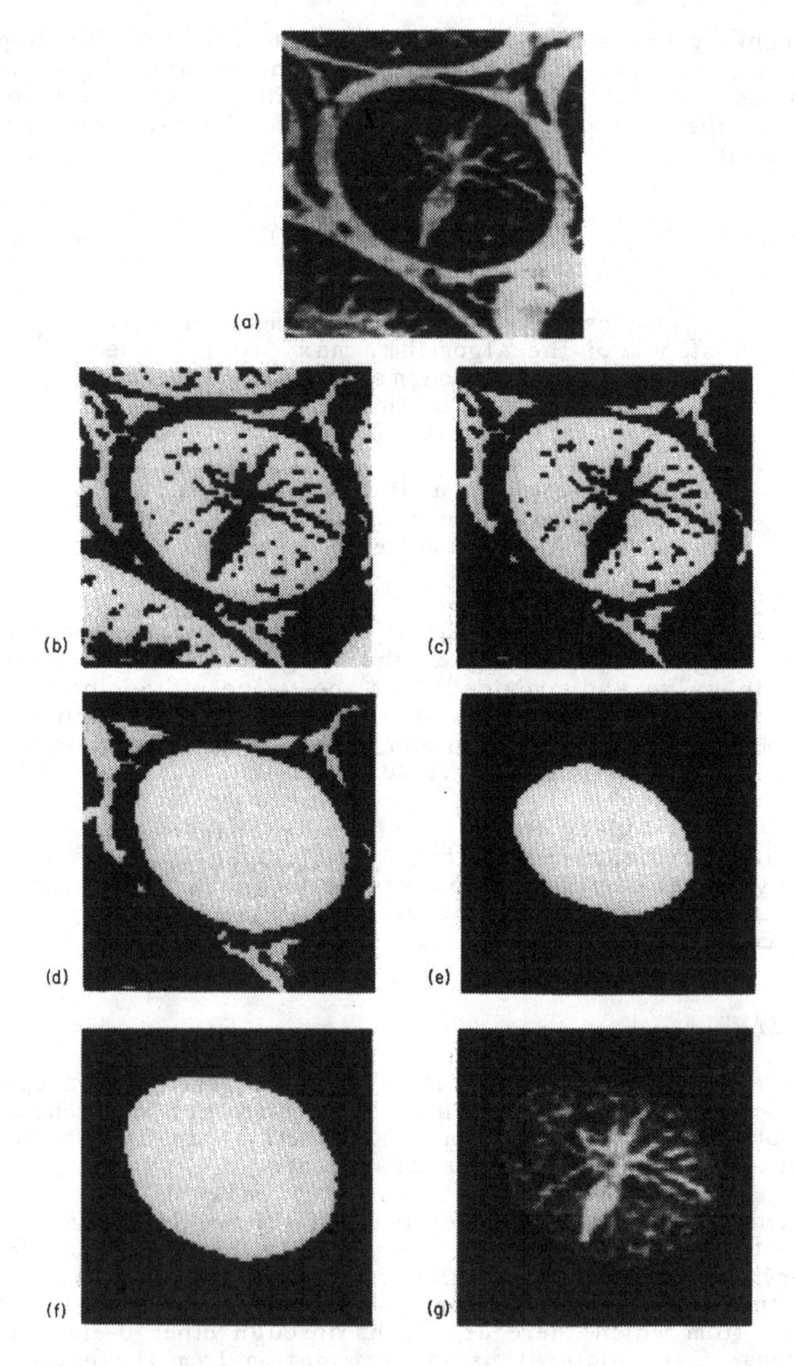

Fig. 5-2 Example of image segmentation by thresholding, propagation, and filtering. See text for details.

tion designed to remove all but the central large object. This object is then used as a propagation source to label the image in (d) producing the result (f). Finally, this resultant mask is ANDed with each of the bit planes in the original image to the graylevel result shown in (g).

This example understates the difficulties normally encountered during segmentation. The successful extraction of the central cell was achieved only because the algorithm used took into account certain important items of knowledge about the input picture, i.e.,

1. The cell to be abstracted does not touch an edge of the picture.

2. It is the largest object in the field of view once edge-connected objects have been removed.

3. It is disconnected from all other tissues.

4. A single density threshold at the histogram trough produces a binary image both satisfying the above conditions and adequately defining the cell shape.

Much more sophistication in the algorithm would be necessary if the tacitly assumed knowledge were neither available nor even applicable.

As stated before, the general task of segmentation is ill-defined. How to identify a "segment" implies some knowledge of the nature of the image and of the way in which it is to be decomposed. On the other hand, there are methods of exposing the data contained in s(i,j) without *a priori* knowledge. These methods are given in Chapter 7.

5.5 POST-PROCESSING

After the application of a boundary finding algorithm, some algorithms will give discontinuous boundaries so that regions are not properly defined. A simple and sometimes satisfactory way of closing boundaries is by expansion of the boundary points followed by reduction or skeletonization. On the other hand, discontinuous boundaries can be taken to indicate "weak" divisions between regions and are, therefore, separating regions which, perhaps, should be merged.

Another effect of augmenting and reducing boundaries

is to eliminate very small regions (i.e., regions whose size is comparable with the discontinuities in the boundaries). The eliminated regions are automatically merged into one or more adjacent regions by this process.

A more parametric approach to region merging can be taken if the regions are labelled with a number representing the value of a parameter characterizing the regions, e.g., the average graylevel. Adjacent regions can then be merged when the gradient of this parameter measured across the boundary separating the regions is less than a selected threshold. This method is particularly effective when applied to regions resulting from the multiple threshold technique described in section 5.4.1, when the retained boundaries will only be those in the high gradient parts of the image.

6. SKELETONIZATION

6.1 INTRODUCTION

Stick-figure representations of objects were fundamental to the cave drawings of primitive man. They are still typical of the first attempts by children when rendering simple pictorial data using pencil and paper. Today we categorize these stick-figure representations as the *skeletons* of objects, or, more strictly, *endoskeletons*, i.e., interior skeletons. There are also *exoskeletons*, i.e., the exterior skeleton of a *collection* of objects, noting that the exoskeleton of a *single* object appears only at infinity. Often the endoskeleton is expressed as the skeleton of the figure (or foreground); the exoskeleton, the skeleton of the ground (background).

The purpose of this chapter is to show how the cellular automaton may be used to compute both the endoskeleton of a single object, the endoskeletons of a collection of objects, or the exoskeleton of a collection of objects. These objects may be presented in a binary data field (either two-dimensional or multi-dimensional) or they may be presented in numerical format, e.g., in the form of a graylevel data field or *image*.

For some time there has been much interest in the formation of the two-dimensional skeleton from both binary and numerical data fields. Recently, due to the stimulus of research in three-dimensional imagery and in time-varying two-dimensional imagery, studies in three-dimensional skeletonization have intensified. In the sections which follow, skeletonizing from binary data fields is first discussed with a description of both two-dimensional and three-dimensional methods. This discussion is followed by sections on two-dimensional graylevel skeletonization.

6.1.1 Early Work in Skeletonization

As with much of the early effort in pattern recognition, the first work in skeletonization of two-dimensional binary fields occurred in the area of applied character recognition. This work is thoroughly reviewed in the first chapter of the book on pattern recognition by Ullmann (1973). Ullmann notes that the first publication on skeletonization is that of Sherman (1960) who developed, at that time, a procedure for thinning the images of individual characters to their endoskeletons. Sherman's method was sequential in that the result of the computation on one element of the binary data field was used in the computation of the value of the next element. Thus Sherman's work is primarily of historical interest in terms of the material covered by this book which concentrates on parallel computation as conducted by the cellular automaton. Sherman used a 3×3 kernel in the Cartesian coordinate system which was scanned across the binary data field. At each position of the kernel, the binary value of the central element was examined. If the value of this element were 0, it would remain unchanged; if 1, it would be changed to 0, unless it was the only 1-element connecting two nonadjacent 1-elements in the neighborhood or if the neighborhood contained only a single 1-element.

Coincident with the work of Sherman (1960), Shelton (1967) filed a patent in 1962 (assigned to IBM) showing how a numerical distance transform of a binary figure could be obtained whose local maxima defined a skeleton. Shelton's idea was later enlarged by Blum (1967) who christened Shelton's transform the "medial axis transform." Although the medial axis transform does not guarantee a connected skeleton, it is significant in being the first truly parallel transform for skeletonization. Later parallel binary methods for skeletonization were developed and are described in Section 6.2.

Other work in the early 1960's, e.g., by Weston (1961), Preston (1961), Golay (1965), also employed sequential skeletonizing algorithms. This fact, however, did not prevent the successful utilization of skeletonizing, especially in the field of character recognition. For example, when an Arabic numeral is skeletonized, it may be relatively easily recognized by measurements made on the graph which is its endoskeleton. Ullmann (1973) provides an example by considering the numerals 6 and 9. In both cases, the graph which comprises the endoskeleton contains a single node of order three and a single end-element. However, in the 6, the end-element is physically located above the node and the node is to the left of the centerline, whereas in the numeral 9, the end-element is physically located below the node and the node is to the right of the centerline. (The numeral 4 is not considered to contain a node of order three because, in most machine designs, the numeral 4 is crossed so that it contains a node of order four as well as either two or more end-elements.)

6.1.2 Later Work on Skeletonization

With the derivation of the Golay hexagonal transform (Golay, 1969) and its reduction to practice in the GLOPR cellular logic machine (Preston, 1971), studies of parallel algorithms for binary skeletonization using subfields in the hexagonal tessellation commenced. A few years later, Golay's basic patent on the hexagonal parallel transform issued (Golay, 1977), assigned to the Perkin-Elmer Corporation. In the same interval, the first extensive theoretical study of connectivity, an essential ingredient in skeletonization, was performed by Rosenfeld (1970) who demonstrated that, unless Golay's subfield method is used, a 5×5 operator is required in order to assure the maintenance of connectivity relationships during skeletonization. This provided the theoretical justification for the use of subfields when employing the 3×3 kernel and led other workers to investigate skeletonization algorithms using various alternative sequences of both 2×2 and 3×3 subarray masks. This work was reported in papers by Levialdi (1972) and by Arcelli, Cordella, and Levialdi (1975).

The extension of this work to studies of three-dimensional skeletonization is a still more recent effort in the field. Although the initial study was done by Arcelli and Levialdi (1973) in the early 1970's, major works in this area were not completed until the 1980's and are those of Srihari (1979), Lobregt, Verbeek, and Groen (1980), Tsao and Fu (1982), and Hafford and Preston (1983). Details are provided below.

6.2 TWO-DIMENSIONAL BINARY METHODS

This section describes and gives examples of parallel algorithms for two-dimensional skeletonization as applied to binary data fields. Three-dimensional skeletonization is discussed in Section 6.3 and methods for the skeletonization of two-dimensional graylevel images are presented in Section 6.4. In the mid-1960's Golay, recognizing the advantages of the hexagonal tessellation as well as understanding the impossibility of performing skeletonization in parallel with a kernel composed of nearest neighbors, invented the idea of subfields (Chapter 2).

Golay's concept was reduced to practice in the GLOPR computer built by Preston (1971) and his co-workers at Perkin-Elmer. Successful skeletonization in the hexagonal tessellation was performed using this instrument in 1968. At approximately the same time Levialdi (1972), working at the Laboratory for Cybernetics in Naples, was investigating reduction algorithms using a 2×2 kernel in the square tessellation.

6.2.1 Connectivity in Two Dimensions

In order to skeletonize a connected region, it is clear that connectivity must be maintained during the skeletonizing process. This can be done by maintaining the local connectivity within the cellular neighborhood. In the Cartesian tessellation, four of the eight neighbors are at unit distance from the central element, while the remaining four are at a distance of $\sqrt{2}$. Based on this difference, there are two degrees of connectivity denoted by "d1-connectedness" and "d2-connectedness." The neighbors exhibiting d1-connectedness are given by the set S'

$$S' = \{(i,j) \mid i \neq j;\ \max(|x_i - x_j| + |y_i - y_j|) \leq 1\} \tag{6.1}$$

while members of the set S" exhibiting d2-connectedness are given by

$$S'' = \{(i,j) \mid i \neq j;\ \max(|x_i - x_j| + |y_i - y_j|) \leq 2\} \tag{6.2}$$

The elements of the set S' are sometimes known as the 4-neighbors and those elements in the set S" as the 8-neighbors. In the hexagonal tessellation, all neighbors are equally distant from the central element so that no distinction is drawn between degrees of connectedness. Further problems arise in the Cartesian tessellation when it is recognized that four of the eight near neighbors connect to the central element at a point which is an infinitesimal. Thus, if the configuration consists of a cross within the 3×3 kernel and it is assumed that the arms of the cross connect to the central element then, when the cross is complemented, the arms of the cross become disconnected. This paradox is sometimes solved by using d1-connectedness for the direct values of the elements in a binary array and d2-connectedness for their complement. An alternate and often more attractive solution is to use the hexagonal tessellation, thus avoiding the connectivity paradox entirely. Section 6.2.3 shows examples of skeletonization in both the Cartesian and the hexagonal tessellations.

6.2.2 Mask Methods

Levialdi used the 2×2 subarray masks shown in Figure 6-1. He solved the fundamental problem that the 2×2 kernel has no central element as follows. Using the Levialdi masks I, the binary result was placed in the upper left (NW) corner of the kernel. Similarly, for the Levialdi masks II, the result was placed in the Southwest (SW) corner; for masks III, SE; masks IV, NE. Conceptually each set of masks was applied in parallel to the binary values of the data contained in all 2×2 subarrays in the two-dimensional binary data field. Levialdi's technique permitted binary 0-elements to be changed to 1-elements which had the rather peculiar effect that a line with a slope of $+45^{\circ}$, when using the Levialdi masks I, translated in the $+135^{\circ}$ direction (towards the Northwest) until reduced to a residue, while a line at -45° would be reduced to a residue at its Northwest end-element. As express by Levialdi (1972),

> "..., this algorithm has the property of being direction oriented, i.e., every object is compressed into one of the four corners of the rectangle circumscribing it."

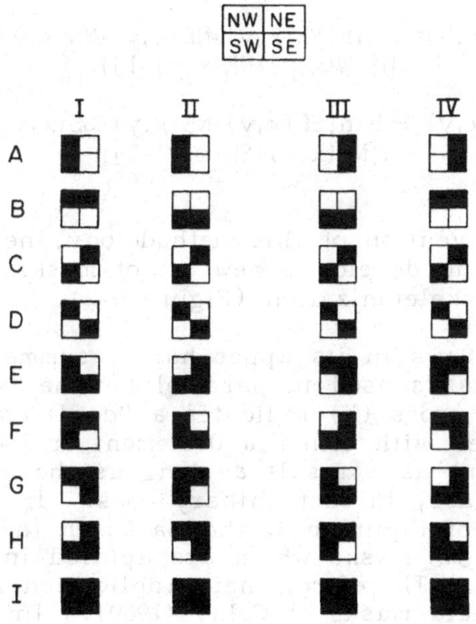

NW NE
SW SE

I II III IV

A
B
C
D
E
F
G
H
I

Fig. 6-1 Levialdi's masks for use in thinning binary images while preserving connectivity. When a local match occurs, the result is output as a 1-element.

Levialdi discovered that his operations could be mathematically expressed in terms of the Heaviside function h(t) denoted by

$$h(t) = \begin{cases} 0 & t \leq 0 \\ 1 & t > 0 \end{cases} \tag{6.3}$$

Using this notation the result of the parallel application of the Levialdi I masks may be expressed by

$$\text{I:} \quad r_L(x,y) = h\{h[SW(x,y)+NW(x,y)+NE(x,y)-1]+ h[NW(x,y)+SE(x,y)-1]\} \tag{6.4}$$

where NW(x,y) is the value of the 1-element in the Northwest position; SW(x,y), Southwest; NE(x,y), Northeast; and SE(x,y), Southeast, and the result is placed in the NW position. Similarly, for the Levialdi masks II,III, and IV

$$\text{II:} \quad r_L(x,y) = h\{h[SW(x,y)+NW(x,y)+SE(x,y)-1]+ h[SW(x,y)+NE(x,y)-1]\}$$

$$\text{III:} \quad r_L(x,y) = h\{h[SE(x,y)+NE(x,y)+NW(x,y)-1]+ h[SW(x,y)+NE(x,y)-1]\}$$

$$\text{IV:} \quad r_L(x,y) = h\{h[SE(x,y)+NE(x,y)+SE(x,y)-1]+ h[NW(x,y)+SE(x,y)-1]\} \tag{6.5}$$

After Levialdi's invention of this methodology, he and his co-workers proceeded to develop a new set of masks which were more suitable for skeletonization (Figure 6-2).

Figure 6-2 shows in its upper half a summary of the I, II, III, and IV masks used in parallel for the 2×2 subarray. In this figure the cross (X) indicates a "don't care" position which may be filled with either a 0-element or 1-element while yielding a 1-element as a result as long as the elements which are shaded black are, in fact, binary ones. In distinction to the upper portion of Figure 6-2, the masks in the lower portion of Figure 6-2 are 3×3 masks which are applied in parallel but not simultaneously. Therefore, their application has some similarity to the subfield masks of Golay (1969). In the lower portion of Figure 6-2, the cross (X) indicates a "don't care" value. The elements shaded black must be 1-elements and the unshaded elements must be 0-elements for the result to be a 1-element.

Since the mask is 3×3 the result is placed in position of the central element. The order in which these masks are applied is: A1, B1, A2, B2, A3, B3, A4, B4. The development of these masks and their application to the thinning of letters of the alphabet were first reported by Arcelli, Cordella, and Levialdi (1975). See Figure 6-3.

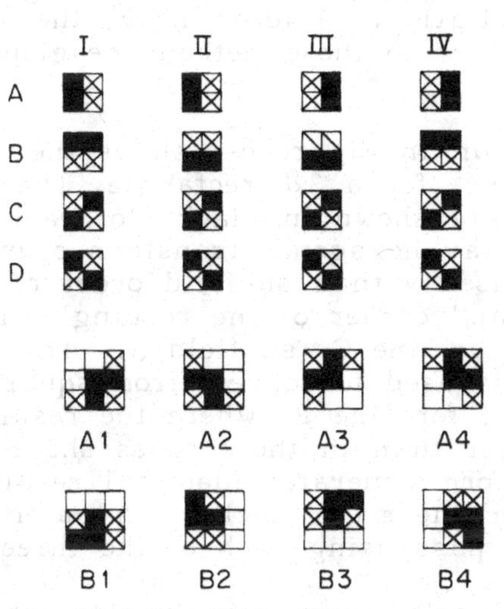

Fig. 6-2 (Upper) summary of the 2×2 masks shown in Figure 6-1. (Lower) summary of the 3×3 masks used in sequence (see text) to obtain the skeletons shown in Figure 6-3.

Fig. 6-3 Skeletons of the letters of the alphabet produced by Arcelli, Cordella, and Levialdi (1975).

6.2.3 Comparison of Masks With Subfields

Figure 6-4 compares the action of the Golay (1969) approach to skeletonization using subfields of three in the hexagonal tessellation with that of Arcelli, Cordella, and Levialdi (1975) using their eight masks (Figure 6-2) in the square tessellation and with that of Preston (1979) using subfields of four in the square tessellation. Interestingly, the method of Golay appears to be superior to those methods developed a full decade later.

Each line shown in Figure 6-4 shows the step-by-step skeletonizing sequence for a 3×8 rectangle. The elements of the final endoskeleton are shown in black (for each step). In lines A through C the Golay hexagonal transform operating in subfields of three is used with a subfield order of 1-3-2. On line A the upper left hand corner of the rectangle lies in subfield 1; line B, subfield 2; line C, subfield 3. The S1 template of Smith (Chapter 2) is used to convert from square to hexagonal coordinates. Except for line B, where the resulting skeleton is one element longer than in those cases shown in lines A and C, the Golay transform generates identical results and produces a final output after one major cycle. (A major cycle is one which includes one pass using each of the three subfields.)

Figure 6-4, line D, shows the skeletonization of the same rectangle using the eight masks of Arcelli, Cordella, and Levialdi. More than twice the number of steps are required using their method to produce the same results as are produced with Golay's method. Finally, lines E and F illustrate skeletonization using subfields of four employing the method of Preston (1979) where line E has the upper left hand corner of the rectangle in subfield 1 and line F has the upper left hand corner of the rectangle in subfield 4. In all cases the subfield order is 1-3-4-2 according to the regime employed in the cellular logic programming language SUPRPIC (Preston et al. 1979). Preston's method produces more variable results than do the earlier methods in that the number of minor cycles required to attain the skeleton depends strongly on the positioning of the rectangle with respect to the subfield lattice. In fact, in line F it is seen that "noise spurs" (Davies, 1981) are produced in the fourth minor cycle. These extend from both the top and bottom of the skeleton as L-shaped protrusions. The elements forming these noise spurs both lie in subfield 4 so that three additional minor cycles would be required before they were removed thus making the entire operation almost as long as that shown in line D. It therefore appears that the hexagonal parallel transform of Golay operating in subfields of three provides not only the most rapid but also the most satisfactory method for skeletonization.

As a final example of the use of skeletonization in the encoding of characters, Figure 6-5 shows the results of encoding Japanese text for minimum entropy facsimile transmission followed by reconstruction by augmentation (courtesy of the Nippon Electric Company, Ltd.).

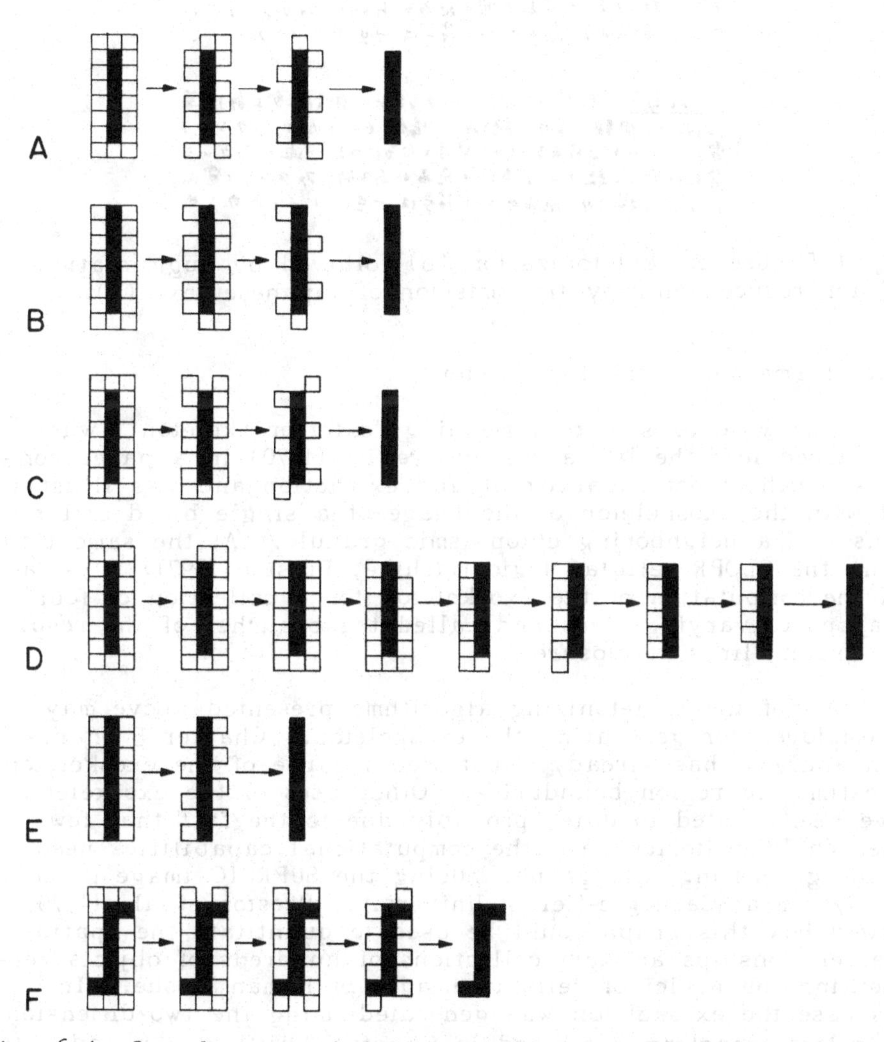

Fig. 6-4 Step-by-step skeletonization using the hexagonal parallel transform of Golay (A–C), the 3×3 masks of Arcelli, Cordella, and Levialdi (D), and the Cartesian parallel transform of Preston (E–F).

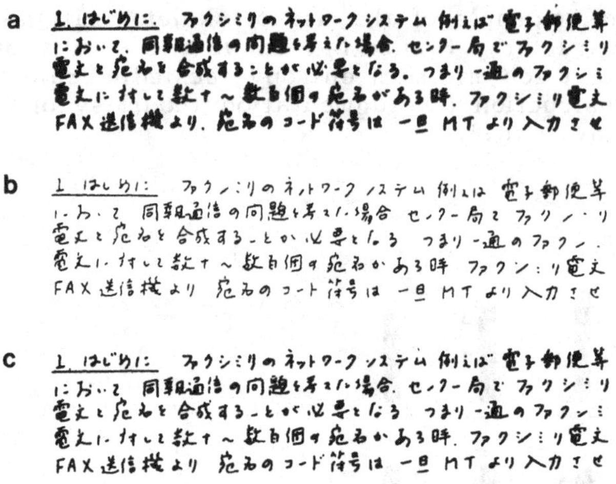

a 1. はじめに ファクシミリのネットワークシステム 例えば電子郵便等において、同報通信の問題を考えた場合、センター局でファクシミリ電文と宛名を合成することが必要となる。つまり一通のファクシミリ電文に対して数十〜数百個の宛名がある時、ファクシミリ電文FAX送信機より、宛名のコード符号は一旦MTより入力させ

b 1. はじめに ファクシミリのネットワークシステム 例えば電子郵便等において 同報通信の問題を考えた場合 センター局でファクシミリ電文と宛名を合成することが必要となる つまり一通のファクシミリ電文に対して数十〜数百個の宛名がある時 ファクシミリ電文FAX送信機より 宛名のコード符号は一旦MTより入力させ

c 1. はじめに ファクシミリのネットワークシステム 例えば電子郵便等において 同報通信の問題を考えた場合、センター局でファクシミリ電文と宛名を合成することが必要となる つまり一通のファクシミリ電文に対して数十〜数百個の宛名がある時、ファクシミリ電文FAX送信機より 宛名のコード符号は一旦MTより入力させ

Fig. 6-5 Use of skeletonization (b) followed by augmentation (c) for reduced entropy transmission of Japanese text (a).

6.2.4 Formation of the Exoskeleton

The word *exoskeleton*, meaning "exterior skeleton," was introduced into the literature by Prewitt (1970) in a paper concerned with object enhancement and extraction and was illustrated with the exoskeleton of the image of a single blood cell nucleus and a neighboring cytoplasmic granule. At the same time, using the GLOPR cellular logic machine, Preston (1971) illustrated the computation of the exoskeleton for a collection of four octagons of varying sizes and called the branches of the resulting graph "lines of closure."

All of the skeletonizing algorithms presented above may be employed for generating the exoskeleton. Chapter 5 on region analysis has already illustrated the use of the exoskeleton in estimating region boundaries. Other uses of the exoskeleton have been limited to date, probably due to the fact that few research laboratories have the computational capabilities needed for generating this graph. Using the SUPRPIC image processing system at Carnegie-Mellon University, Preston et al. (1979) showed how this graph could be used to quantitate the spatial interrelationships amongst collections of hundreds of objects representing the nuclei of cells in images of human tissue. In this case the exoskeleton was generated using the two-dimensional logical transform described in Chapter 2 with $\Xi = 6$ and a crossing number of four ($X = 4$) using subfields of four with the subfield order 1-3-4-2. Examples, taken from the later work of Preston (1981), are shown in Figure 6-6.

Figure 6-6A shows a field of residues and the corresponding exoskeleton. Where the branches of the exoskeleton join to form nodes, the crossing number may be either 6 or 8 in the square tessellation. Several measures may be made from the exoskeleton which include a count of the number of nodes of different order, a histogram of the sizes of the tiles (polygons) defined by the branches of the exoskeleton, or a histogram of the length of the exoskeleton branches. Figure 6-6B shows how this histogram is formed by isolating the exoskeleton branches by deleting 1-elements at the nodes (except for nodes at branches where they touch the borders of the field.) The branches then may then be reduced to residues, and, in the process, the residue histogram (Chapter 7) is generated. Figure 6-6D shows the specific set of histogram branches which fall in the fourth bin of the particular histogram generated for use in the analysis of the lengths of the exoskeleton branches shown in Figure 6-6C.

In order to better understand the usefulness of the exoskeleton branch-length histogram in the analysis of topology, consider the N×N checkerboard as an example of a data field containing a completely regular topology. The branches of the graph which is the exoskeleton of the residues of the checkerboard squares are identical to the borders of the checkerboard squares themselves. There are 2N branches per residue and each polygon defined by the exoskeleton branches is a square. The branches of the exoskeleton are all identical in length so that the entropy of the histogram is zero. For a data field containing objects or residues of objects in an irregular grouping, it is found that the number of exoskeleton branches per polygon increase (Figure 6-6) and the entropy of the histogram of the exoskeleton branch lengths increases proportionally with the degree of irregularity. Thus exoskeleton measurements may be used to determine both the complexity and the variability of the structural organization of collections of objects in the plane. This general field of analysis is related to the polygons of Voronoi (1908) and is reviewed in an article by Ahuja (1982).

6.3 THREE-DIMENSIONAL SKELETONIZATION

Only a few research centers in the world are currently working on skeletonization in spaces of greater than two dimensions. Research is concentrated at the University of Maryland, Carnegie-Mellon University, Purdue University, the State University of New York at Buffalo, the University of Delft, the Yokohashi University. Some of the research at these universities is represented in papers by Srihari (1979) Morganthaler and Rosenfeld (1980), Tsao and Fu (1982), Lobregt, Verbeek, and Groen (1980), Toriwaki (1982), and Hafford and Preston (1983) respectively. The first publication in this field appears to be that

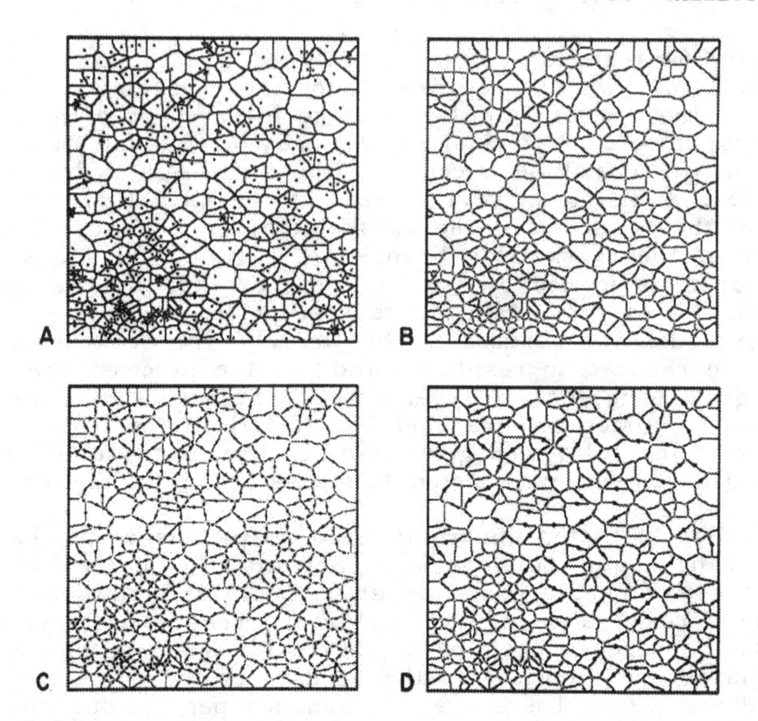

Fig. 6-6 Examples of exoskeleton analysis showing (A) exoskeleton formed from residues, (B) nodes removed, (C) residues of the exoskeleton branches, (D) branches in the fourth bin of the exoskeleton branch histogram ORed with their residues.

of Arcelli and Levialdi (1973) who developed a method of object counting by reduction to residues in three dimensions. This work is an extension of the earlier work of Levialdi et al., in two dimensions cited in Section 6.2.2.

6.3.1 Connectivity in Three Dimensions

In addition to d1-connectedness and d2-connectedness as defined by equations (6.1) and (6.2) above, the kernel of the Cartesian tessellation in three-dimensions encompasses three separate sets having d1-, d2-, and d3-connectedness. These three sets may be represented by the general equation

$$S^n = \{(i,j,k) \mid i \neq j \neq k;\ \max(|x_i - x_j| + |y_i - y_j| + |z_i - z_j| \leq n\} \tag{6.6}$$

When n = 1, we have the three-dimensional 8-neighborhood; n= 2, the 18-neighborhood; n = 3, the 26-neighborhood. Once again there is a connectivity paradox, but only for the 26-neighborhood. This complication may be avoided by using the hexahedral tessellation and the tetradecahedral neighborhood in which all elements are equidistant from the central element (Chapter 3).

Skeletonization of binary data fields is more difficult in three dimensions since, as discussed in Chapter 3, there is no longer any meaning to the concept of *crossing number*. Instead, the skeletonizing program must examine the effect of the removal of each individual 1-element on the connectivity of the region being skeletonized. This may be done by computing the connectivity number CN which, for a three-dimensional solid which is not penetrated by tunnels and does not contain cavities, is given by

$$CN = F + E - N = 2 \tag{6.7}$$

where F is the number of faces, E is the number of edges, and N is the number of nodes of the closed netted surface of the solid. Closed netted surfaces for the tetradecahedron, the 2×2×2 cube and the 3×3×3 cube are shown in Figure 6-7. As presented in Chapter 3, the tetradecahedron has 14 faces, 24 edges, and 12 nodes (vertices). Values for F, E, and N are shown for the 2×2×2 cube and the 3×3×3 cubes in Figure 6-7. In all cases CN = 2. In general, as was proven in the last century by Euler and Poincare (see Hilbert and Cohn-Vossen, 1932) three-dimensional, closed, netted surfaces of solid objects have a connectivity number of two. If a cavity is added within such an object, its connectivity number increases by two for each cavity; if a hole or tunnel exists through the object, its connectivity number is decreased by two. By preserving the connectivity number during the skeletonization operation, it is assured that the original topology of the object is maintained.

6.3.2 Present Status

In the work of Arcelli and Levialdi (1973), a 2×2×2 kernel was employed with a parallel reduction algorithm which may be expressed in terms of the Heaviside function given in equation (6.3). The mathematical expression for one of these reduction algorithms is as follows

$$\begin{aligned} r_L(x,y) = h[&h(BNW+BNE+FSW-2)+h(BNE+FNE+BSW-2)+h(BNW+FNE+BSE-2) \\ &+h(BNE+FSW-2)+h(BNE+BSE+FNW-2)] \end{aligned} \tag{6.8}$$

where BNE, BNW, BSW, BSE represent the NW, NE, SW, and SE corners of the *back* plane of the 2×2×2 cube, while FNW, FNE, FSW, FSE represent the corresponding elements in the *front* plane of the cube. Since there is no central element in the kernel, the result must be placed in one of the eight corner positions of the cube. For the algorithm given in equation (6.9) the result is placed in the back plane of the cube in the NE position. There are, of course, seven similar equations which the user must use if it is desired to place the result in one of the other corner positions of the 2×2×2 cube.

Lobregt et al. (1980) extended the work of Arcelli and Levialdi to the 3×3×3 cube. They decomposed this cube into eight partially overlapping 2×2×2 cubes. They developed an algorithm which conserved the connectivity number (CN) of the 3×3×3 cube by determining the contribution to CN of each 2×2×2 cube. These contributions are tabulated for all twenty-two orientation-independent configurations of the 2×2×2 cube in Table 6-1. By addressing this table eight times for each 3×3×3 cube and summing, the total contribution to the connectivity number may be determined. In this manner, a decision may be made as to whether to remove a particular voxel. This may be done for all voxels in parallel. Later Srihari and Srisuresh (1983) developed a connectivity preserving algorithm combining those of Lobregt et al. (1980) and Toriwaki (1982). Figure 6-8 gives an example.

A parallel algorithm requiring only one table lookup operation per voxel was developed by Hafford and Preston (1983) using the hexahedral tessellation introduced in Chapter 3. An

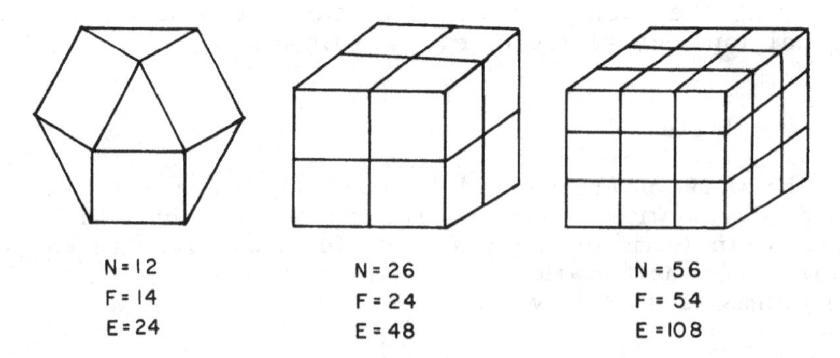

Fig. 6-7 The tetradecahedron, 2×2×2 cube, and 3×3×3 cube with values given for the number of nodes (N), faces (F), and edges (E.)

Table 6-1 Contribútion to Three-Dimensional Connectivity (n_i) of the 3×3×3 Cube by all Rotationally Symmetric 2×2×2 Sub-Cubes

BSE	BSW	BNE	BNW	FSE	FSW	FNE	FNW	n_i
0	0	0	0	0	0	0	1	1/4
0	0	0	0	0	0	1	1	0
0	0	0	0	1	0	0	1	-1/2
1	0	0	0	0	0	0	1	-3/2
0	0	0	0	0	1	1	1	-1/4
0	1	0	0	0	0	1	1	-3/4
0	0	0	1	0	1	0	0	-1/4
0	0	0	0	1	1	1	1	0
0	0	1	0	0	1	1	1	-1/2
0	0	0	1	0	1	1	1	-1/2
1	0	0	0	0	1	1	1	0
1	1	0	0	0	0	1	1	0
0	1	1	0	1	0	0	1	1
1	1	1	0	1	0	0	1	-3/4
1	0	1	1	1	1	0	0	1/4
1	1	1	1	1	0	0	0	-1/4
0	1	1	1	1	1	1	0	1/2
1	1	1	1	0	1	1	0	1/2
1	1	1	1	1	1	0	0	0
1	1	1	1	1	1	1	0	1/4
1	1	1	1	1	1	1	1	0

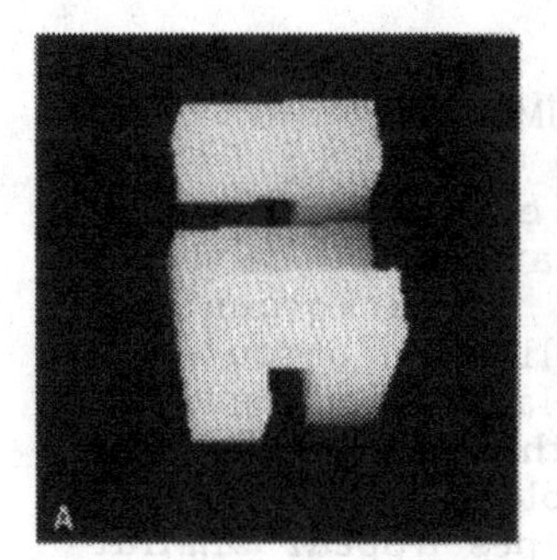

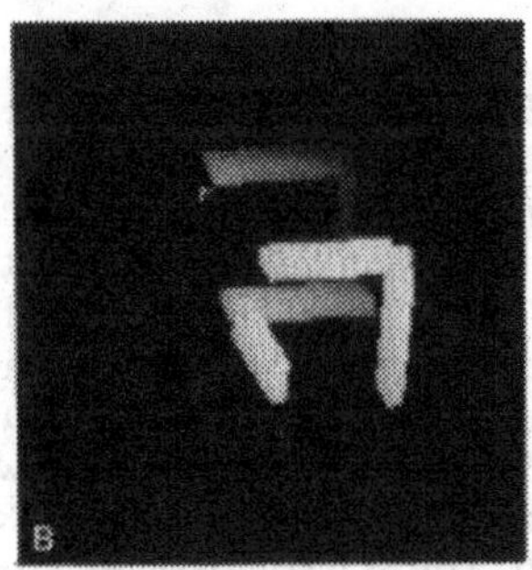

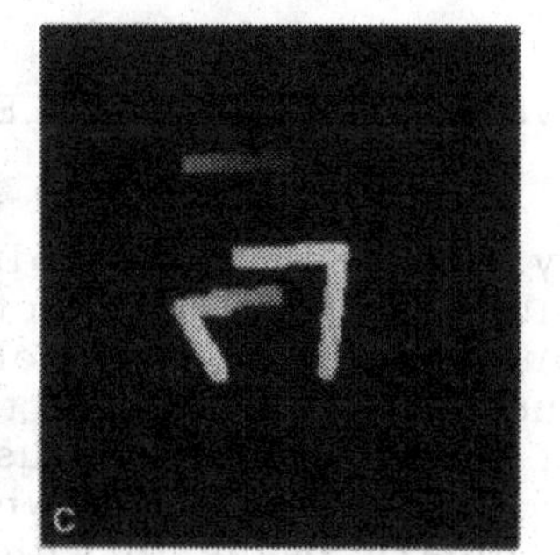

Fig. 6-8 Example of three-dimensional skeletonization due to Srihari (1979).

elaborate procedure, using a hierarchy of Ξ values starting at $\Xi = 6$ and progressing upward, if necessary, to $\Xi = 8$ is employed with $X_0 = X_1 = 4$. A flow chart is given in Figure 6-9. This hierarchical procedure is necessary due to a deficiency of the tetradecahedral kernel. Although the kernel is complete in the sense that all straight lines passing through the central element have $X_1 = 4$ and $X_0 = 2$, it is incomplete in terms of planes passing through the central element which lie along the cubic directions of the hexahedral tessellation. There are three of these planes through the central element as well as the four hexagonal planes. A 1-element lying in a cubic plane of 1-elements has eight neighbors in that plane, but only four of these are members of the set of elements defined by the kernel. Examination of the geometry shows that these are the 4-neighbors in the cubic plane so that $X_1 = 8$. Thus, the algorithm mistakes such 1-elements for elements which lie at a nodal point in the skeleton. This deficiency is overcome by the algorithm charted in Figure 6-9. The algorithm tests periodically for voxels which exhibit high values of the crossing number. This is followed by selective augmentation of those areas in the binary field where such voxels exist. Using this method, satisfactory skeletons have been produced as exemplified by those shown in Figure 6-10 (see also Chapter 9).

A final example of three-dimensional skeletonization is given from the work of Tsao and Fu (1982). Figure 6-11 shows three-dimensional objects in the form of block letters of the Roman alphabet skeletonized using various algorithms. The algorithms are not given in detail by the authors, but they are connectivity-preserving and are based upon the use of six subfields in the Cartesian tessellation. These authors must also employ what they call "checking planes" because six subfields are only a subset of the eight which exist in this tessellation.

6.4 GRAYLEVEL SKELETONIZATION ALGORITHMS

Not every object in a graylevel image can be adequately represented by a binary image. Segmentation may be possible so that the object can be lifted out of its background, but subsequent thresholding at any level will often obscure the detailed shape of the object. A particularly good example of this problem is illustrated in Fig. 6-12 which shows images of some chromosomes which cannot be thresholded so as to yield a binary image in which the chromosome shapes appear similar to the "gray shapes" seen in the original images. The problem here is to devise techniques in which all the shape information in the graylevel image is used in some way to compute a skeleton. Two alternative approaches are described in the following sections.

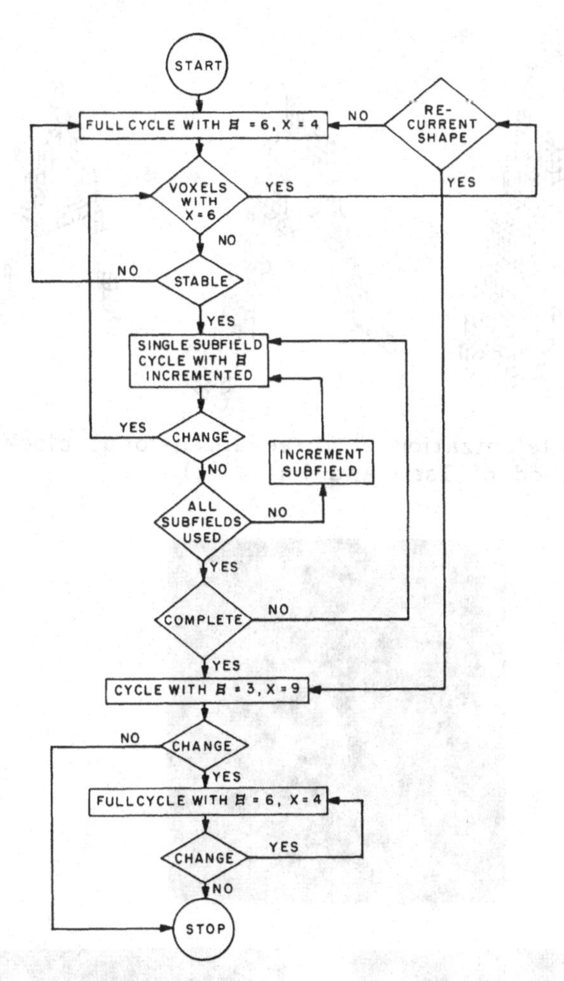

Fig. 6-9 Flow chart of the algorithm used by Hafford and Preston (1983) for three-dimensional skeletonization in the tetradecahedral tessellation.

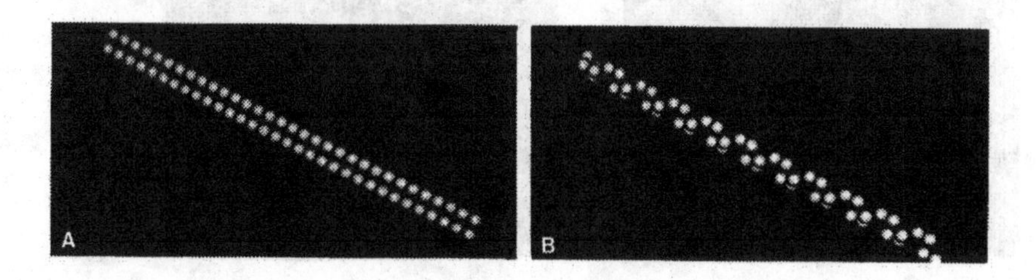

Fig. 6-10 Example of skeletonization of a double strand of voxels using the algorithm charted in Fig. 6-9.

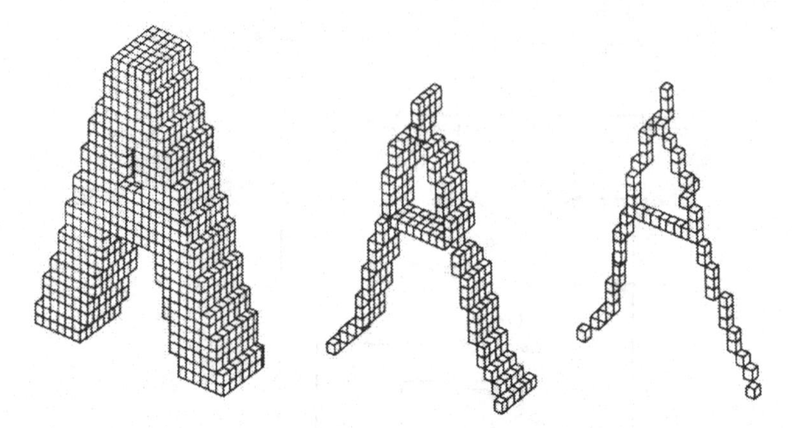

Fig. 6-11 Skeletonization of three-dimensional block letters using the method of Tsao and Fu (1982).

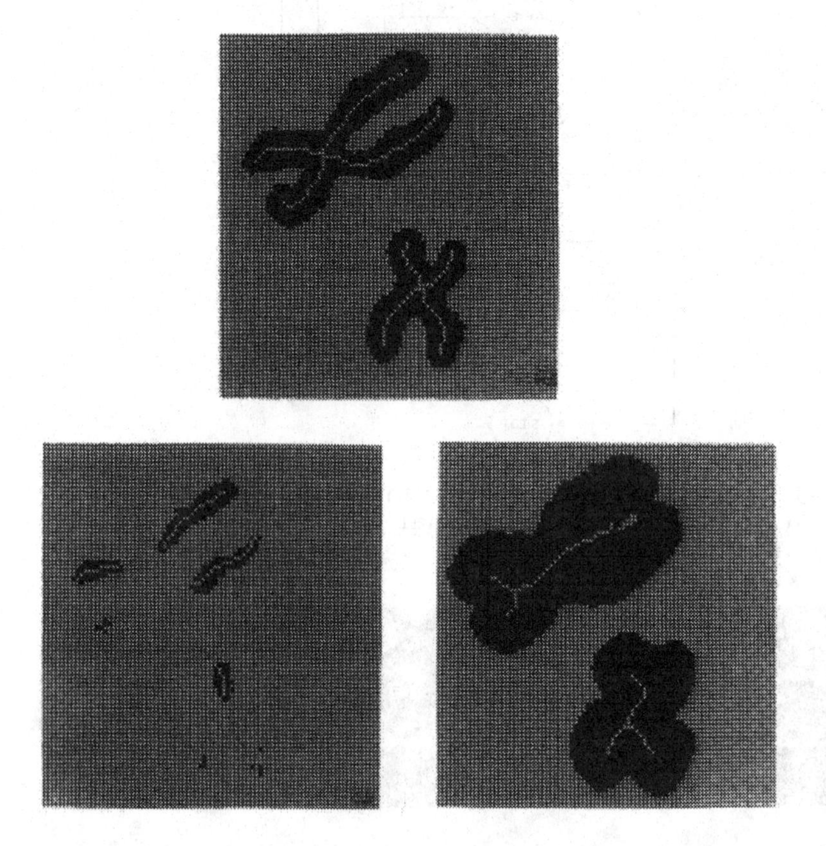

Fig. 6-12 Chromosome images used to illustrate the problems inherent in strictly binary skeletonization.

6.4.1 Min/Max Skeletonization

Goetcherian (1980) has shown that many algorithms developed for binary images can be adapted for use with graylevel images by substituting MIN and MAX functions for AND and OR functions and by treating the logical inverse of a binary image as being equivalent in graylevel images to one minus the image. Applying this technique to the skeletonization algorithm due to Arcelli et al. (1975), which is described in section 6.2.2, a procedure was evolved which, for the first mask, translates as follows:

(1) Extract from the image I0 the 1-elements having three 0-elements in positions 1, 2 and 8 (numbering clockwise around the 3x3 neighborhood). Store the result in I1. This is equivalent to: Extract from the image I0 the elements having three neighbors less than themselves (in gray value) in positions 1, 2 and 8. Store the result in I1.

(2) Extract from I0 the 1-elements having two 1-elements in positions 4 and 6. Store the result in I2. This is equivalent to: Extract from I0 the elements having two elements equal to or greater than themselves in positions 4 and 6. Store the result in I2.

(3) Perform the logical AND operation between the contents of I1 and I2. Subtract the result from I0. This is equivalent to: Replace the gray value of those pixels in I0 for which both the above conditions are satisfied, by the maximum of the values of the neighbors in positions 1, 2 and 8.

In this description, which follows Goetcherian's convention, it is assumed that dark pixels are represented by high graylevel values and that the object to be skeletonized is dark in a lighter background. Equivalent results can be obtained with these conventions of image brightness reversed. The results of applying the complete algorithm to the chromosomes shown in Fig. 6-12 is shown in Fig. 6-13. Note that this process produces a "gray" skeleton, i.e. one in which elements of the skeleton are not binary valued but have graylevels distributed in some relationship to the original image. Broadly speaking, dark objects produce dark skeletons, and vice versa,

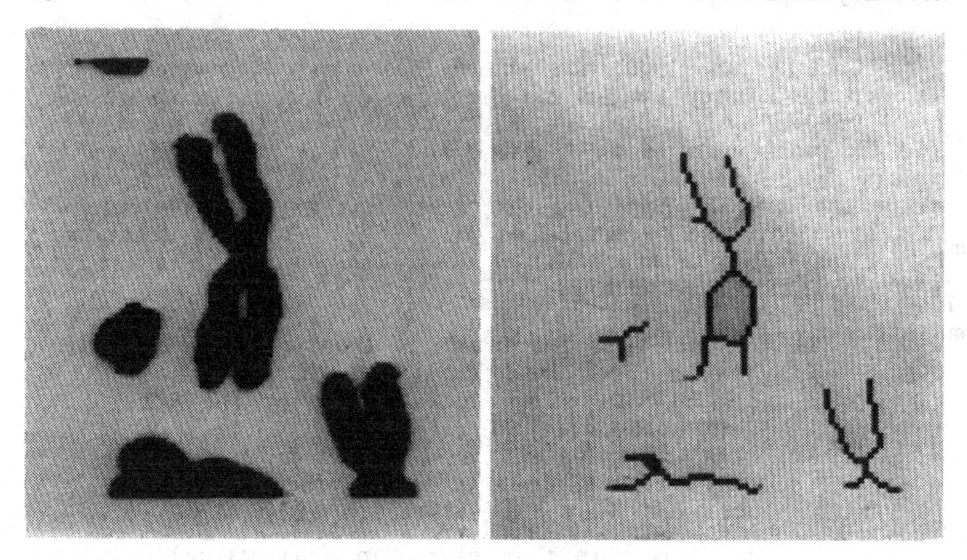

Fig. 6-13 Example of graylevel skeletonization of the chromosome images shown in Figure 6-12 using the algorithm of Goetcherian (1980).

as can be deduced by studying the algorithm quoted above. Hilditch (1969) published a modification of this chromosome skeletonizing algorithm and has implemented it successfully on the CLIP4 image processor. Details are given in Chapter 9.

7. CELLULAR FILTERING

7.1 INTRODUCTION

This chapter begins with a review of traditional linear filtering using numerical methods. A comparison is then made with logical filtering with demonstrations of the use of logical filtering for feature extraction. The problem of detecting signals in noise is next discussed, followed by a section on correlation and convolution using the cellular automaton.

An important concept in linear filter theory is that of *matched filtering*. As far as can be determined the first use of the phrase "matched filter" was in a paper by Van Vleck and Middleton (1946), whose work was influenced by earlier work, undertaken for the United States National Defense Research Committee during the early 1940's, and published in a classified wartime report by Wiener (1942). This was later published in the open literature as the now famous book *Extrapolation, Interpolation, and Smoothing of Stationary Time Series* (Wiener, 1949). Wiener provided the solution to the optimum linear filtering problem as related to the design of signal processing systems. Wiener, in turn, acknowledged the parallel and independent work of Kolmogoroff (1941) and Kosulajeff (1941) in Russia who were pursuing similar lines in studying signal estimation and prediction. In their excellent book *Radar Signals* Cook and Bernfeld (1967) reference other parallel efforts in this field by Huttman (1940) and Cauer (1950) in Germany and Sproule and Hughes (1948) in Great Britain. Particularly noteworthy are the contributions of North (1943) in the United States and, in fact, the matched filter is often referred to as a "North filter" as well as the more common

"Wiener filter." The efforts referred to by Cook and Bernfeld were closely allied to theoretical developments in the field of pulse-compression radar during World War II. These authors summarize these wartime investigations as follows:

> "The development of the matched-filter concept represented an effort to define, independent of practical limitations, theoretical performance criteria for pulse-radar systems. The fact that such widely separated scientists demonstrated the insight to arrive at a more or less identical solution to a common problem is remarkable enough in itself."

The interested reader who wishes to pursue the origins of the matched-filter concept more rigorously should review the commentary on Wiener's work by Levinson (1947) and the well-known monograph *Probability and Information Theory, with Applications to Radar* by Woodward (1953).

7.1.1 Mathematical Formulation

In general terms the matched filter for a given signal s(x,y) is that filter whose frequency-domain representation "matches" that of the signal, i.e., is identical in amplitude and conjugate in phase. This is the frequency domain representation corresponding to the given signal which will detect that signal optimally in white Gaussian noise. Given a signal s(x,y) which is bounded in x and y and whose Fourier transform is $S(\omega_x,\omega_y)$, then the filter which is matched to s(x,y) is given by

$$H(\omega_x,\omega_y) = S^*(\omega_x,\omega_y) \tag{7.1}$$

where ω_x and ω_y are the radian spatial frequencies. In order to find the space-domain representation, h(x,y), the inverse Fourier transform is utilized to yield

$$\begin{aligned} h(x,y) &= k \iint S^*(\omega_x,\omega_y)\exp j(x\omega_x+y\omega_y)d\omega_x d\omega_y \\ &= s(-x,-y) \end{aligned} \tag{7.2}$$

where k is a constant of proportionality. As can be seen,

h(x,y) is simply the space-domain inverse of s(x,y). It follows, therefore, that the response of the matched filter to a unit impulse is the space-domain inverse of the signal to which the filter is matched.

Using the Wiener-Khintchine theorem, it can be written that

$$(1/2\pi)\iint S^*(\omega_x,\omega_y)H(\omega_x,\omega_y)\exp j(x\omega_x+y\omega_y)d\omega_x d\omega_y$$

$$= \iint h(x',y')s(x'+x,y'+y)dx'dy' \qquad (7.3)$$

The latter half of the above equation is clearly recognized as a correlation function and explains why use of the matched filter is often called "correlation detection." Writing the correlation coefficient C as

$$C = \left|\iint h(x',y')s(x',y')dx'dy'\right|^2$$

$$\div \iint h^2(x',y')dx'dy' \iint s^2(x',y')dx'dy' \qquad (7.4)$$

it is seen that, by substituting equation (7.1) in equation (7.4), we obtain C = 1. Thus, the matched filter maximizes the correlation coefficient.

The correlation coefficient can also be thought of as the cosine of the angle between the functions s(x,y) and h(x,y) when these functions are considered as vectors. In fact, the matched filter provides *coherent* integration over the signal domain while elsewhere the integration is *incoherent*. Such a filter, therefore, is most useful when it is employed to detect a signal in a background of random noise when the properties of the signal are uniquely defined. For example, if the signal processing task was to locate all the occurrences of the letter m on this page, a matched filter given by equation (7.1) could be utilized. This would ensure the maximum likelihood of success if it is assumed that the noise caused by non-uniform inking of this page could be characterized as a classical random process.

7.1.2 Generalization

Other signal detection tasks, however, such as that implied by the request, "Bring me a book," require a broadened approach to filtering. Clearly if the size, shape, orientation, structure, and color of the book requested were known, then a matched filter could be employed to locate it. However, if the question implied a desire for any book rather than a particular book, the signal space must be searched for a region where the general qualities or features of the entity *book* are present.

This chapter takes a rather broad look at the concept of filtering for signal detection as implemented by the cellular automaton where the purpose of filtering is to detect, classify, and/or identify a particular spatial signal or configuration, or some ensemble of spatial signals or configurations. In addressing such a problem a number of filters may have to be utilized. The implementation of these filters is likely to depart significantly from the classical concept of a matched filter, and, in fact, many aspects of this chapter touch on the more general fields of study known as *feature extraction* and *pattern recognition*, but all within the context of filtering. The emphasis of the discussion is on two-dimensional patterns with some, but not all, applications emphasizing image analysis.

7.1.3 Pattern Recognition

In general, the goal of pattern recognition in image analysis is to design filters which are capable of extracting measurements or features which can be used to separate a set of images or portions of a single image into unambiguous and meaningful subsets. The number of subsets present in pictorial data may vary enormously. There may be several billion subsets (as with pictures of the faces of all current inhabitants of Earth), many thousands of subsets (as with images of the leaf structures of all species of plants), a few tens (as with images of the major classes of human white blood cells), or as few as two (as with images of man-made objects versus images of objects which are not man-made). Useful pattern recognition requires accurate measures of meaningful features in the field(s) in which members of subsets of interest are to be located, classified, and/or identified. Measures whose values cluster in a subset specific manner in measurement space are the measures desirable in the performance of these tasks.

Much has been written about design of clustering algorithms. This book does not treat this subject due to the belief that poor measures, i.e., measures whose values provide ambiguous subset delineation, cannot be clustered by definition. Rather, the filter designer must devise more suitable measures or devise new measures of new features which inherently produce the clustering desired. It is the design and evaluation of such filters as implemented by the cellular automaton that is the subject of this chapter.

7.2 BINARY PATTERN PRIMITIVES

As mentioned in Chapter 1, one of the most significant early efforts in applying the cellular automaton to pattern recognition was that of Unger (1958) at Bell Telephone Laboratories. Unger employed nine processing elements simultaneously in his cellular automaton. Chapter 1 also shows the work of Dineen (1955) who employed logical filtering for noise removal in character recognition. In both cases a binary image was employed and each point in the field was observed through a 3×3 window.

Other workers used the 3×3 window as a neighborhood for what was called "mask-matching." This technique is directly related to matched filtering by correlation detection and was used initially for finding of edges in binary images. Nadler (1965) used the four masks shown in Figure 7-1 for the purpose of locating vertical, horizontal, northwest-southeast, and northeast-southwest edges, respectively. Genchi et al. (1968) at the Toshiba Electric Company Ltd. used the seven masks shown in Figure 7-2 for filtering binary images. Correlation detection, as represented by equation (7.3), was employed and all four of Nadler's matched filters were utilized simultaneously. The filter which produced the highest output for each point in the binary image was then used to score the edge direction at that point. However, if the magnitude of the highest output was below a given threshold, then a decision was made that the image point addressed was not on an edge. Finally, the sign of the highest output was used to determine whether the edge represented a black-white or white-black transition.

Besides finding edges by the use of multiple matched filters, other workers used the output of correlation detectors to encode binary images of characters to extract features for character recognition. Genchi's masks corresponded to empty

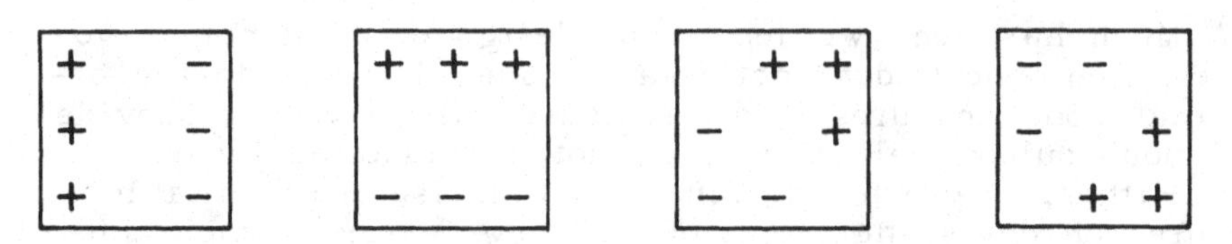

Fig. 7-1 Nadler's 3×3 masks for use in character recognition by matched filtering.

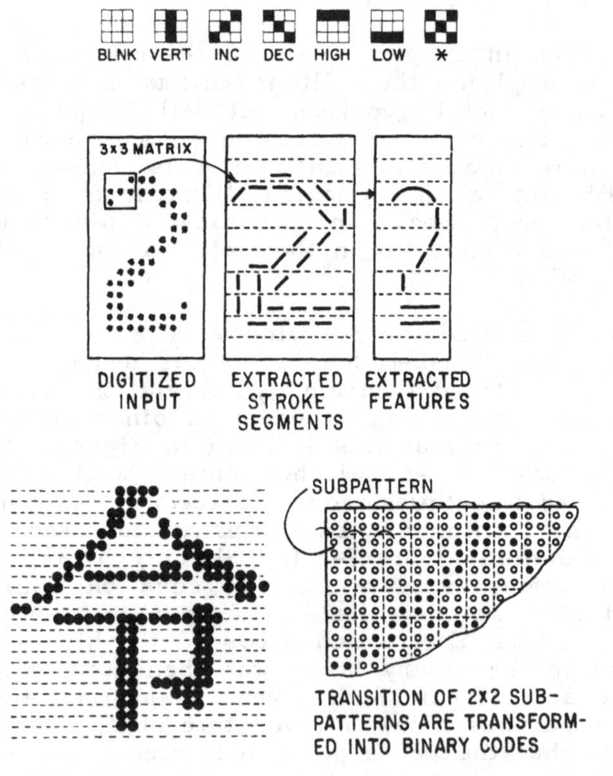

Fig. 7-2 Masks used by Toshiba Electric Co. and the Nippon Electric Co. in matched filtering with results shown for encoding Kanji characters.

regions (BLNK), verticals (VERT), northeast-southwest (INC), northwest-southeast (DEC), high and low horizontals (HIGH, LOW) and the cross (*). Not only did Genchi et al. use these seven idealized masks but actually encoded all 512 possible binary patterns in the 3×3 window in accordance with the closest correspondence of each to one of the seven masks given. They used this technique to extract what were called "stroke segments" from each character (Figure 7-2). Using the stroke segment representation, features were extracted and each character identified. Similar work was performed by Ohmori, et al. at the Nippon Electric Company Ltd. (1972) as is also illustrated in Figure 7-2.

Another set of binary pattern primitives are those of Golay (1969). As explained in Chapter 2, these primitives are based on the hexagonal tessellation and encode all 64 possible binary codes in the six-element hexagonal neighborhood into the 14 primitives which are orientation independent. This feature of Golay's primitives immediately implies that they are not intended for character recognition where orientation plays a crucial role. Instead, the binary pattern primitives of Golay are designed for extracting basic features from the binary images, e.g., edges, interior points, exterior points, corners, end-elements, and filaments irrespective of orientation. An illustration taken from the early work of Preston (1971) is provided in Figure 7-3 showing the extraction of both the edges and corners from the input signal. Golay primitive pattern number 2 is employed for the detection of corners and the simultaneous detection of patterns numbers 3, 4, or 5 for edges. A more general display illustrating all of the Golay primitives is shown in Figure 7-4.

7.3 GENERALIZED LOGICAL FILTERING

Mask-matching operations using correlation of a binary input signal with one or more binary pattern primitives provides single or multiple matched filtering for the purpose of testing for the existence of certain features of the signal in a single iteration. Once the test is performed and the features extracted, then the signal may be recognized or classified, e.g., as in character recognition. It is assumed that the user knows in advance the features to be detected and thus the proper masks for performing matched filtering. In some cases the user has no such *a priori* information. This section shows how logical filtering, operating upon singly-thresholded signals and using multiple iterations, may be used to explore the structure of an initially unknown signal using an orderly progression of measurement techniques. The methods described here are broader than traditional matched

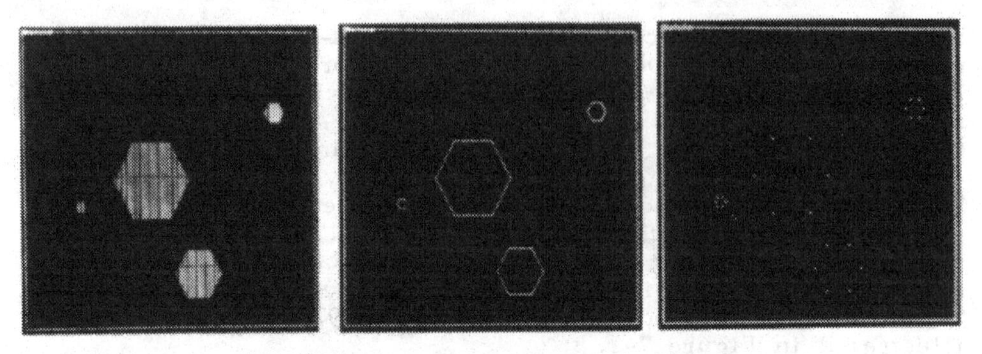

Fig. 7-3 Use of the Golay primitives in the hexagonal tessellation to locate both edges and corners.

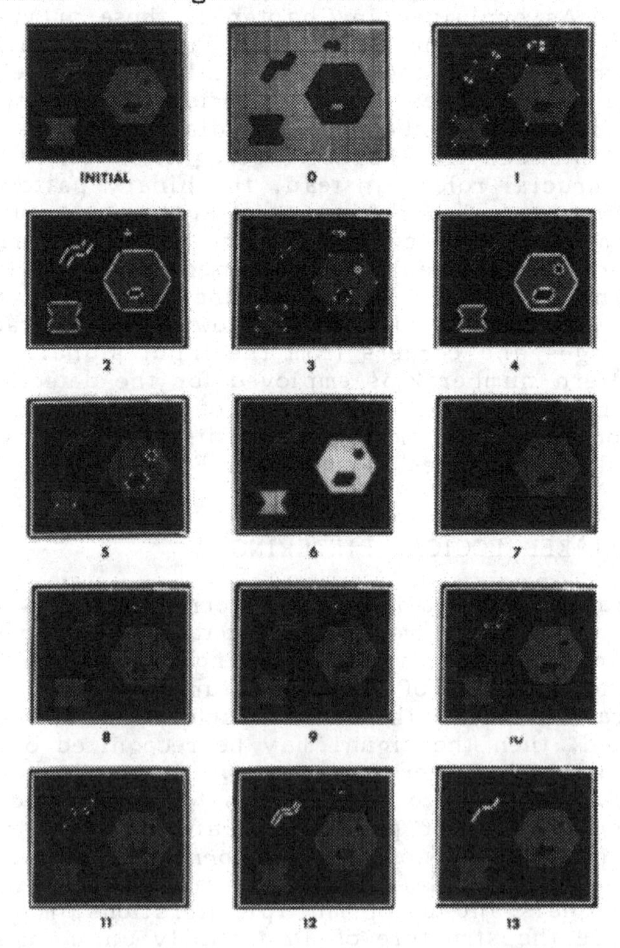

Fig. 7-4 Synthetic images showing location of all picture elements corresponding to each of Golay's primitives.

filtering. The approach used does not require *a priori* information on either the detailed structure of the signal or on the components of the signal.

For purposes of illustration the two-dimensional signal fields shown in Figure 7-5 are employed in this section. These signals have been deliberately stylized for the purpose of demonstrating the utility of logical filtering in many different situations. The following section then compares logical filtering with more traditional matched-filtering methods.

Each of the fields in Figure 7-5 contains eight separate regions each containing the same signal structure. If the individual signal structure is expressed by the equation

$$s'(x,y) = B + s(x,y) \tag{7.5}$$

then each s'(x,y) shown in Figure 7-5 is arranged to have the same bias B and the same signal energy E as given by

$$E = \iint |s(x,y)|^2 \tag{7.6}$$

Specifically, each signal field was digitized over a 512×512 array at eight bits per pixel with B = 128. For each signal structure within the signal field E = 65536 (corresponding to an rms signal level of eight over an area equal to 1024 elements). These parameters were deliberately chosen so that the contrast of each signal structure was close to the threshold of human vision. (Contrast is defined as the rms signal divided by the bias and the threshold for human vision is at a contrast equal to about 4%.) This was done to emphasize the point that the decisions made visually about the presence or absence of a signal or signal feature in a signal field are often misleading, as is illustrated in the section which follows. Thresholded versions of each signal field shown in Figure 7-5 are provided in Figure 7-6.

In Figure 7-5a each individual signal structure is a 32×32 square; Figure 7-5b, a 4×256 rectangle; Figure 7-5c, a horizontal cross-section of the multifrequency test pattern given in Chapter 2 consisting of 16 cycles of a linear frequency sweep whose frequency varies from zero to a frequency of 64 cycles across the 512-element span of the field and having a vertical extent of four elements.

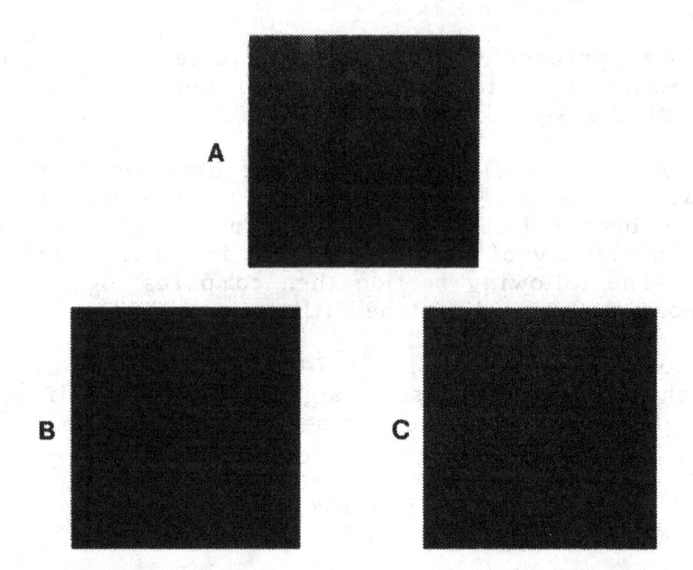

Fig. 7-5 Three types of signals are shown at low contrast above: (A) 32×32 squares, (B) 256×4 rectangles, and (C) 256×4 frequency sweeps.

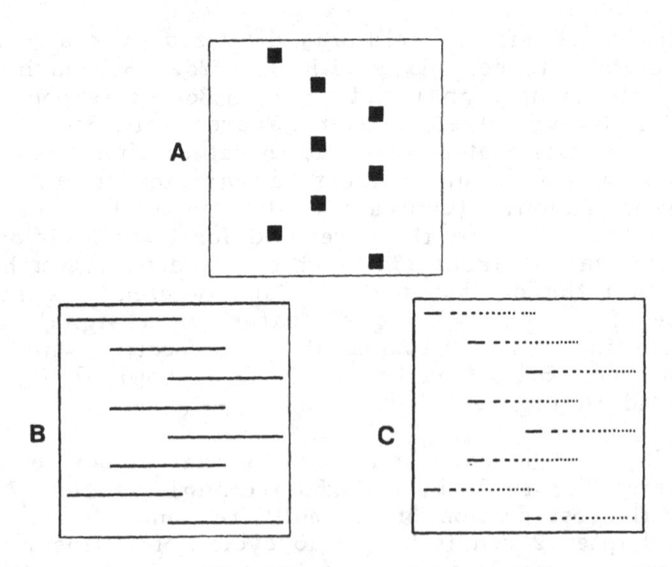

Fig. 7-6 The signal fields shown in Figure 7-5 may be thresholded at their mid-range to yield the results above.

7.3.1 Exploration of the Signal Field

Measurements may be made on any signal field by using filters which are matched to known features. From the results of these measurements s(x,y) may in many cases be identified. In other cases the signal field must first be partitioned into regions (Chapter 5) before the measurements are conducted.

In order to use logical filtering to extract features from the signal fields shown in Figure 7-5, thresholding is performed. The result of such a thresholding operation is shown in Figure 7-6. Utilizing a single threshold set at the mid-range of the values of s'(x,y), all measurements upon the thresholded signal $s_L(x,y)$ are performed by counting. Defining those picture elements which are black in Figure 7-6 as 1-elements, counting yields measures of the number of 1-elements in s'(x,y which are above the mid-range value. Let each measure be called the "total area" of all the signal structures present in the signal field. A description of the use of this measurement and other measurements is given in the sub-sections which follow.

7.3.1.1 *Area*

When a count is made of the number of 1-elements in the three fields shown in Figure 7-6, the results are 8192, 8192, and 4576. The investigator immediately notes that these numbers are extraordinarily low in comparison with the total number of elements in each field (262144). From this one measurement alone the investigator knows that each s'(x,y) must be strongly modal. If desired, this could be pursued further. For example, one could test each s'(x,y) by thresholding again with a small positive displacement of the threshold, say 20%, from the mid-range value and performing a Boolean exclusive-or with the previously obtained result. If this computation is carried out for the first two cases illustrated in the Figure 7-5, the result of the computation would be null. This would tell the investigator that the corresponding signal fields were bimodal with a small number of elements all having the same value with respect to a more-or-less constant background value. In the third case illustrated in Figure 7-5, the result of the computation would be a number smaller than the original result, indicating signal structure(s) whose values fluctuated slightly but were still significantly different from a more-or-less constant background value. Thus by area measurements alone the investigator, in this case, has learned a great deal about a signal field under study.

7.3.1.2 *Edge Measurements*

The next most frequently used measure is that obtained from an evaluation of the edge content of the logical signal. Using the two-dimensional logical transform described in Chapter 2 with values of $\Xi = 9$ and $X = 2$ for one iteration followed by counting, the numbers obtained for Figure 7-6 are 992, 4128, and 2800, respectively. From this and the results from Section 7.3.1.1 the area/edge ratios may be computed as 8.2, 2.0, and 1.6. This immediately indicates to the investigator a marked difference between the contents of the three signal fields. However, insufficient information is available at this point in the investigation to make any definite determination of the identity of the signal(s) present. For example an area of 8192 and an area/edge ratio of 8.0 could imply a long rectangle about 7 elements wide, a few shorter and somewhat wider rectangles, several circles of diameter 32, or eight squares 32×32 (the correct answer). Similarly an area of 8192 and an area/edge ratio of 2.0 could imply a multiplicity of lines four elements wide (the correct answer) but also could imply many squares each having dimensions of approximately 7×7. In the final case (area/edge = 1.6) the signal configuration might consist of many long lines about three elements wide, a large number of small circles, or many small (5×5) squares. Clearly, at this point, further investigations are required to delineate the structures present in the signal fields.

7.3.1.3 *Octagonal Convex Hull*

At this point the investigator could readily determine the orientation of the signal structures by finding whether their boundaries followed the primary axes of the tessellation. To perform this study the octagonal convex hull is generated by iterating the logical transform using $\Xi = 4$ and $X = 9$. A Boolean exclusive-or is then taken with the initial signal field. In all three cases considered here the result is null. This immediately indicates to the investigator that the boundaries of all structures in the signal fields under study lie along the principal directions of the tessellation. This rules out the possibility of there being circles, or other structures with curved borders, in the signal field. It is now evident that the signal field contains convex structures having borders aligned with the principal axes of the tessellation.

7.3.1.4 *The Residue Histogram*

Next the investigator might decide to determine the number of contiguous structures present in the signal fields under study. For this purpose the appropriate logical transform uses $\Xi = 8$ and $X = 4$ with multiple iterations carried out in subfields. A subfield sequence of 1-3-4-2 is elected to preserve symmetry. After each iteration residues are counted and the counts are graphed as a function of the number of iterations. The graph is the *residue histogram*. The residue histograms for the signal fields shown in Figure 7-6 are presented in Figure 7-7. The residue histogram of Figure 7-6a shows that there are eight objects in the signal field each of identical size. All are reduced to residues after 16 iterations. Since the total area is 8196, the area of each signal structure is 1024. It is also known from above that the area/edge ratio is 8.0 and that the edges are aligned along the principal axes of the tessellation. Since only the 32x32 square has these properties and reduces to a residue in 16 iterations using $\Xi = 8$, the investigator has discovered that the signal field contains eight such squares.

Similar information is obtained from the residue histogram for Figure 7-6b. There are eight objects which reduced to residues simultaneously at 128 iterations. These objects, therefore, must be long lines (extended rectangles) of dimensions 4x256. The residue histogram for Figure 7-6c indicates 128 objects of different sizes. About 75% of the objects reduce to residues in the first few iterations, but all are not reduced until 20 iterations.

7.3.1.5 *Exoskeleton Analyses*

At this point in the study the investigator knows the identities but not the relative positions of the objects in two of the signal fields. At the same time only the number and boundary orientation of 128 objects in the third signal field is known. Also, since all objects were reduced to residues,

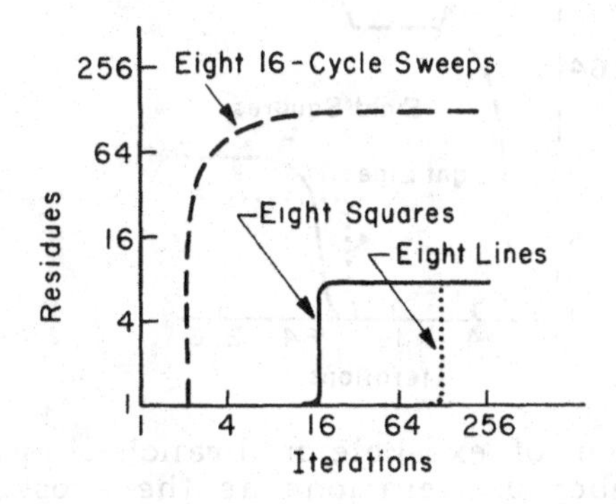

Fig. 7-7 Residue histograms of the fields in Figure 7-6.

the investigator also knows that there are no objects containing voids. (If objects containing voids had been present the region labeling methods described in Chapter 5 could be employed to obtain the connectivity number.)

To determine more about the relative position of the objects in each signal field an exoskeleton analysis may be performed (Chapter 6). Results are presented in the graph shown in Figure 7-8 and the actual exoskeletons are shown in Figure 7-9. At each iteration in the generation of the exoskeleton the number of exoskeleton branches is determined by reducing them to residues and counting the residues. Thus Figure 7-8 shows the number of exoskeleton branches as a function of the number of iterations.

The most informative of these graphs is the one relating to the signal field shown in Figure 7-6b. Here no branches form in the exoskeleton until the 32nd iteration at which point 11 branches form simultaneously. This indicates that the eight lines which exist in this signal field are parallel and equally spaced. No other structural positioning would produce such a result.

Information conveyed by the graph relating to the signal field contained in Figure 7-6a is somewhat less clear. Nine branches form at about 50 iterations while additional branches (up to a total of 12) form through the 90th iteration. This indicates that some of the eight squares are clustered while other members of the set are somewhat farther apart. Clearly, the squares are not uniformly distributed over the 512x512 signal field. Not much more than that can be determined at this time.

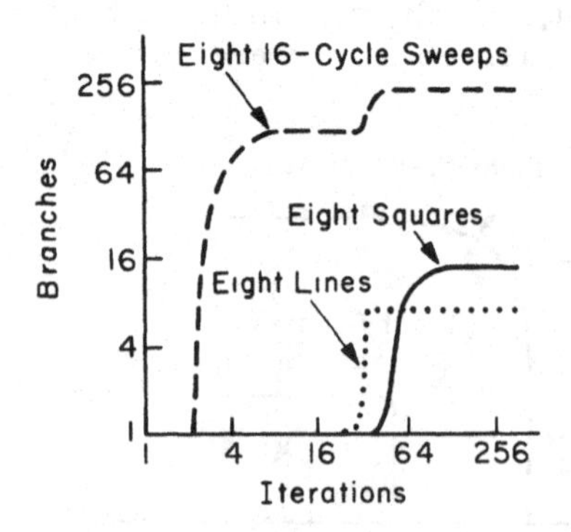

Fig. 7-8 The number of exoskeleton branches is given as a function of the number of iterations as the exoskeletons of the signal fields shown in Figure 7-6 are computed.

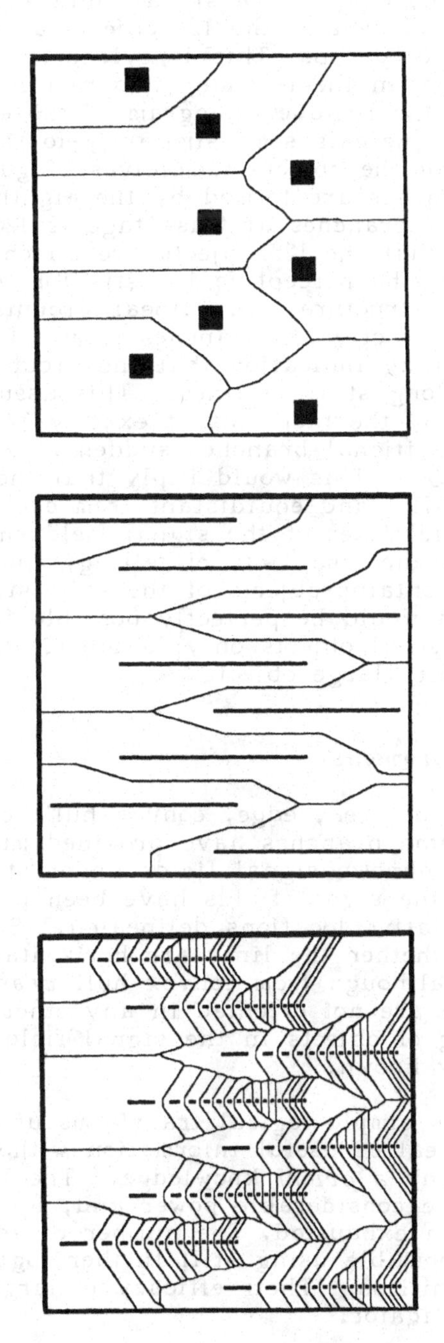

Fig. 7-9 Exoskeletons of the signal fields given in Figure 7-6.

Finally, in the case of the signal field shown in Figure 7-6c, it becomes obvious that the 128 objects are collected into a small number of groups. This is evident, since there are two clear stages in the residue histogram of the exoskeleton branches. As the residue histogram of the exoskeleton branches develops, there is some similarity to the direct residue histogram of the objects themselves (Figure 7-6c). A multiplicity of branches are formed by the eighth iteration. The total number of branches at this stage is 120 which strongly indicates that the 128 objects are bunched together in groups. In fact, the perceptive investigator would note that, if the signal structures were linear groupings of 16 objects each, then exactly 120 branches should form. There is, therefore, a strong indication that the groups of objects are associated in long straight lines. This assumption is further reinforced by the fact that at exactly 32 iterations a large number of additional branches suddenly forms, yielding a final total of 290. This would imply that the linear groupings are parallel and equidistant from each other, much as the eight parallel lines of the signal field shown in Figure 7-6b. There is no clear way of telling, however, whether each grouping contains objects of the same size or of different sizes. It would be perfectly possible for one linear grouping to be of small objects only; another, intermediate size objects; another, large objects.

7.3.1.6 *Other Measurements*

Measurements of area, edge, convex hull, object residues, and exoskeleton branches have provided much information on the three unknown signal fields under study. The objects contained in the signal fields have been partially identified and their relative locations delineated. Still missing is information on whether the lines are horizontal, vertical, or at 45 degrees (although from convex hull measurements it is known that they are not oriented in any other direction). The exact grouping of objects in the signal field of Figure 7-6c is only partly known.

The relatively simple logical transforms utilized have provided a great deal of useful information without the necessity of obtaining any *a priori* knowledge. The logical transforms employed have considerable power and, by no means, has this power been exhausted. The number of additional measurement methods possible using still further logical transform are almost limitless. Their efficacy is largely in the hands of the investigator.

If information on orientation is required, the investigator can make use of the dual-point kernels that are described in Chapter 2. Using these kernels it is possible to ascertain whether the lines discovered in Figure 7-6b are horizontal, vertical, or at a 45 degree angle to the major axes, since these kernels propagate boundaries only when these boundries exist at specific orientations. If the investigator desires further information on the groupings of objects in Figure 7-6c, it is possible to proceed using logical transforms that cause nearby objects to merge before the exoskeleton analysis is performed. When this operation is carried out, it can be determined that each of the linear groupings already found in Figure 7-6c contains both small and large objects. Furthermore, again using dual-point kernels, orientation information may be obtained. In fact, by observing the size distribution of the objects contained in each grouping it would be possible to discover that each signal structure was a linear frequency sweep.

7.4 NOISE IMMUNITY

Before leaving the topic of the logical transform for matched filtering and feature extraction, immunity to noise in comparison with the traditional linear matched filter should be presented. Section 7.3 has pointed out the advantages of the logical transform in extracting a multiplicity of features from an unknown signal field for the purpose of analyzing the structure of the signals present. The examples provided in Section 7.3 were relatively stylized and purposefully designed to illustrate the use of the logical transform for feature extraction and measurement for a variety of inputs.

This section treats the problem of extracting information from such signals when they are severely corrupted by noise. In this realm of signal analysis the matched filter is the optimum linear detector and it is assumed in this section that *a priori* information is available to the user. Let it be assumed that the signal fields shown in Figure 7-5 are received via a transmission channel in which thermal noise over the full bandwidth recorded in the 512×512 field has an rms value four times that of the signal amplitude, i.e., the signal is 12dB below the noise level. The result is shown in Figure 7-10.

The signals shown in Figure 7-5 were deliberately selected so that their energy, as given by equation (7.6) were equalized. This implies that they are equally detectable under the matched-filtering regime expressed in equations (7.1) through (7.3). The random noise added to these signals (Figure 7-10) has an rms value of 32 units which produces the re-

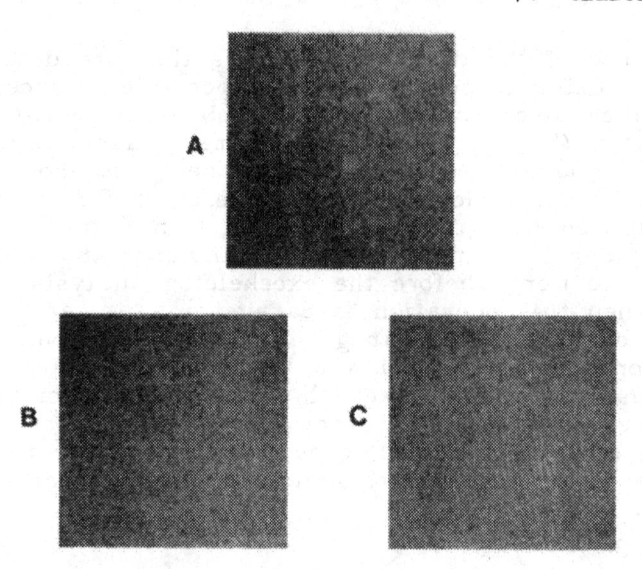

Fig. 7-10 Appearance of the signal fields shown in Fig. 7-5 when noise is added at 12dB above the rms signal level.

sultant signal-to-noise ratio of -12dB. It is interesting to observe the ability of the human visual system to detect these signals. Some of the 32×32 squares are detectable but, even if the human observer rotates Figure 7-10 by 90° and looks at a glancing angle to the plane of the page, the other signals are difficult, if not impossible, to locate.

7.4.1 Linear Matched-Filtering

The spectrum of the noise added to the signal to obtain Figure 7-10 covers the entire frequency range recordable in both the x and y directions, i.e., from one to 256 cycles across the aperture in both x and y. The two-dimensional space-bandwidth product (*extent* or *etendue* in optical signal processing parlance) covers both negative and positive frequencies from -256 to +256 cycles across the aperture and is equal to $512^2 = 262144$. Matched filtering or correlation detection involves integration which effectively reduces the bandwidth (but not the space) covered by the noise.

In order to perform correlation detection, the matched filter given by equation (7.1) may be used to detect the signals present in Figure 7-10. As given by equation (7.3) this operation is equivalent to auto-correlation. This requires multiplication, shifting, and integration. Since the integration is taken over 1024 elements in each of the cases studied, there is an improvement in signal-to-noise ratio which is given by the expression

$$S/N = 10 \log_{10}(1024) = 30\text{dB} \tag{7.7}$$

so that the resultant signal-to-noise ratio is 18dB. Figure 7-11 furnishes an illustration of this signal-to-noise improvement. Figure 7-11a shows the probability density function of the noise (- - -) and of the noise plus signal(---) before correlation detection, i.e., when the signal-to-noise ratio is -12dB. Since the signal is pure, i.e., is not noisy in and of itself, addition of signal to the noise produces a right-shift in the probability density function by 0.25σ where σ is the standard deviation of the noise. In the case illustrated in Figure 7-11a the rms noise is 12dB above the signal level. Ideally, matched filtering narrows the noise bandwidth without affecting the bandwidth occupied by the signal and, therefore, reduces the noise energy without affecting the signal energy to produce the result illustrated in Figure 7-10e. Here the signal is now 18dB above the rms noise.

Figure 7-12 (upper portion) shows the post-correlation results for each of the signal fields shown in Figure 7-10. Each field is shown in the original 512×512 format even though, due to bandwidth reduction, far fewer elements are required to satisfy the sampling theorem. In the case of the squares, the cutoff frequency in x is the same as in y, i.e., 512/64 = 16 cycles across the aperture. Thus the sampling theorem requires only 1024 samples over the 512×512 field arranged in a 32×32 matrix. In the case of the rectangles, in the x direction the cutoff frequency is 2 cycles across the aperture and 128 cycles across the aperture in the y direction. This also leads to 1024 samples in the 512×512 field but now the matrix is 4×256. The case of the frequency sweep is different in that the bandwidth of the correlation kernel is not determined by its span in x but rather by the number of cycles of the highest frequency in the frequency sweep. In this particular case the period of the highest frequency is eight elements so that the output bandwidth in the x direction is 64 cycles across the aperture. In the y direction the result for the frequency sweeps is identical to that for the rectangles yielding a total number of samples equal to 32768 in a 128×256 matrix over the 512×512 field. Thus it would be expected for the frequency sweeps that, although the likelihood of an error per sample is the same as that for the squares and rectangles, more false positives occur due to the increased number of samples required. This is in fact the case as is shown in both Figure-11 and in the first

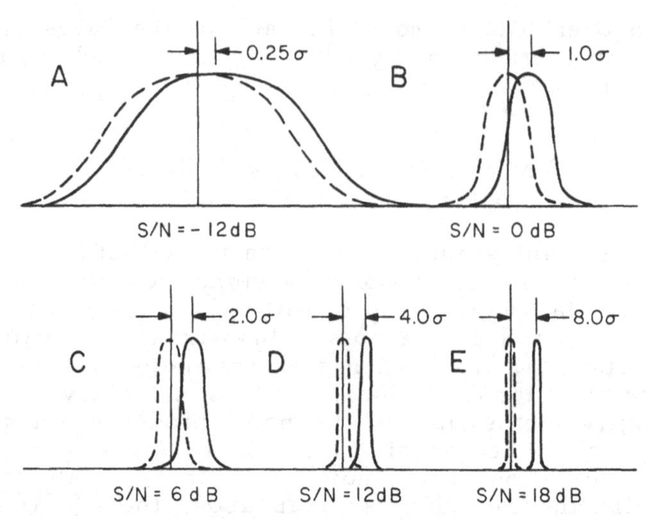

Fig. 7-11 Signal to noise probability density relationships for various signal-to-noise ratios.

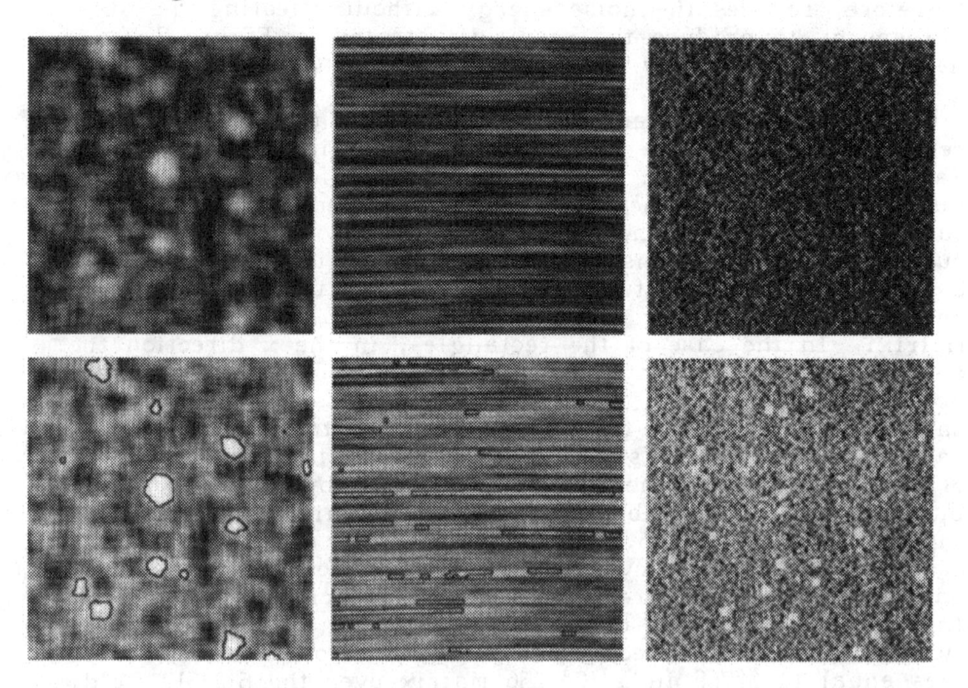

Fig. 7-12 Filtering of the contents of the three data fields shown in Figure 7-11 using the corresponding matched filters: (above) the direct graylevel outputs of the filters and (below) after equiprobability thresholding.

line of Table 7.1. In computing the numbers given in Table 7.1 multiple false positives in the matrix position corresponding to a single sample are, of course, counted as a single error.

7.4.2 Signal Detection Using the Logical Transform

When using linear matched filtering as demonstrated in Section 7.4.1, *a priori* information was assumed to be available on the structure of the signals to be detected in a background of random noise. The problem addressed in this section is to determine to what extent the logical transform may be used for detecting a signal in noise when no *a priori* information is available. In other words the problem is to determine the noise immunity of the logical transform. It is possible to analyze the noise immunity of this transform for signal structures imbedded in random noise by computing the likelihood of various values of ξ as a function of threshold. Figure 7-13 shows Figure 7-10a thresholded at four levels selected at equally spaced intervals on either side of the mean of the probability distribution function. Despite the fact that the signals are 12dB below noise, it is evident that the signals have a marked effect not only on the number of 1-elements which appear in those regions where the signal is present but also on the specific 3×3 neighborhood patterns. The average number of elements above threshold may be derived from the normal distribution. Furthermore the likelihood of all values of ξ may be computed as a function of threshold. Such a plot is provided in Figure 7-14.

Table 7-1 Tabulation of Results for the Cases Shown in Figures 7-12, 7-15, and 7-16.

FILTER TYPE	FALSE POSITIVES			FALSE NEGATIVES		
	SQUARES	LINES	SWEEPS	SQUARES	LINES	SWEEPS
WIENER	6	16	25	0	0	2
LOGICAL	12	12	12	0	3	5
WIENER PLUS LOGICAL	0	1	5	0	2	4

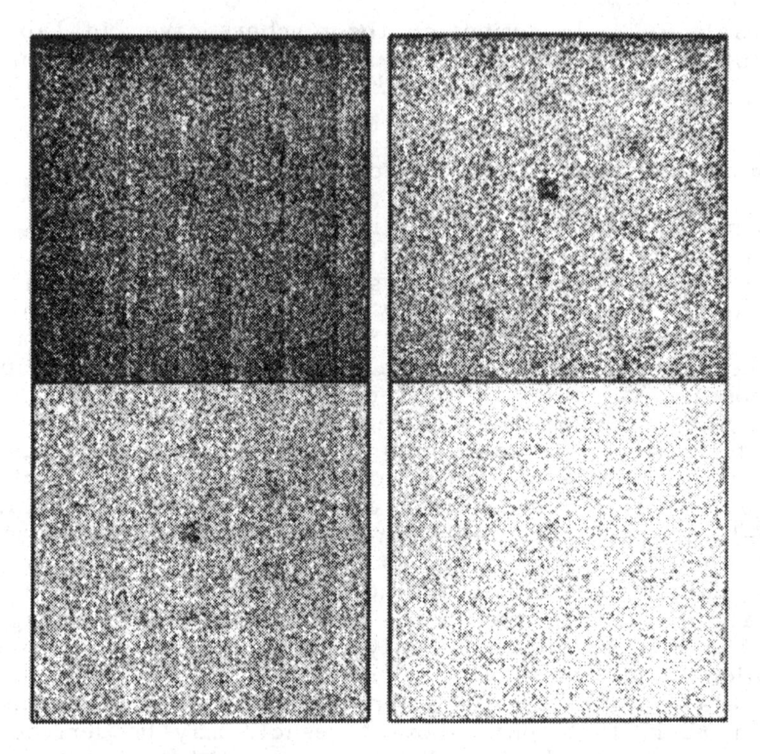

Fig. 7-13 Figure 7-10a thresholded at four levels over the 99 percent range of its probability distribution function.

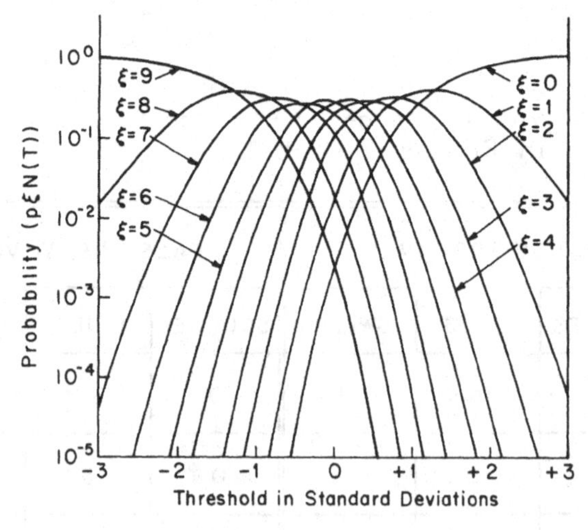

Fig. 7-14 Probability graphs for several values of ξ of $p\xi N(T)$ for white Gaussian noise plotted as a function of the threshold.

The likely number of neighborhoods exhibiting a given value of ξ is a countable quantity which, in turn, varies statistically according to the Poisson distribution. Figure 7-14 shows that this number changes rapidly with threshold, i.e., changes significantly, at a given threshold, in regions where a signal is present. The logical transform therefore offers the user a powerful method of signal detection without the *a priori* information on signal structure which would be required by traditional matched filtering. Using the logical transform it is possible to perform signal detection by designing a transformation or series of transformations sensitive to the spatial structure of the thresholded signal and also sensitive to the number of elements which occur above threshold in the neighborhoods where the signal is present.

Note that Figure 7-14 indicates that the characteristic of $p\xi 4$ at a threshold of $\sigma = 0.25$ and the characteristic of $p\xi 5$ at a threshold of $\sigma = -0.25$ are points where the probability is rapidly changing. Each of the signal fields shown in Figure 7-10 was thresholded at these two levels separately. The resultant two binary images were logically transformed utilizing $\Xi = 4$ for one and $\Xi = 5$ for the other for a multiplicity of cycles. The results were then ORed and the presence or absence of 1-elements in each of the appropriate sampling zones was scored in order to determine false positives and false negatives. The results are shown in the second line of Table 7-1. Note that 12 false positives occurred in the noise field alone, which explains why the number of false positives is identical for each signal field. As can be seen there were no false negatives for the signal field containing squares, three for the rectangles, and five for the frequency sweeps. Results are shown in Figure 7-15 demonstrating the capability of the logical transform to detect the presence of weak signals in noise without *a priori* information.

7.4.3 Post-Detection Processing

Since the spatial structures in those regions where false positives occur at the output of the Wiener filter differ from those regions where true positives occur, it is possible by post-detection processing using the logical transform to improve the overall performance of the filter. As can be seen from Table 7-1, the total number of errors (false positives plus false negatives) generated after Wiener filtering was 6 for the squares, 16 for the rectangles, and 27 for the frequency sweeps. By simply utilizing a logical transform which differentiates between large and small groups of 1-elements (see Figure 7-16), it is possible to decrease markedly the number of false positives while making small increases in the number of false nega-

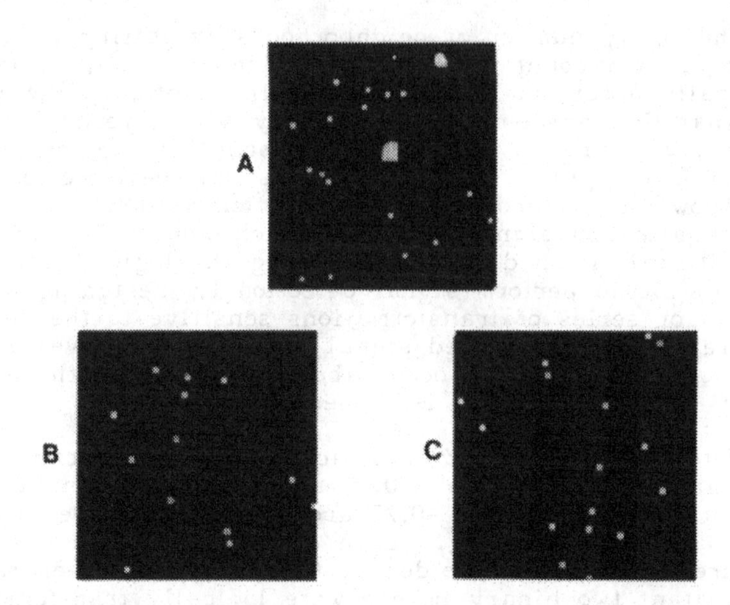

Fig. 7-15 Cellular logic transform used to detect presence of squares, rectangles, and frequency sweeps in noise using the binary images generated from Fig. 7-10 at $\sigma=+.25$ and $\sigma=-.25$.

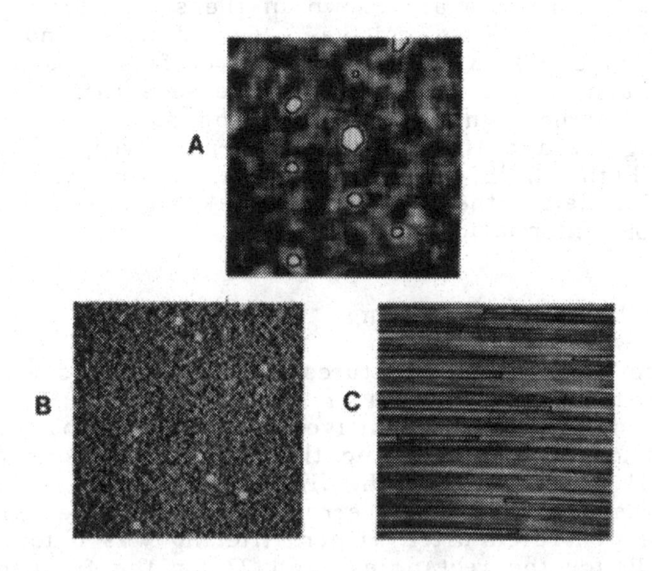

Fig. 7-16 Post-detection processing using the logical transform of the results of equiprobability thresholding (Figure 7-12) improves performance considerably.

tives. As shown in the last line of Table 7-1, this method leads to a final result where there are zero errors for the squares, while for the rectangles the total errors may be reduced to 3, and, for the frequency sweeps, to 9. The errors are reduced primarily by decreasing the number of false positives at the expense of increasing the number of false negatives. Whether this is an improvement depends on the penalties paid for missing some of the signals versus the penalties paid for falsely indicating that a signal is present in an area which actually contains no signal. In the case of the squares it is clear that perfect performance may be achieved. However, in the case of the rectangles, when 25% are missed, a single false detection results. The worst performance is in the case of the frequency sweeps where, when 50% are missed, there are still five false detections.

It is evident from this discussion that the logical transform can be used not only for the direct detection of signals in noise but also can be employed in conjunction with linear matched filtering to perform useful post-detection operations. In many cases, especially those where the signal extent is considerably smaller than the total area of the signal field, the logical transform carries out the very essential operation of removing some of the false positives which occur due to those occasional small but insignificant regions where noise exceeds the detection threshold.

7.5 NUMERICAL FILTERING

Although it is possible to extract valuable information from graylevel images by first thresholding them and then applying logical filters to the resulting binary images, the cellular automaton can also be used to apply numerical local-neighborhood operators to the graylevel images themselves. Techniques used for performing arithmetic operations in cellular automata are described in detail in Chapter 4 and the concept of a local neighborhood operator in which the graylevel of a transformed pixel is a weighted sum of the graylevels of pixels surrounding the original pixel is introduced in Chapter 5. In the following sections these methods and concepts are utilized to derive procedures for performing discrete correlation and convolution and for constructing a wide range of graylevel filters.

7.5.1 Convolution and Correlation

If $f(x,y)$ and $g(x,y)$ are functions of the discrete variables x and y, then the convolution and correlation of these functions are defined by

$$f(x,y) * g(x,y) = \sum_{m=o}^{M-1} \sum_{n=o}^{N-1} f(m,n)g(x-m,y-n) \tag{7.8}$$

and

$$f(x,y) \circ g(x,y) = \sum_{m=o}^{M-1} \sum_{n=o}^{N-1} f(m,n)g(x+m,y+n) \tag{7.9}$$

respectively. The functions f(x,y) and g(x,y) are assumed to be discrete arrays of size A×B and C×D respectively, which are periodic in the x and y directions with periods M and N, and in which

$$M \geq (A + C - 1)$$
$$N \geq (B + D - 1) \tag{7.10}$$

Both functions are assumed to be zero outside the regions A×B and C×D, i.e. they are "embedded" in infinite planes of zeros. These boundary conditions prevent interference between successive periods of the periodic functions and allow the resulting function to be interpreted as a single period of the discrete, two-dimensional convolution (or correlation).

A particularly simple form of these discrete operations occurs when g(x,y) represents a graylevel image of size C×D (assumed embedded in zeros as before) and f(x,y) is a 3×3 local neighborhood operator. In the case of correlation, the process reduces to forming at all pixels in g(x,y) the weighted sums of all 3×3 groups of pixels in g(x,y), the weights being given by the nine values of f(x,y). Specifically

$$f(x,y) \circ g(x,y) = \sum_{m=o}^{2} \sum_{n=o}^{2} f(m,n)g(x+m,y+n) \tag{7.11}$$

Convolution is identical to correlation except that one of the functions is reflected in both x and y as expressed by

$$f(x,y) * g(x,y) = \sum_{m=o}^{2} \sum_{n=o}^{2} f(m,n)g(x-m,y-n) \tag{7.12}$$

A cellular automaton, in which each cell has access to data in all cells in its local neighborhood, may therefore perform both of these operations in parallel without the need to shift one function with respect to the other. If a convolution (or correlation) window larger than 3×3 is required, elements which are not adjacent to the central elements must be weighted and summed. In this case one or another of the functions must be shifted with respect to the other by a distance sufficient to bring every element into the local neighborhood of all non-zero elements of the convolving function. Figure 7-17 shows how a function can be correlated with a 5×5 window by using four diagonal shift operations and employing four different correlation kernels f_{NW}, f_{NE}, f_{SW} and f_{SE} at each shifted position. The required function is obtained by summing, on an element-by-element basis, the four sub-results from each shifted correlation with a typical example given in Figure 7-18.

In order to perform equiprobability thresholding of the result of the correlation c(x,y) to detect in the original image g(x,y) a signal such as f(x,y), some normalisation must be performed. A new result n(x,y) is formed in which each pixel has a value equal to the average of the squares of the values of f(x,y) taken over windows of dimensions A×B surrounding each pixel. The result c(x,y) is also squared and divided by n(x,y), element by element. Maxima in the final result of this operation correspond to regions where the required signal(s) have been detected. The averaging can be regarded as a correlation (or convolution) of g(x,y) with a window in which all the elements are ones. It is advisable to mask out regions of very low intensity since this can give spurious results because of rounding errors in the division process.

Figure 7-18 shows a sequence of images in which signal structures with an intensity distribution equivalent to the correlation kernel given in Figure 7-17 are detected as described above. Note that a similar distribution has apparently been detected at the end of the image of the key.

7.5.2 Filters for Noise-Removal

Instead of performing correlation detection, it may be required to improve the detectability of a signal within a noisy background by other means, either to enhance its visibility or else to render it more amenable to detection by subsequent processes. As discussed in Section 7.4 this is useful when *a priori* information on the exact signal structure is unavailable.

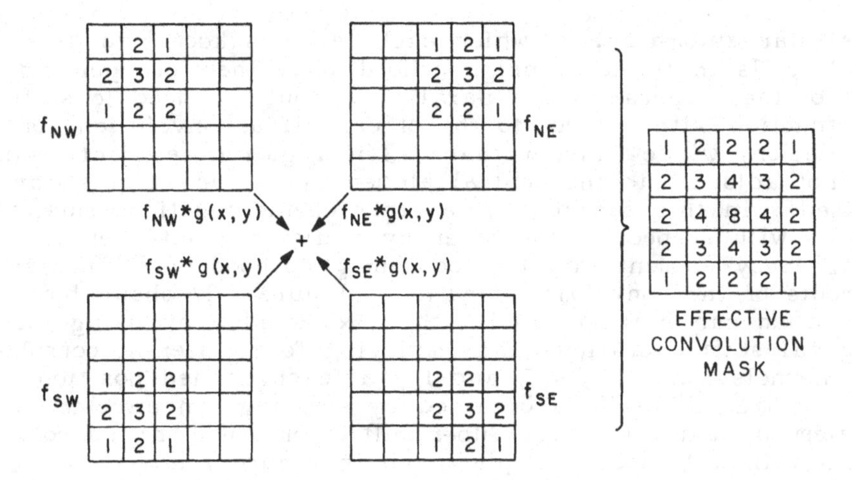

Fig. 7-17 Four 3×3 weighting matrices may be combined directly by the cellular automaton to produce the effect of a 5×5 matrix correlation operation.

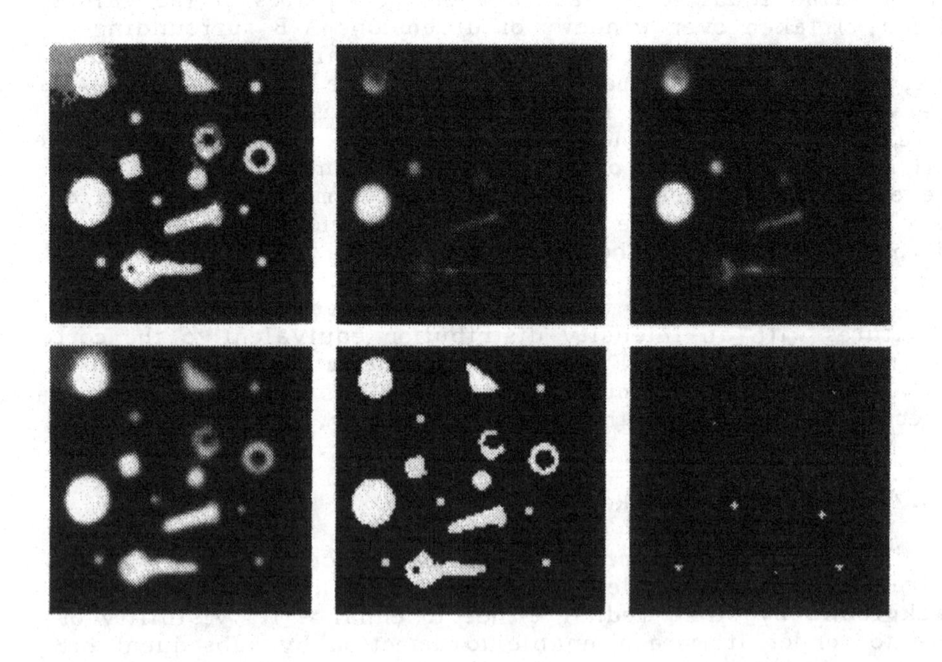

Fig. 7-18 Example of the application of the correlation matrix given in Figure 7-17.

7.5.2.1 *Averaging*

One way to reduce the noise amplitude in a noisy image is to use low-pass filtering, in the case when the difference between the spatial statistics of the noise and those the signal reside in a frequency shift. For example, for two adjacent elements, it is more probable that their graylevels will have similar values in a noise-free image especially when the objects of interest are large compared with the element spacing. Replacing element graylevels by the average graylevel in the local neighborhood will therefore have little effect on the signal except to blur the edges of objects in the image. On the other hand, if the spatial bandwidth of noise in the image is comparable with the reciprocal of the element spacing, adjacent noise elements will usually exhibit differing graylevels. Thus the noise amplitude will be substantially reduced with respect to the signal amplitude by averaging. Similar arguments apply for averaging windows larger than the cellular neighborhood and the limit for this method of noise removal will be set by the acceptable level of blurring, i.e., by the minimum size of significant object detail which must be retained. Averaging windows may be equally weighted or, more usually, weighted more strongly in the center than at the extremes. The larger windows (greater than 3×3) are implemented by shift and add sequences of 3×3 windows as before. Figure 7-19 shows the effect of a 3×3 and 5×5 averaging windows on a noisy image.

7.5.2.2 *Augment/Reduce Operators*

In a binary image, noise can be removed by sequences of logical operations, i.e., by logical filtering. In a reduction operation ($\Xi = 8$) object elements are replaced by background elements if they have a background element in the local neighborhood. Conversely, augmentation ($\Xi = 0$) causes background elements to be replaced by object elements if there are object elements in their neighborhood. A sequence of Q cycles with $\Xi = 8$ followed by Q cycles with $\Xi = 0$ removes "noise objects" of maximum dimension 2Q. The reverse procedure removes "background noise" regions 2Q in size. Edges of objects are made less ragged by this process, which can be regarded as being comparable to edge sharpening. These operations are called "ouverture" and "fermeture" by their inventor, Serra (1982). Their graylevel equivalent is, for reduction, to assign to each element a graylevel value equal to the minimum graylevel value in its local neighborhood (augmentation uses the maximum graylevel). As expected it is found that noise and small detail are eliminated but edges of large objects are sharpened as shown in Figure 7-20.

Fig. 7-19 The effect on an image containing broadband additive noise of averaging over 3×3 (left) and 5×5 (right) windows.

Fig. 7-20 Example of high frequency noise removal and edge sharpening by max-min filtering.

7.5.2.3 *Rank Filtering*

A third technique for noise elimination involves rank filtering. In principle the elements in the local neighborhood are first ordered with respect to their graylevel values and a rank assigned to each. A certain function of the rank is then calculated to give the result. Finally, the central element is assigned a new graylevel equal to that of the element having the rank selected. The four most useful rank filters are the maximum, minimum, median and extremum. The maximum and minimum are described in the previous section. The median operation selects a new graylevel value having as many elements with values below as there are above in the local neighborhood. Extremum filtering is a combination of a maximum and minimum filtering in which the value of the central element is replaced by either the maximum or the minimum in its local neighborhood, selecting whichever is nearest to its original value. The median and extremum filters are illustrated in Figure 7-21.

Rank-order filters present an interesting challenge to those who seek algorithms of high efficiency. Danielsson (1981) has reviewed the better known algorithms and suggested another which is particularly suitable for implementation in cellular automata.

Fig. 7-21 Examples of median (left) and extremum (right) filtering.

If the gray levels in a defined neighborhood of a particular pixel are sorted in order of increasing intensity, the application of a filter of rank-order r causes the original pixel gray level to be replaced by the value which is the rth in the ordered list. In principle, this operation can be applied simultaneously at every pixel in an image, but the process of sorting in a parallel mode is difficult to envisage and also liable to involve undesirably large amounts of image memory. The method proposed by Danielsson involves the inspection of the gray levels of the neighboring pixels on a bit-by-bit basis. The algorithm can best be explained by means of an example.

Suppose the 3×3 neighborhood of a particular pixel contains the gray levels G1...G9 where G5 is the value of the particular pixel and the remaining eight values refer to the local neighbors. In this example, the actual values are assumed to be between 0 and 15 and can, therefore, be represented in four bits (Table 7-2).

Table 7-2 First Stage Danielsson Ranking

i	MOST SIGNIFICANT BIT B1	B2	B3	LEAST SIGNIFICANT BIT B4	Gi	RANK-ORDER r
1	1	0	0	1	G1	5
2	0	0	1	0	G2	1
3	0	1	0	1	G3	3
4	1	1	1	0	G4	9
5	1	0	1	0	G5	6
6	0	1	0	1	G6	4
7	1	0	1	1	G7	7
8	0	1	0	0	G8	2
9	1	1	0	1	G9	8

If the median is required, then the value sought is that of the pixel whose rank order is five. The process starts by dividing the G values into two groups on the basis of the first bit B1. In the high value group are the set (G1, G4, G5, G7, G9) and in the low group the set (G2, G3, G6, G8). It is clear that the median must be in the high value group and that the set (G2, G3, G6, G8) can, therefore, be discarded. A counter Q is incremented by four, being the number of elements in the low value group. The reduced list of candidates for rank four is shown in Table 7-3.

Table 7-3 Second Stage Danielsson Ranking

i	B1	B2	B3	B4	Gi	RANK-ORDER r
1	1	0	0	1	G1	5
4	1	1	1	0	G4	9
5	1	0	1	0	G5	6
7	1	0	1	1	G7	7
9	1	1	0	1	G9	8

The second bit B2 can now be examined and the candidates divided again into high and low value sets. At this stage, the high value set is (G4, G9) and the low value set (G1, G5, G7). The counter Q is incremented by three (the number of values in the low value set) and reaches a value (seven) which is greater than the desired rank. This implies that the median value is now in the low value set and (G4, G9) can be discarded. The new candidate table is shown in Table 7-4.

Table 7-4 Third Stage Danielsson Ranking

I	B1	B2	B3	B4	Gi	RANK-ORDER r
1	1	0	0	1	G1	5
5	1	0	1	0	G5	6
7	1	0	1	1	G7	7

Proceeding as before on the basis of B3, the high value set is (G5, G7) and the low value set (G1). Since the value of Q is now greater than the rank desired, it is known that the median is one of the seven elements indicated by the value of Q. Specifically, it is further indicated as being one of the values of the members of the set (G1, G5, G7). The object is, therefore, to remove the higher values from the table, leaving the set containing the median still in the table, reducing Q by the number of elements removed. The final value of Q is, in fact, five, which is the required result, indicating that the median is G1.

The entire process is summarized in the flow chart of Figure 7-22. The gray values are assumed to have bits (B1,...,Bk, ...,BN) and the neighborhood has elements (G1,...,Gj,...,GN). B1 is the set of most significant bits. The required rank-order is r. The quantity is the set of candidate elements.

In the form given, this algorithm does not obviously lend itself to implementation in a cellular automaton. It must be

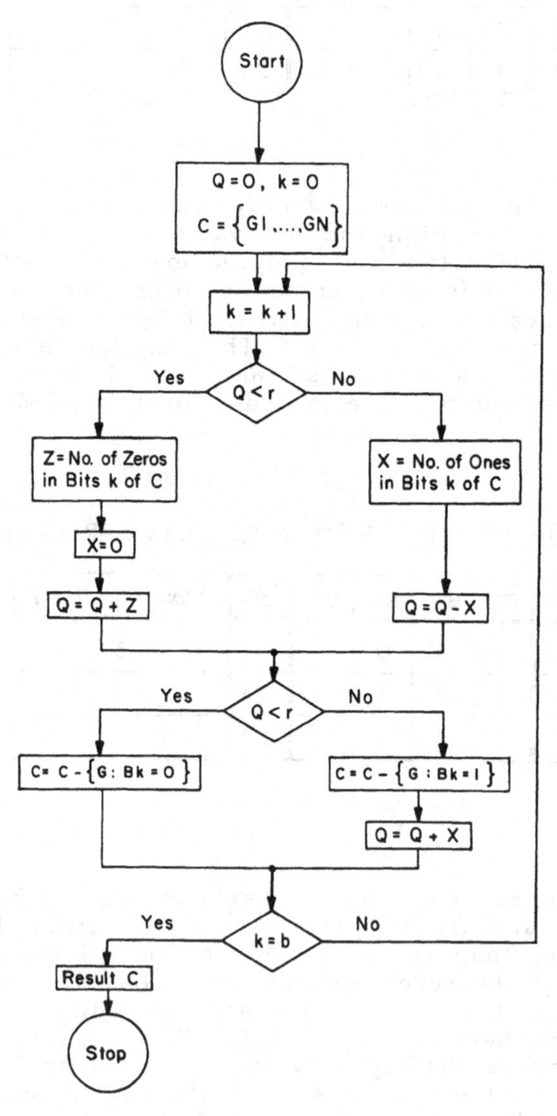

Fig. 7-22 Flow chart for rank-order filter.

remembered that conditional branches will usually be obeyed in both directions at different elements of the array. A parallel version of the algorithm must, therefore, be adapted from the serial version. A set of counters and registers is associated with each element of the array, as follows

$$\begin{aligned} Q &= \{Q1,\ldots,Qn\} \\ Z &= \{Z1,\ldots,Zn\} \\ X &= \{X1,\ldots,Xn\} \\ P &= \{P1,\ldots,PN\} \\ K &= \{K1,\ldots,KN\} \end{aligned} \tag{7.13}$$

where $n = \lceil \mathrm{Log}\ N \rceil$ the first stage of the parallel algorithm involves shifting the most significant bit of each of the neighbors into P1 of the central element, the next bit into P2, and so on. In the second stage, the Z bit stack counter is incremented by the number of zero elements in P which correspond to one elements in K, the candidates register, i.e., Z is incremented by $PJ'.KJ$ for $J = 1$ to N. At the same time, X is incremented by $PJ.KJ$). Q is then added to Z. A flag F is then set at one where $Q < r$ (and zero elsewhere). Referring to Figure 7-22, it can be seen that, when F is one, the left branch is taken and the count X is no longer required. F' can, therefore, be ANDed with all the elements of X to zero the X counter. Similarly, if F is zero, the right branch is followed and the Z count can be zeroed by ANDing the elements of Z with F. The value of Z is then added to Q and X subtracted from Q.

The new values of Q are next tested and the flag again set at one where $Q < r$ and zero elsewhere. F' is ANDed with X which is now added to Q. It only remains to update the candidate list K by eliminating the elements represented by an I in P (where F = 0) or by a 0 in P (where F = 1). This can be effected by the simple operation

$$KJ = KJ.\mathrm{AND}.(F.\mathrm{EXOR}.P) \tag{7.14}$$

The process is illustrated in the flow diagram in Figure 7-23. The carry function implies that bit-stack subtraction is performed retaining only the carry plane at each bit. After the final (most significant) bit, the carry will contain ones where the result is negative. The AND operations between the flag plane F and a bit-stack (such as X) are performed between F and each bit plan in turn. The operations + and - refer to arithmetic addition and subtraction of bit-stacks.

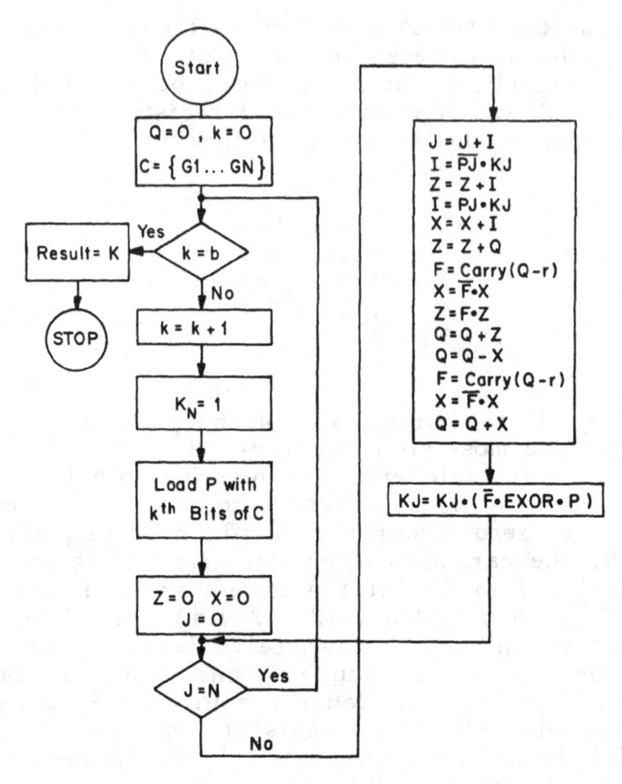

Fig. 7-23 Flow chart for parallel implementation of rank-order filtering.

The cost of this process can be estimated (in comparison with, say, forming the average of a neighborhood of N elements) by analyzing the flow diagram. Apart from the shifting operations (common to both processes), the rank-order filter demands

5bN	n-bit additions
bN	n-bit stack incrementations
2bN	n-bit carry preserving additions
3bN	n-bit stack ANDing operations

which is equal to order 11bNn operations. Note that a simple averaging process would require N additions on a bit-stack of size (n + b) bits, which is obviously a much less expensive process than rank-order filtering.

8. SCIENTIFIC APPLICATIONS

8.1 INTRODUCTION

This chapter describes applications of cellular automata to data which do not necessarily represent optical images, even though images can be used to represent the data, e.g., an electrostatic field. The array should therefore be regarded not as part of a vision system but rather as a computation and *visualizing* system. It will be shown that this approach to computation stimulates the conception of unusual computational algorithms which are not only capable of efficient implementation by the cellular automaton but also, in some cases, more efficient than when implemented on conventional serial machines.

As a simplification to this discussion, it will be assumed that the array dimensions of the cellular automaton always match the dimensions of the data array being considered. In reality, this often will not be the case and techniques will need adjusting to take the mismatch into account. It can also be assumed that such adjustments are likely to be fairly minor; the basic algorithmic structure of the problem solution can be expected to be unaffected.

The order of the sections in this chapter is arbitrary since no obvious ordering suggested itself. Also, the descriptions of algorithms are somewhat imprecise although results presented are from working programs which have been run on CLIP4. The imprecision in the descriptions is intended to avoid the temptation to "tailor" the algorithms to this one machine.

8.2 MINIMUM PATH DETERMINATION

The general problem is to join two points in an array by a path satisfying certain minimization criteria and other limiting constraints. For example, the path between two given points might be required to avoid a set of obstructions and to have the minimum unobstructed path length.

Applications are to be found in many fields, including image processing in which the joining of points is often a stage in more complex tasks such as image segmentation. In industrial technology, routing problems play an important part in printed circuit board and integrated circuit design. Three examples will be considered here in order to illustrate a range of different approaches from which the most suitable might be selected for a specific application.

8.2.1 Simple Point Joining

If there is a straight unobstructed path between the two points to be joined, then an extremely simple algorithm can be employed; see, for example, Stamopoulos (1975). In Figure 8-1, the two points are labeled X and Y and the four array sectors surrounding each point are Xa, Xb, Xc, Xd and Ya, Yb, Yc and Yd, respectively. The steps in the algorithm are as follows

1. X and Y are represented as 1-elements in two arrays Ix and Iy otherwise containing only zeros.

2. Ix is directionally propagated into one of the four array sectors. After propagation the result is ANDed with Iy to determine whether the sector includes the point Y.

3. When a direction causing an inclusion of Iy is found, Iy is then propagated in the reverse direction.

4. The two results (from Ix and Iy) are ANDed together to form the rectangle XEYF.

5. The rectangle is skeletonized using a standard skeletonizing algorithm (Chapter 5), taking care to prevent the elimination of X and Y during the process.

6. The final result is the minimum path between X and Y.

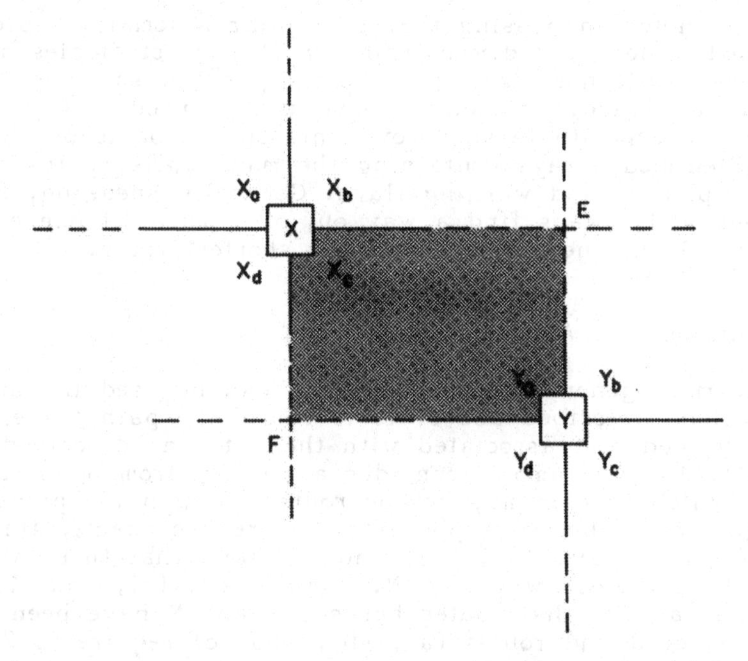

Fig. 8-1 A path connecting points X and Y is found by generating the rectangle XEYF and skeletonizing.

This algorithm does not yield a straight, minimum path. Better results can be obtained by using more array directions, ANDing more sectors. For an in-depth investigation of the minimum path problem the reader is referred to Lee (1961).

8.2.2 Maze Problems

A somewhat artificial path problem concerns the solution of a maze. If the path area in a maze is exactly one element wide at all points, then the path will appear as a tree-like structure when removed from the array displaying both the walls and paths of the maze. The branches of the tree will terminate at the maze "dead-ends," except for one branch which finishes at the goal or exit of the maze.

Solving the maze presents no difficulties since all that is required is to apply a logical transform which recursively removes line ends while re-establishing the ends forming the maze entrance and the maze exit after each iteration. When the process reaches a stable state, the residual path is the required path. Detection of line ends is a trivial task using local neighborhood operators (Chapter 2).

It is noted in passing that a cellular automaton could also be used to test and demonstrate *serial* maze strategies although these will not result in efficient parallel solutions. In the serial approach, an isolated element is caused to traverse the maze, making single step movements on the basis of the local neighborhood arrays containing the maze walls, paths previously explored, and warning flags. Generally speaking, such techniques will always find a way out of a maze, if one exists, but will not necessarily find the shortest route.

8.2.3 Minimum Cost Algorithm

The more general approach, which can be used in both the situations described above, is to relate the path traversed to some defined cost associated with the path, as described by Lee (1961). As an example, consider a journey from point X to point Y, which may or may not be routed through the point N (see Figure 8-2). Let us suppose that there are exactly three routes into N, namely (a), (b), and (c), and that there are exactly three routes away from N, namely (d), (e), and (f). Let us also assume that routes between X and N have been previously explored and routes (a), (b), and (c) require 5, 2,

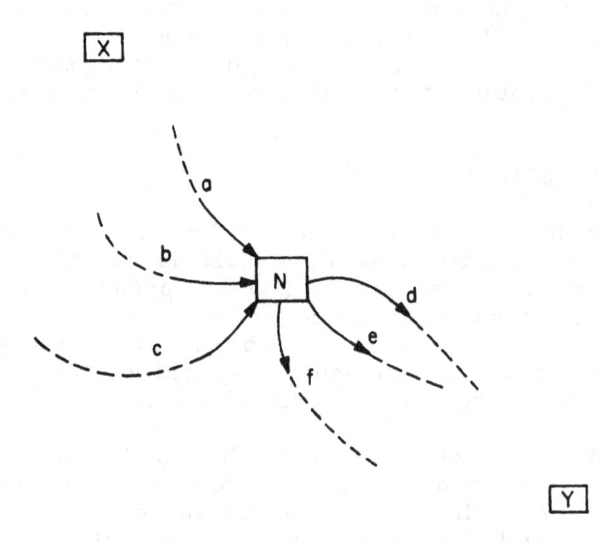

Fig. 8-2 Configuration for determining the minimum cost of taking a path through N along incoming routes (a),(b),(c) and outgoing routes (d),(e),(f).

and 3 units of fuel, respectively. The next stage in the operation will be to set off simultaneously in the three directions (d), (e), and (f), measuring fuel consumption along each path, until further nodes are found. However, before leaving node N, the node is tagged with the symbol (b) as representing the path which could be taken back to X with minimum fuel consumption. Once Y has been reached, a reverse path can be traced, using the tags to select the minimum path each time a node is crossed. The retraced path is then the path between X and Y allowing minimum fuel consumption.

When the cost associated with a route is merely the integrated path length, the cellular automaton provides a very neat solution to the problem. In this case, the contents of an array I_x which is initially empty except for a 1-element at X, are augmented and the result NANDed with a second array I_o which contains 1-elements representing path obstructions. Thus

$$I_x \leftarrow E(I_x) \cdot I_o' \tag{8.1}$$

where $E(I_x)$ is the result of augmenting I_x. If this process is repeated indefinitely, I_x will eventually touch the point Y held in another otherwise empty array I_y. This can be detected by examining $I_x \cdot I_y$ which will then be non-empty.

Prior to each cycle of augmentation, the 3×3 local neighborhood of each zero element of I_x is inspected and a label stored to indicate the position of 1-elements in the neighborhood. Conveniently, the labels can be the numbers 0 to 7 (requiring three bits of storage) for the eight neighbors, arbitrarily selecting the lowest number when more than one of the neighbors is a 1-element.

Back-tracking involves shifting a marker in a sequence of directions indicated by the stored labels, ORing the new position into an array after each step so as to form the complete path. It will be noted that this technique does not seem to take full advantage of the parallel processing capability of the array, particularly during back-tracking when only one processing element is in use at each step.

Figure 8-3 provides illustrations of both point joining and minimum path construction, using CLIP4.

8.3 FIELD CALCULATIONS

Cellular automata provide a convenient method for calcula-

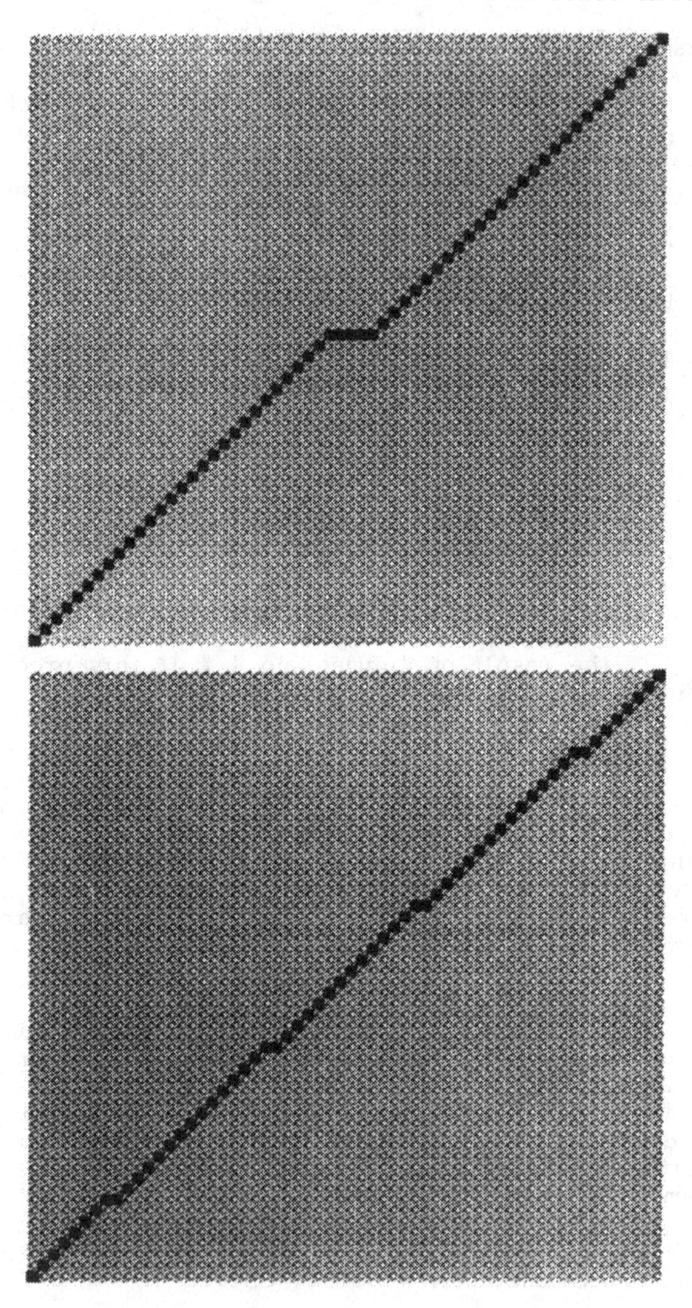

Fig. 8-3 The minimum path between two points is obtained using the Stamopoulos algorithm as shown in A. The minimum unobstructed path may also be calculated by the Lee algorithm (B).

ting potential fields by relaxation. Suppose an electrode structure is represented by a two-dimensional numerical array, each element representing the potential at one point on a particular electrode (which will be assumed to be held constant). The problem is to calculate the resultant potentials at all other points (elements) in the plane.

The method is iterative and consists of replacing the potential in each element by the average of that of itself and of its eight neighbors at each iteration. The electrode elements are restored to their original constant potentials after each iteration. In practice, it is computationally much more efficient to ignore the central element when averaging since this simplifies the required division to a division by 8 (which can be accomplished by shifting) rather than by 9.

The principle is illustrated in a one-dimensional example in Figure 8-4 which represents a six-element linear array, converging at the eleventh iteration. Although somewhat outside the scope of this chapter, it is interesting to note that the stable situation reached at the end of the process is the result of a wave of activity passing from the left to the right, i.e., from high to low potential. The speed of transmission has been increased by rounding up the averages to the nearest integer. A further speed increase can be obtained by starting waves of activity from both the left and the right, which is achieved by setting the intermediate points at an initial value approximately half way between the extremes. In this case, oscillations set in after an initially rapid convergence (three iterations), the oscillations being due to a combination of the effects of two opposing waves and rounding errors (see Figure 8-5). The stability of solutions obtained by these matters is a matter for further investigation. Figure 8-6 shows an example of a field calculation using these techniques.

8.4 DYNAMIC MODELING

Computer simulation of physical processes is a familiar practice which is of particular value when the processes are either slow or physically difficult to realize due to high cost, danger, inaccessibility, or some other constraining factor. The success or failure of a simulation will depend on how closely the model matches the original process, thus depending on the available knowledge of the physical laws governing the process and the magnitudes of the relevant parameters, and on the accuracy with which these laws and parameters can be represented in the computer domain. The penalty for too low an accuracy is a model whose behavior does not resemble that of the simulated physical process, whereas an unnecessarily high

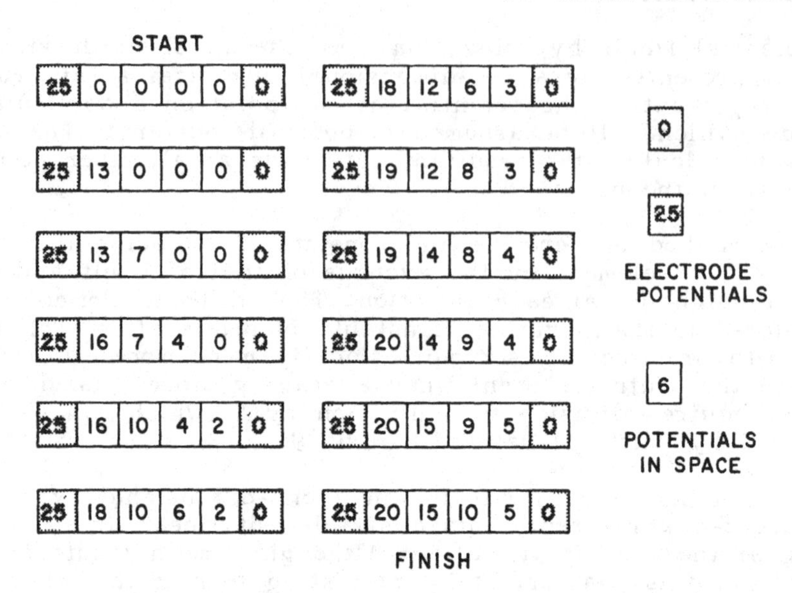

Fig. 8-4 The cellular automaton may be programmed to calculate electric field potential. Steps of a one-dimensional example using left-to-right propagation are shown above.

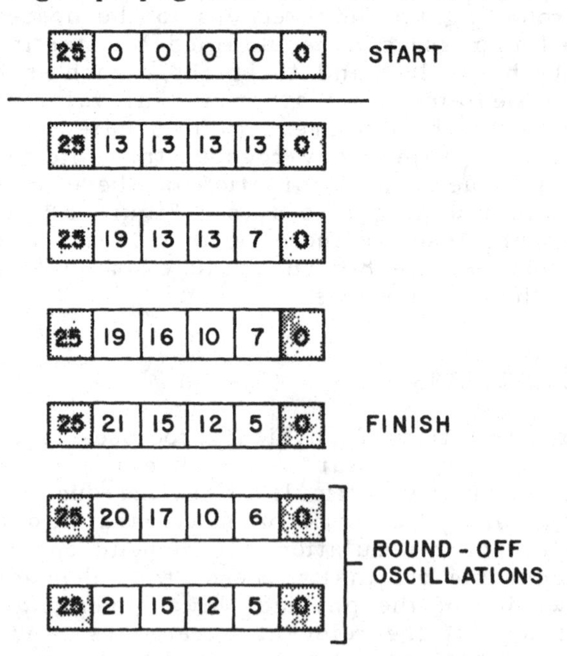

Fig. 8-5 Calculations of electric field potential may also be carried out by simultaneous propagation from the left and the right.

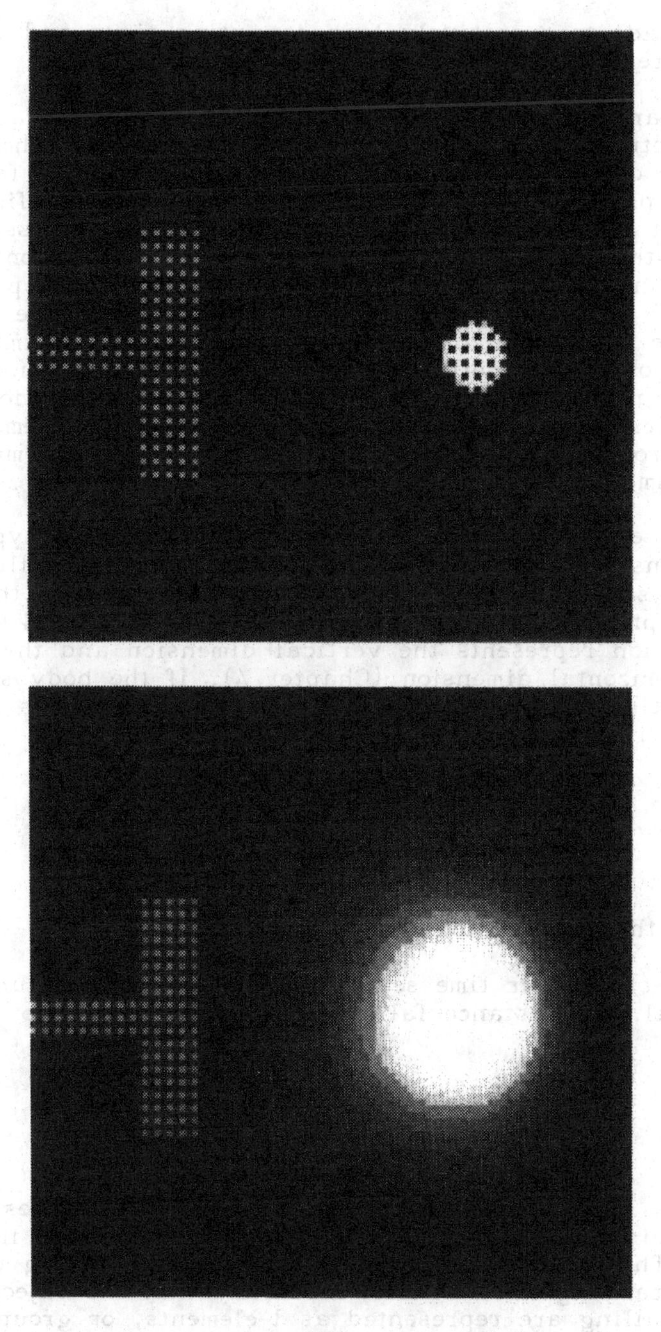

Fig. 8-6 Electric field calculations showing (A) the electrode structure and (B) the resulting electric field.

accuracy leads to excessively long computing times and wasteful use of resources.

Cellular automata are most likely to be effective as modeling architectures when the process being modeled is inherently of the same dimensionality as the modeling array. In fact, although few physical processes are two-dimensional (confined to a plane), it is often a satisfactory approximation to assume uniformity in the third dimension. The plane of calculation is taken to be a cross section through the three-dimensional process in which all relevant parameters are independent of the position of the plane in the third dimension. Subject to this limitation, it is then possible to define the interaction between points in the modeled region in terms of local neighborhood functions. Analytic problems arise due to discretization of the mathematics but these problems are not specific to the cellular automaton but are common to all computer models.

As an example to illustrate an approach to this type of problem, consider an object falling freely under the influence of gravity (see Figure 8-7). At the beginning ($t = 0$), the object is represented by an isolated 1-element at $s(i,j,k)$ where the y direction represents the vertical dimension and the x direction horizontal dimension (Chapter 4). If the body starts with an initial velocity equal to u, then the distance s fallen in a time t is given by

$$s = ut + \tfrac{1}{2}gt^2 \tag{8.2}$$

where g is the acceleration due to gravity.

If we choose our time scale so that, in the first unit time interval, the distance fallen is one row (from u to u-1), then

$$1 = 0 + \tfrac{1}{2}g \tag{8.3}$$

yielding $g = 2$. At subsequent times, the total distances fallen, measured in units of the array spacing, are given in Table 8-1. This process could be implemented by setting up the simple program segment as follows. Assume that the objects which are falling are represented as 1-elements, or groups of 1-elements, in image plane I_o. Also assume that a set of bit planes $V_1, V_2, \ldots, V_v$ is associated with the binary image plane.

Table 8-1 Position Table for Free-Falling Body
(See text for explanation of symbols.)

t	0	1	2	3	4	5	6	7	8
s	0	1	4	9	16	25	36	49	64
y	u	u-1	u-4	u-9	u-16	u-25	u-36	u-49	u-64
Δy	0	1	3	5	7	9	11	13	15

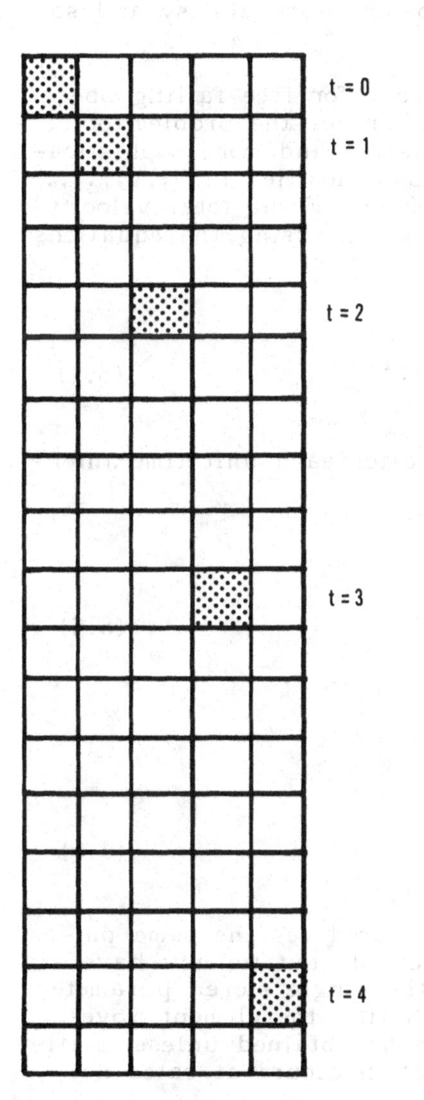

Fig. 8-7 Time-sequential position of a free-falling body having a non-zero left-right initial velocity.

Stored in this bit stack above each element in I_o is a binary number which will be the number of downward shifts to be applied to that element in the next unit time interval; V_1 the least significant bit plane. After each interval, all bit stacks are incremented by two units (Table 8-1).

To produce the correct shift for every element in any particular interval, V_1 is ANDed with I_o and with the bit stack, and the resulting v+1 planes shifted down by one unit, ORing the shifted elements into the original planes and NANDing the shifted elements from the original planes. The process is repeated with V_2, but this time giving two downward shifts, and so on for $V_3,\ldots,V_v$.

While this approach is satisfactory for free falling objects, it does not offer a simple solution for the problem of colliding or bouncing objects. An alternative and more sophisticated technique is to store instantaneous velocities in $V_1, V_2, \ldots, V_v$. These can be the x and y components of the total velocity if two-dimensional motion is to be modeled. Using the equations

$$v = u + gt$$

$$s = ut + \tfrac{1}{2}gt^2 \qquad (8.4)$$

where v is the final velocity, then, after each unit time interval,

$$\Delta u = g$$

$$\Delta s = u + \tfrac{1}{2}g \qquad (8.5)$$

and, for g = 2 (as above)

$$\Delta u = 2$$

$$\Delta s = u + 1 \qquad (8.6)$$

It can readily be seen that this produces the same pattern of motion as did the previous method, but we now have a way of dealing with bouncing since the single stored parameter u can be systematically adjusted each time the element moves. Satisfactory precision is not likely to be obtained unless smaller time intervals are used so that motion occurs at rates not greater than one array element per unit of time. Note that bouncing involves (1) testing for an obstruction, held as a binary image in another array, (2) reversing the shift direction, and (3) reversing the acceleration (implying decrementing rather than incrementing the stored variables).

We note here that the local neighborhood operator concept is used to detect obstructions. This could be extended for the calculation of elastic collisions between moving bodies of known mass, two velocity components and the bodies' masses being stored in $V_1, V_2, \ldots, V_v$. Ultimately, modeling of this type is likely to be difficult primarily because of the discrete representation.

8.5 LAY PLANNING

In certain industries, it is sometimes necessary to solve a problem which has much in common with a jigsaw puzzle. Whenever a collection of two-dimensional shapes is to be cut out from a uniform sheet of material, there is a need to fit together the shapes so as to use the material economically. A particular instance occurs in the clothing trade where the problem is simplified (for automatic solution) by the requirement that the orientation of the pattern parts is fixed in relation to the pattern on the cloth.

The solution to this problem involves a sequence of searches for the optimum position of the next part to be located in a partially completed plan and consists of a constrained exhaustive search moderated by heuristics. A working algorithm has been described by one of the authors in Duff (1980). Typical results are reproduced here in Figure 8-8. Although it is not worth reproducing the complete algorithm in this chapter, it is of interest to review the principles involved. At each stage in the process, the following procedures are carried out

1. A pattern piece is selected as the next candidate for insertion in the plan.

2. The plan is explored to determine possible insertion positions for the new piece.

3. The positions are evaluated against certain criteria.

4. (a) The new piece is moved into position or

   (b) The piece is set to one side for later consideration.

Fig. 8-8 In lay planning the cellular automaton is used to calculate the positioning of the clothing pattern parts (A) which yields the optimum use of the material as shown in (B).

In the published solution, possible insertion areas are found by a form of autocorrelation. The incomplete plan is scanned over a copy of itself, adopting a scanning path which corresponds with the edge of the new piece to be inserted. The copy is ORed into the plan at every step in the scan and the remaining blank areas are candidate insertion locations for the point on the edge of the new piece at which the scan path started. The uppermost of the most left of these points is selected. Step 3 is performed by measuring the overlap of the new piece with what has been referred to as the "committed area." This is the area of cloth bounded by the left edge of the cloth and its upper and lower edges, closed by a vertical line just touching the extreme right hand side of the elements in the partially completed plan. If the new piece is contained within the committed area, it is inserted. Otherwise the area of intersection between the new piece and the committed area is measured and stored. The next new piece (choosing pieces in order of decreasing area) is tried and inserted if completely contained in the committed area and also of larger overlap area than the previous piece. The aim is eventually to insert the piece whose overlap area is a maximum, out of all those pieces remaining to be inserted.

Even from this brief description, it can be seen that many of the operations involved are eminently suitable for implementation by the cellular automaton. The example illustrated was computed in a few seconds on CLIP4.

8.6 UNCONVENTIONAL MATHEMATICS

The natural way to represent data in a two-dimensional cellular automaton is, of course, as a two-dimensional array. It follows that certain mathematical operations can be approached in quite unconventional ways once the data have been transformed into an array format; it is also possible that the data may already be in such a format if they have themselves been produced in or by the array.

8.6.1 Location of Extrema

Consider, for example, the sequence of numbers in Figure 8-9 shown plotted as linear columns. If these numbers are integers d_0 to d_{11}, from left to right, then the largest can be found by the simple program which is outlined on the following page.

START: $j = 0$

$d_{max} = d_j$

LOOP: $j = j + 1$

if $d_{max} > d_j$, go to TEST

$d_{max} = d_j$

TEST: if $j \neq 11$, go to LOOP

END

However, if the data already appears as the array shown in Figure 8-9, one method of finding the maximum at C would be to lower a line AB until the line intersects the data. A faster method would be

1. PROPAGATE a copy of the data left and right.

2. EXTRACT the top edge of the rectangular expanded area.

This short program (only two operations in CLIP4) produces a line at the height of the maximum of the data, corresponding to the line ACB. The next step would depend on what facilities exist for reading data from the array; for example, some cellular automata might enable coordinates of points to be extracted. Note also that the minimum, or lowest integer, could be found by the equivalent process of propagating the white regions in the figure before extracting the top edge.

Taking the same data plotted in binary coded columns in a bit plane (Figure 8-10a), an alternative approach can be used to find maxima and minima in the graph or histogram expressing the data. The array is shifted one column to the right (Figure 8-10b) and the new array subtracted from the old, as explained in Chapter 4. The difference array is shown in Figure 8-10c. It will be noted that the negative results produce 1-elements, shown shaded, in the top row. If the top row only is transferred into an empty array, then elements X and Y in Figure 8-10d are defined by

X $\equiv$ any 0-element with a 1-element to its immediate right

Y $\equiv$ any 1-element with a 0-element to its immediate right

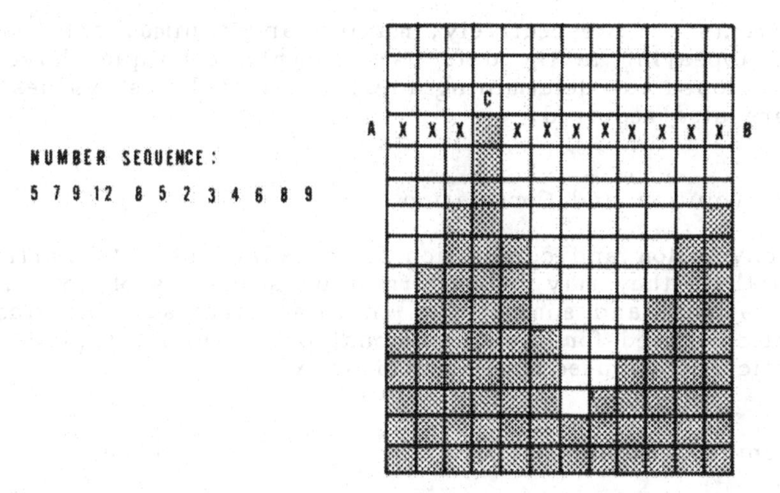

Fig. 8-9 The maximum value of a series of numbers can be determined from the bar-graph representation of the numbers in a cellular array.

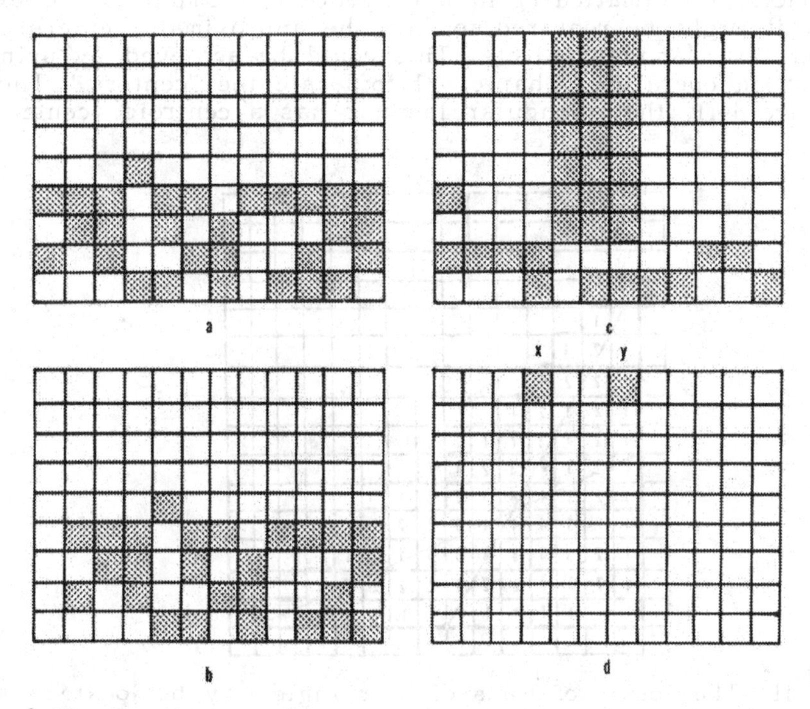

Fig. 8-10 Binary coded data arranged in the cellular array as shown above can also be transformed so as to locate extrema. (See text for details.)

The elements are, respectively, maxima and minima, only one of each appearing in the data used. Simple techniques have been developed for determining the highest or lowest values if these are also required.

8.6.2 Convolution and Correlation

Convolution and correlation fit cellular automata particularly well as they may be performed by sequences of shifts, multiplications, and summations. In these processes, bit stack arithmetic is used for two-dimensional operations; bit plane arithmetic can be used one-dimensionally.

8.6.3 Center of Mass

Another aspect of the unconventional use of cellular automata for arithmetic operations is that such an array will sometimes provide a non-arithmetic method for performing an operation which would normally, although artificially, be carried out strictly mathematically in a conventional computer. For example, it might be required to find the approximate centers of mass of some laminar objects. This could be achieved by using a reduction operation (Chapter 2) to locate the "centers." Thus in Figure 8-11, the triangular laminar has a centroid (center

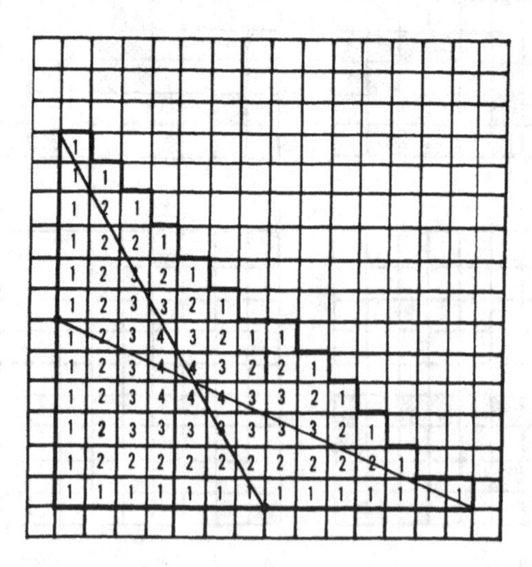

Fig. 8-11 The center of mass of a triangle may be located using reduction with $\Xi = 7$ until the iteration just before the step when erasure occurs. The remaining 1-elements can then be reduced to a residue using subfields.

of mass) marked by the intersection of the two lines joining the apexes to the midpoints of edges. The numbers in Figure 8-11 indicate the index of the cycle of reduction needed to eliminate each element using $\Xi = 7$. It can be seen that, after three cycles, a small group of elements remains (labeled "4") which would disappear after one more cycle and which surround the centroid.

8.6.4 Line Length

The discrete nature of images represented in an array implies certain unusual properties. Consider the three line segments shown in Figure 8-12. The lines shown shaded are the minimum 8-connected lines joining the end elements. The dotted elements must be added to produce 4-connected lines. Table 8-2

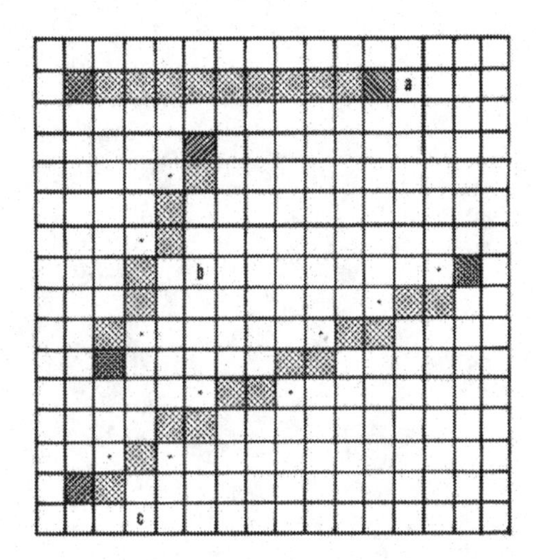

Fig. 8-12 Line length may be estimated by taking the geometric mean of the number of elements in the 4-connected path and the 8-connected path between two points.

Table 8-2 Length Table for 4-Connected, 8-Connected, and True Measures

LENGTH				
LINE NO.	4-CONN.	8-CONN.	TRUE	GEOM. MEAN
a	11	11	11	11.00
b	11	8	9	9.38
c	21	14	16	17.15

shows the true line lengths (measuring to the end-points of the end elements) and the lengths obtained by counting the elements in the 8- and 4-connected representations. The geometric mean is also tabulated.

It has been found empirically that the geometric mean of the number of elements in the 8- and 4-connected discrete representations is a good approximation to the line length in continuous space. Furthermore, the length of a line and the area of a laminar can both be found by counting the elements of the array. This points to the importance of devising efficient counting mechanisms in the array, a feature which sometimes tends to be neglected.

9. BIOMEDICAL APPLICATIONS

9.1 INTRODUCTION

Without question, the most important and extensive commercial application of cellular logic techniques is in biomedicine. The diff-series of cellular logic machines initially developed by Perkin-Elmer and currently under development, manufacture, and distribution by the Coulter Biomedical Research Corporation are employed worldwide. During periods of peak operation, these machines both screen and analyze the images of the red cells, white cells, and platelets of approximately 50,000 individuals daily. All employ the GLOPR cellular logic machine described in Chapter 10. At the time of writing, if these machines were in use simultaneously, they would be capable collectively of reviewing at low resolution the images of about 30 million red cells per hour, analyzing simultaneously 30,000 high-resolution white cell images, while also measuring platelet characteristics. This is the largest automatic image analysis effort anywhere in the world and exceeds by an order of magnitude the image *sensing* capability of LANDSAT.

As yet, the array automaton is not used commercially due to its relatively recent reduction to practice. More than one CLIP4 machine is currently in use with, at the time of writing, a single MPP in operation (see Chapter 11). Interesting research in biomedical image processing has been conducted, not only with these machines, but also using the earlier CLIP3. This work is reported in this chapter. The chapter begins with a discussion of nuclei finding and counting followed by a description of the use of both GLOPR and CLIP3 for the analysis of chromosomes. This is followed by an illustration of the application of exoskeleton analysis to the quantitative description of tissue architecture as displayed in images of kidney sections. Next, the combination of both endoskeleton and exoskeleton analysis to the description of the structural properties of human white cells is treated, followed by sections on the application of CLIP4 to computed tomography and on the three-dimensional skeletonization of neurons.

9.2 NUCLEI FINDING AND COUNTING

The earliest application of cellular logic transforms, using a 2×1 kernel, was that reported by Mansberg and Segarra (1962). The biomedical specimens used (brain tissue sections) provided an ideal input signal in that objects of interest could be made to appear at high contrast by means of suitable staining and illumination techniques. As mentioned in Chapter 1, an intercept association method was used, based on the still earlier work of Causley and Young (1955) at University College London, in counting red blood cells. These latter workers avoided digital techniques and employed a television microscope and analog circuitry.

Object counting in high-contrast biomedical imagery is a relatively simple matter using a single threshold followed by a cellular logic operation which reduces each object to its residue. Total residues may be counted. Also, by dividing the image into sub-images, the local density of the residues may be determined so as to measure both the average number of objects per unit area and their variance.

Figure 9-1 is an example of a more difficult case where the objects of interest (the cell nuclei) appear at a variable contrast against a complex background. The image shows the tubule structure of the normal cortical region in the kidney, along with several glomeruli. In some regions, such as along the boundaries of the tubules, cell nuclei are clearly defined. They are the round, darkly-stained objects and are about eight micrometers in diameter. In other regions, such as in the glomeruli, the cell nuclei appear in dense clusters. In still other areas, they are weakly stained and scarcely visible. If one were to graph a single row of data in the digitized image, the results would be as shown in Figure 9-2. Each mode in this graph is associated potentially with a single nucleus and, if a count of these nuclei is desired, it is clear that many threshold levels must be examined. There is clearly no single threshold at which the multilevel image may be converted into a bilevel image wherein all cell nuclei would be clearly defined.

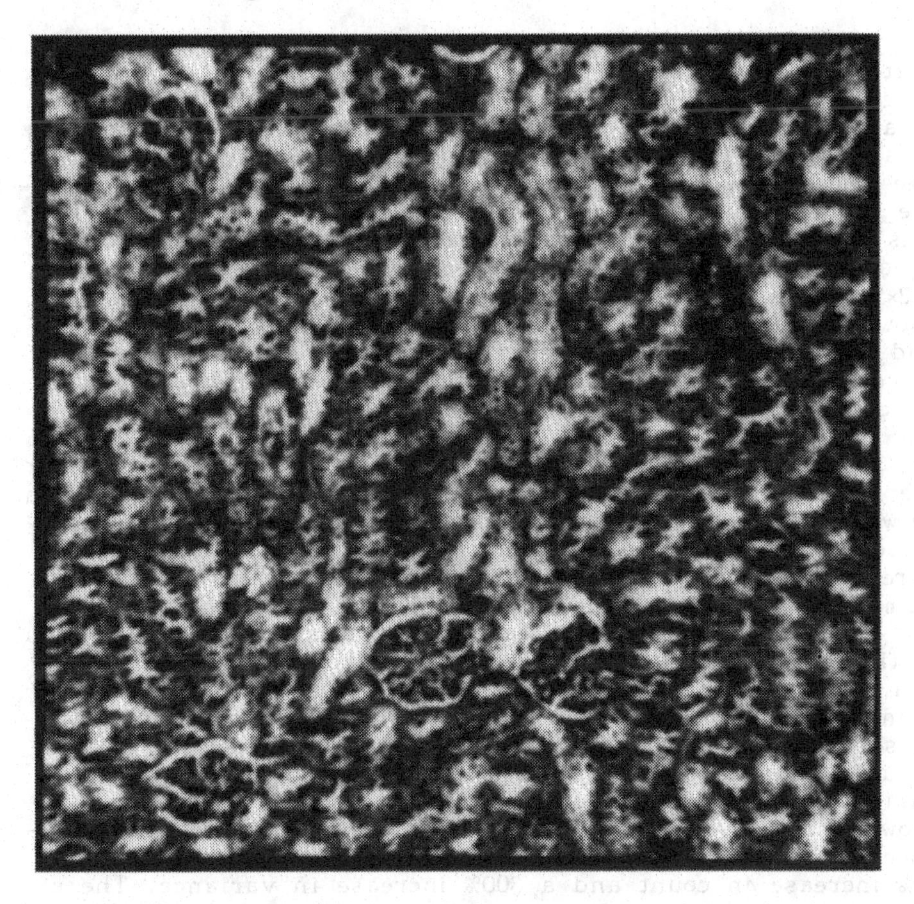

Fig. 9-1 Graylevel image of a 0.8×0.8 millimeter area of normal human kidney tissue.

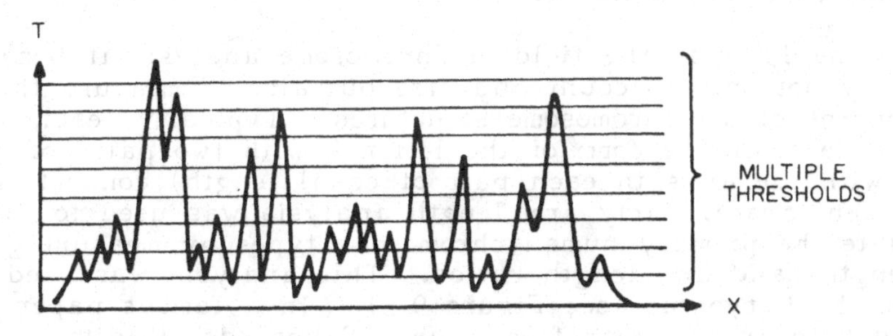

Fig. 9-2 Graph of a portion of single row of data from the negative of an image such as that shown above. Each mode corresponds to the nucleus of one cell.

In order to handle object counting in images such as that shown in Figure 9-1, multilevel thresholding is employed. At each threshold, a cellular logic operator is used to reduce to a residue any contiguous cluster of 1-elements corresponding to an object as large as about ten micrometers in diameter. The results are ORed into an accumulator and, to remove multiple counts, clusters of residues are merged (Figure 9-3). (In this figure, both single and merged residues are shown as 3×3 squares due to the difficulty of observing single points in a 512×512 array.) Results may be evaluated from the display shown in Figure 9-4. A visual comparison between Figures 9-1 and 9-4 indicates good agreement between the actual locations of cell nuclei and those regions indicated as being cell nuclei by cellular logic image processing.

This analytical method was employed in an experiment for the purpose of determining whether it is possible to distinguish between images of the cortical region of a normal kidney with those of diseased kidneys (Preston, 1981). Tissue sections were gathered from four normal rabbit kidneys, as well as from four kidneys in which pyelonephritis had been induced. The cellular logic computer was used to analyze images of each section and to score the locations of approximately 30 thousand cell nuclei. It was found that not only the absolute nuclei counts, but also their variances, were specific to kidney disease. Normal rabbit kidney tissue measurements were closely grouped near a total count of 3,000 per 512×512 field with a variance over 64×64 windows of less than 700. Abnormals showed not only increased cell counts, but also increased variances. Typically, an image of abnormal tissue displayed a 30% increase in count and a 300% increase in variance. The effect of disease in creating *architectural disarray* in the kidney was thus confirmed.

9.3 ENDOSKELETON ANALYSIS

In the 1960s in the field of chromosome analysis it became necessary not only to count and size but also to measure the arm lengths of all chromosome structures. Typically, each chromosome is in the form of the letter X with two pairs of arms (with the arms in each pair of equal length) connecting at the centromere. Early arm length analysis was used to differentiate the primary human chromosome types by measuring arm lengths and arm length ratios. This analysis was used to form the karyotype (see Figure 9-5). In a famous paper entitled "Linear Skeletons from Square Cupboards" Hilditch (1969) gave some examples of the problems involved in finding the edges of chromosome arms prior to arm-length measurement. The authors are fortunate to have obtained the results

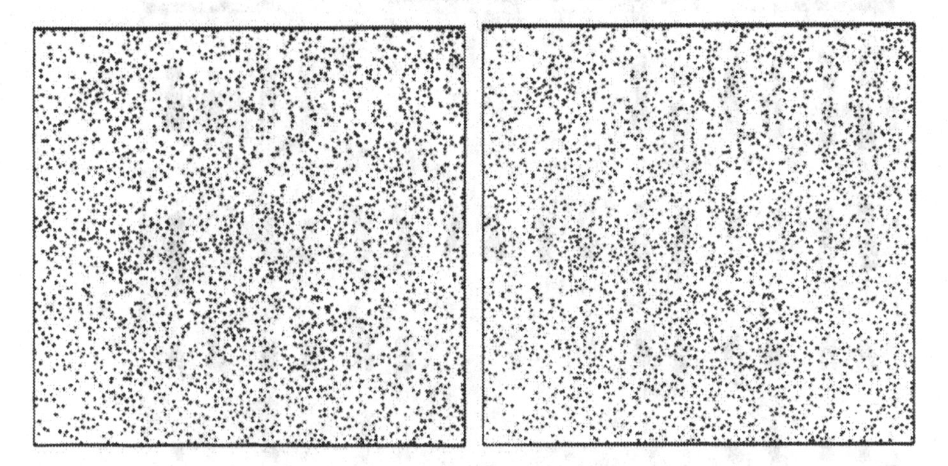

Fig. 9-3 Residues (left) and merged residues (right) obtained from a multithreshold cellular logic transform executed on the original kidney image data.

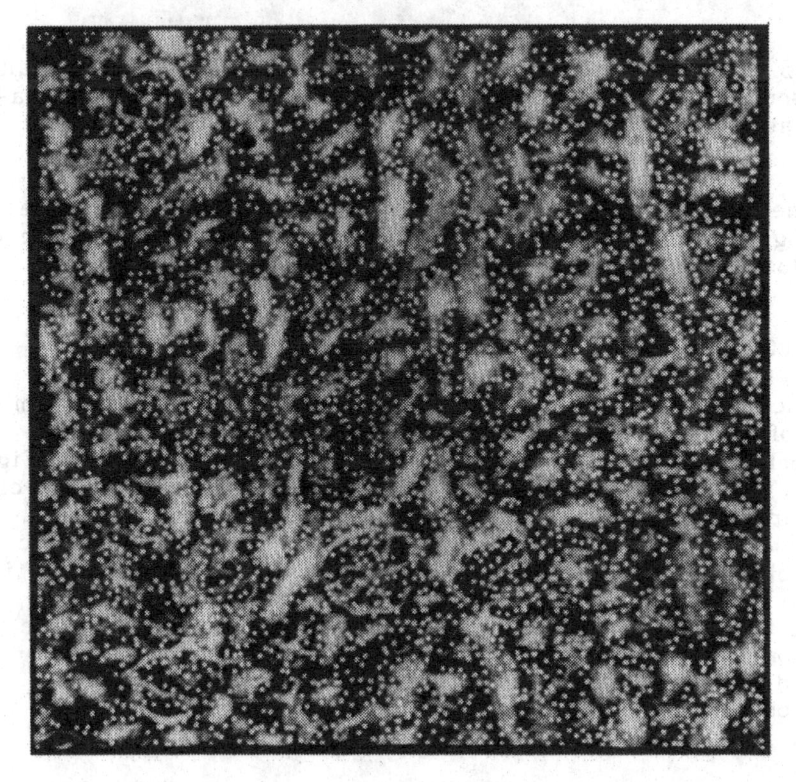

Fig. 9-4 The original kidney tissue image marked with the merged residues illustrated above.

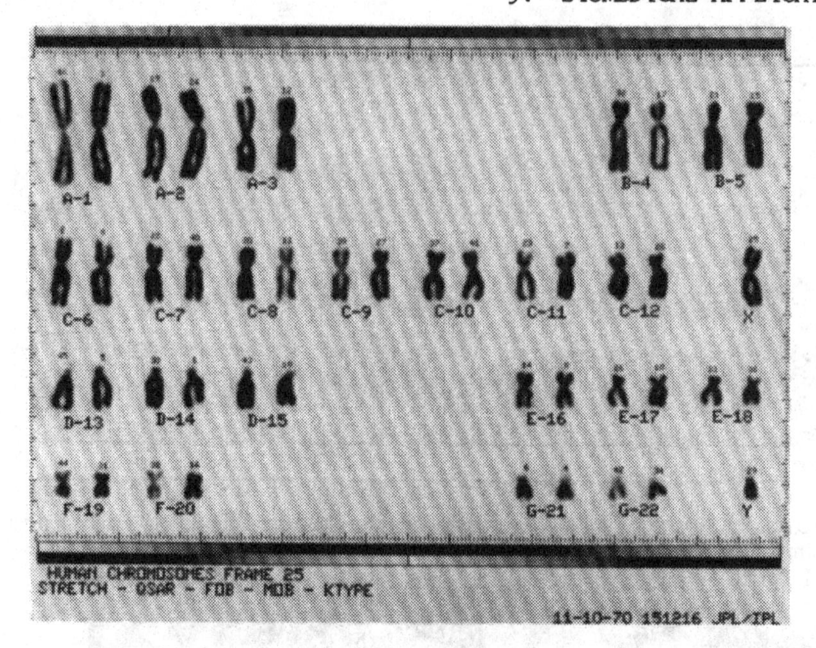

Fig. 9-5 Computer generated karyotype of a cluster of human chromosomes. (Courtesy, K. Castleman, Jet Propulsion Laboratory, Pasadena, California.)

of further, unpublished experiments by Hilditch, who, in the following paragraphs and figures, describes work on the use of the endoskeleton in chromosome analysis.

9.3.1 GLOL Program for Chromosome Analysis

Since Hilditch refers to the chromosome analysis program of one of the authors (Preston, 1973), this section presents that work first. A flow chart of the program is shown in Figure 9-6. The first part of the program digitizes the image of a cluster of chromosomes using an electo-optical microscope scanner as the input device. A histogram of the photometric values obtained in one scan is generated and then smoothed in order to locate the two histogram modes which represent the average transmission of both the background and that of the chromosomes. The minimum between these two modes is then employed as the threshold used to form a binary image. All chromosome image elements are scored as binary 1s; background, binary 0s.

The binary image is stored in image memory A of the GLOPR cellular logic machine and the next program step cop-

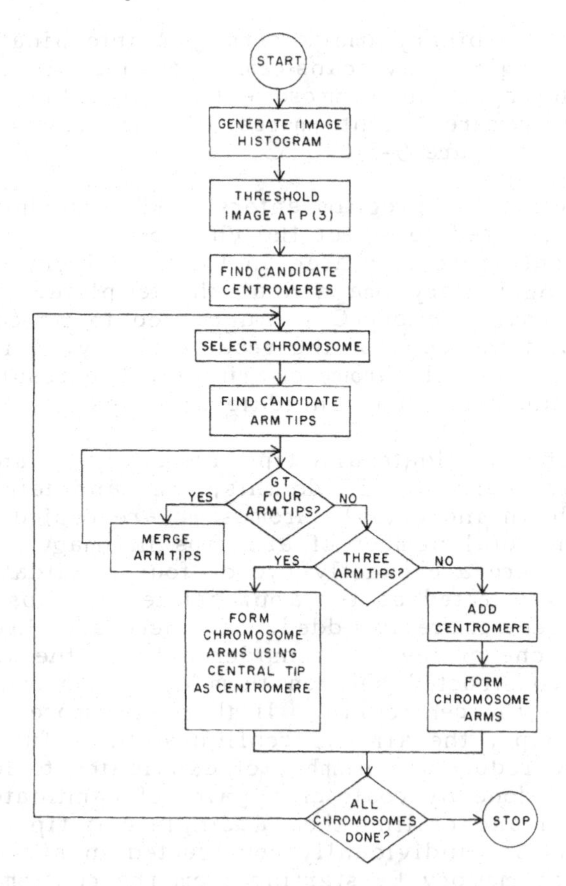

Fig. 9-6 Flow chart of the cellular logic chromosome analysis program used by Preston (1973).

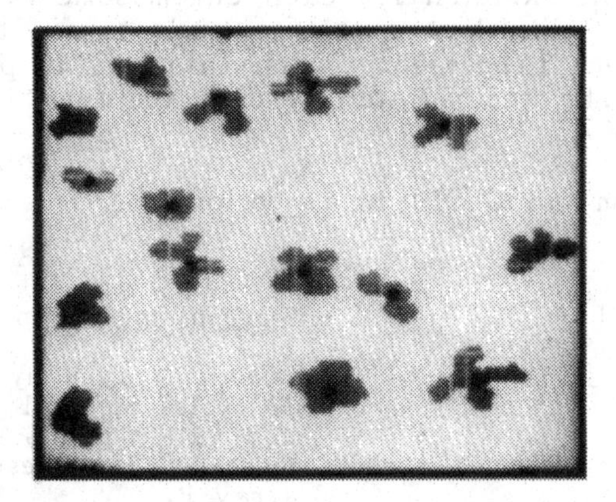

Fig. 9-7 Cluster of chromosomes with centromeres (shown as bright points) estimated by reduction to residues.

ies the contents of binary image memory A into binary image memory B. A single Golay transform is then applied for the purpose of reducing each chromosome to its residue and thus locating its centromere. Centromeres are then saved in binary image memory B (Figure 9-7).

At this point, the program enters a RUN, during which the PICK command is used to select the chromosomes one at a time. The selected centromere is placed in binary memory C and then propagated using memory image A as the template. Next, the chromosome in image memory C is augmented to its convex hull (Figure 9-8) and the edge of the resultant polygon is intersected with the original chromosome image. The result is then reduced to residues to form candidate arm tips.

Finally, the candidate arm tips are analyzed and the chromosome arms are defined. To do this, the candidate arm tips associated with an individual chromosome are copied into image memory D. The total number of arm tips in image memory C is computed. If there are exactly two or four candidates, they are immediately selected as the appropriate arm tips and the corresponding centromere is added. If there are three candidate arm tips, one of them is considered to be the centromere and arms are constructed between them using the central candidate arm tip as the centromere. If there are more than four candidate arm tips, the arm tip configuration is further processed so as to reduce the number of candidates to four or fewer. This is done by coalescing pairs of candidate arm tips by augmentation and calling them a single arm tip. In all cases, the arms are individually constructed in a sequential manner in image memory E, starting from the centromere and joining one arm tip to it at a time.

The resultant skeleton for each chromosome is placed in image memory F. Finally, a test is made to see that all chromosomes have been processed. When this test is positive, the program stops. The accumulated results are then in image memory F (Figure 9-9).

The program (in GLOL) is as follows:

HISTOGRAM H	Store histogram in array H.
SMOOTH P,H,32,5	Smooth array H by at most 32 cycles or until five peak and valley points have been located. Store results in array P.

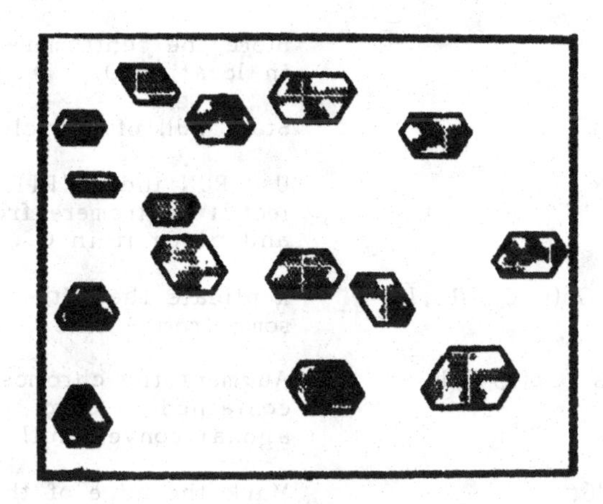

Fig. 9-8 Chromosomes and their convex hulls.

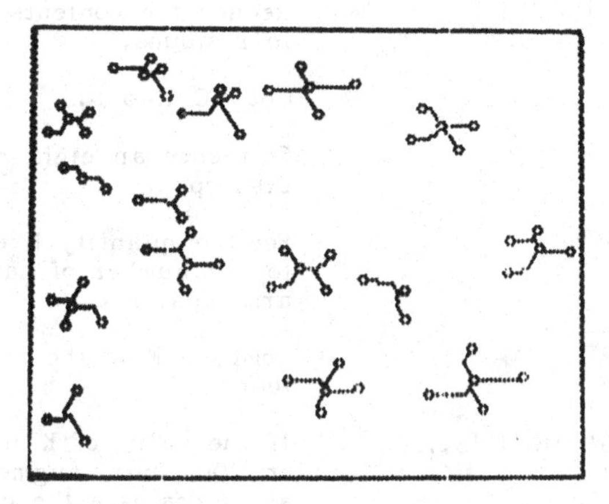

Fig. 9-9 Chromosomes and their endoskeletons found using the cellular logic program of Preston (1973).

ACQUIRE A,P(3)	Threshold the image at the third value stored in array P and store in A.
B=A	Copy the contents of A into B.
B=M[G'(B)B]1-4,I,3	Reduce each chromosome to its residue (centromere).

COUNT B,Q	Store the centromere count in location Q.
RUN Q	Start RUN of Q cycles.
PICK C,B,RI	Use RUN index (RI) to select a centromere from B and place it in C.
C=M[G(C)A+G'(C)C]1-13,I,	Replicate the *RI*th chromosome from A into C.
C=M[G(C)+G'(C)]13-5,I,	Augment the chromosome contained in C to its hexagonal convex hull.
C=M[G'(C)]6,,	Mark the edge of the convex hull in C.
C=C*A	Intersect the contents of C with the chromosomes in A.
C=M[G'(C)C]1-4,13	Reduce the contents of C to residues.
D=C	Copy C into D.
DO I/1,8	Commence an eight cycle DO-loop.
COUNT C,K	Set the quantity K equal to the number of candidate arm tips in C.
IF K.GT.4	Compare K to the number four.
C=M[G(D)+G'(D)D]1-13,,	If the value of K is greater than four, augment so as to coalesce the contents of D (otherwise jump to ELSE).
C=D	Copy the contents of D into C.
C=M[G'(C)C]1-4,I,3	Reduce the contents of C to residues.
ELSE	Take no action if K is equal to or less than four.
STOP	Terminate the conditional statement and return.
END LOOP	Terminate the DO-loop.

IF K.EQ.3	Compare the number of candidate arm tips (K) with the number three and jump to ELSE if the value of K is not equal to three.
C=M[G(C)C']1,,	Replace each arm tip in C with an empty hexagon.
C=M[G(C)+G'(C)C]1-5/7-13,8,	Augment and coalesce the hexagons in C.
C=M[G'(C)C]1-4,I,3	Skeletonize the chromosome arm contained in C.
F=F+C	Add the contents of C to F.
ELSE	Take the conditional action.
PICK D,B,RI	Place the centromere of the chromosome being processed in D.
DO I/1,4	Commence a four-cycle DO-loop.
PICK E,C,I	Select the *I*th arm tip from C and place it in E.
E=E+B	Add the contents of B to E thus creating a pair of residues consisting of the centromere and the selected arm tip.
E=M[G(E)E']1,,	Replace each residue in E with an empty hexagon.
E=M[G(E)+G'(E)E]1-5/7-13,8,	Augment and coalesce the hexagons in E.
E=M[G'(E)E]1-4,I,3	Skeletonize the chromosome arm contained in E.
F=F+E	Add the contents of E to F.
STOP	Terminate the conditional statement and return.
END LOOP	Terminate the DO-loop.
END RUN	End Program.

The above program ran on the original Perkin-Elmer research GLOPR and took about five minutes to go from the input image to the final result shown in Figure 9-9. Although never run on the diff3 GLOPR, the run time would have been about 3 seconds. In the following section, Hilditch comments on run times using the CLIP3 in the CLIP4 simulator mode.

9.3.2 Chromosome Analysis Using CLIP3/CIMU

The work described in this section was carried out on the CLIP3/CIMU system constructed prior to the advent of CLIP4 which, essentially, was a CLIP4 simulator. The operations possible with this system are described by Duff (1974). CLIP3 CIMU employed the 16×12 CLIP3 cellular automaton and was capable of executing Boolean operations and functions over a simulated 96×96 CLIP4 array. A useful feature of this system was that neighbors of points could have not only their original values but also those obtained as a result of the application of some operation. In particular, when using a procedure which requires propagation across the array for the purpose of finding a connected component, execution was possible using a single machine operation. Also, the CLIP3/CIMU system could execute basic branch instructions and test the state(s) of the array(s) emulated.

If an array of chromosomes were ideal, namely with no overlaps, then a relatively straight-forward analysis would be possible. Although images can sometimes be found in which most chromosomes are ideal, such cases are few and far between and it is not possible to limit one's analysis to these. Any viable system must, therefore, be able of recognizing and handling such complexities as bent, twisted, touching, and overlapping chromosomes. The analysis required for such configurations, using general purpose computers, has been found to be too slow to be efficient. For example, skeletonization, although promising (Hilditch and Rutovitz, 1969) is a time-consuming aid to chromosome recognition. It was, therefore, decided to implement skeletonization on the CLIP3/CIMU system to assess what improvement in execution time would be possible using an array automaton.

9.3.2.1 *Experiment I*

First the algorithm for chromosome skeletonization described by Preston (1973) for use with GLOPR was implemented (see Section 9.3.1). This algorithm was designed for use on chromosome images recorded at a much lower resolution than that which we have found necessary. Also it produces good skeletons only in straight-forward cases. For these reasons, the algorithm was implemented directly without alteration, although basic differences in the design of GLOPR and the CLIP machines meant that the resulting program was far from optimal. The program took between 5 and 10 minutes to produce results for a 96×96 bit field containing images of 10 to 15 chromosomes, depending on their number and size. This would correspond to an anticipated maximum time of 0.1 to 0.2 seconds on CLIP4 and compares with the time (5 minutes) reported by Preston (1973) for a similar field of chromosomes using GLOPR.

9.3.2.2 *Experiment II*

The next experiment was to apply a straight-forward thinning algorithm to produce skeletons from binary images of chromosomes at the resolution to which we were generally accustomed, i.e., with points of the digitized image taken at about 0.25 micrometer intervals on the original substrate. At this resolution, the 96×96 matrix covers a 12×12 micrometer area. Since in a chromosome cluster suitable for analysis the largest chromosome would not measure more than 10 micrometers, this should cover any single chromosome as well as almost all configurations consisting of touching or overlapping chromosomes which are likely to be encountered.

The thinning algorithm used was based on the masks developed by Levialdi (see Chapter 6). The exact algorithm which we used was a slight modification of Levialdi's in which the asymmetry inherent in the original masks was removed. For example, mask (B) was replaced by the two masks (B1) and (B2) shown in Figure 9-10, with a 1-element being changed to a 0-element if it satisfies either mask. This variant of the Levialdi algorithm, although slower than the original, was found to be insensitive to slight irregularities in the boundary which, using the original algorithm, produced unwanted extensions of the skeleton. Some results obtained are shown in Figure 9-11. These results illustrate the disadvantage inherent in the use of binary images, namely sensitivity to the threshold level, since the chromosome arms which one is trying to distinguish tend to lie closely together with no gap between them at lower threshold levels.

9.3.2.3 *Experiment III*

The final experiment was the implementation of a density sensitive algorithm such as had been found most satisfactory (but very slow) in previous work using general purpose computers. This uses the graylevel image acquired from the input

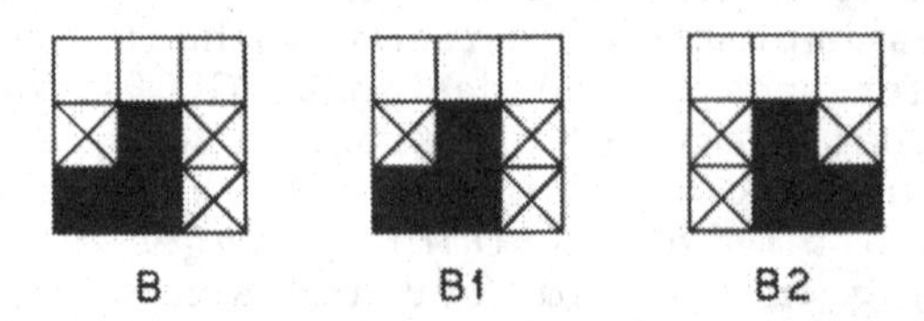

Fig. 9-10 Original Levialdi mask (B) modified by replacement with masks B1 and B2.

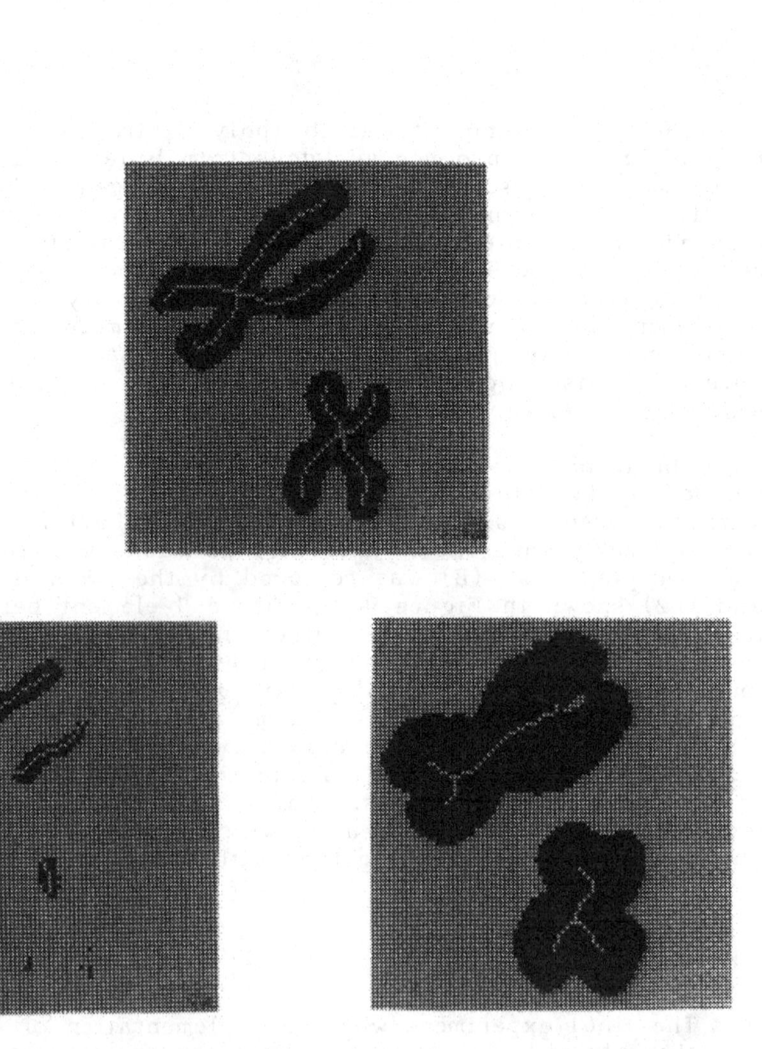

Fig. 9-11 Illustrations of the extreme variability in the computed endoskeleton as a function of threshold.

device using eight density levels. The image is first smoothed by setting each point to the average of its eight neighbors. Next, thinning is employed, using the algorithm described above, on a binary picture consisting of all above-background points. There is an added condition, namely, that only points that are one density level above background in the graylevel image may be deleted. This continues until no further points can be deleted. Then, points at the next density level become available for deletion and so on. The lower density troughs lying between the higher density arms are gradually eroded. The final skeleton lies along the arm ridges. This effect is enhanced by reverting to thinning at a lower density level whenever any such points are uncovered. The process is prevented from removing genuine holes, i.e., depressions completely surrounded by chromosome arms, in the following way. Whenever the level at which thinning is taking place is raised, if any points remain with a density more than a specified number of levels below this thinning level, then these points are deleted. The results of applying this algorithm to some chromosomes is shown in Figure 9-12. The times taken for these examples were between 5 and 15 minutes. These correspond to an estimated time of 0.1 to 0.3 seconds for the CLIP4 system showing a considerable improvement over our previous work on a general purpose computer where times were two orders of magnitude slower.

9.3.3 Summary of Chromosome Skeletonization

This section has shown that chromosome skeletonization may be performed by means of both cellular logic techniques on binary images and by means of cellular integer arithmetic on graylevel images. A straightforward thinning algorithm for binary images and a density thinning algorithm for graylevel images have been illustrated as applied to chromosome images using a resolution of 0.25 micrometer spacing between points in digitizing the image data. Similar results may be expected for other examples of both binary and graylevel imagery when it is desired to obtain the endoskeleton of the basic image components.

9.4 EXOSKELETON ANALYSIS

Useful biomedical applications of exoskeleton analysis using the cellular automaton have been found in studies of image data gathered from tissue sections. Continuing with the example given in Section 9.2, Figure 9-13 shows the results of using a high threshold to find the relatively light areas in the image shown in Figure 9-1 for the purpose of locating vessels.

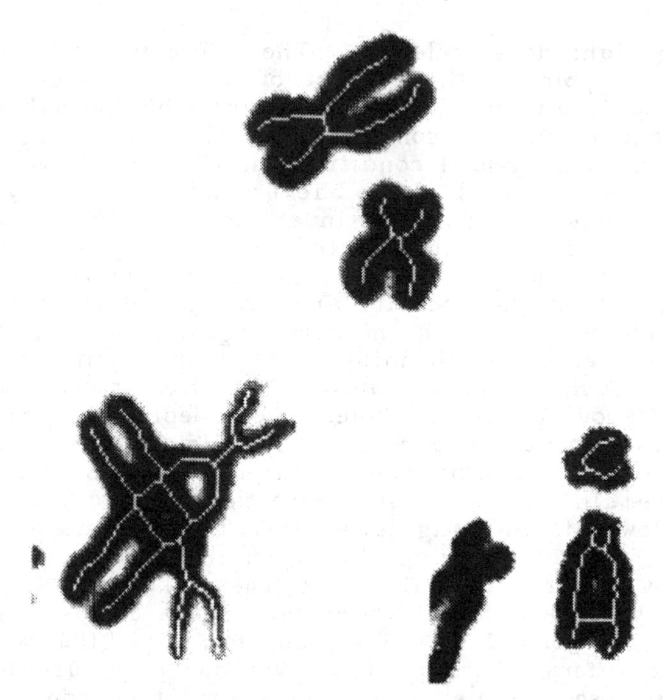

Fig. 9-12 Three chromosome images processed using the eight-density-level skeletonizing algorithm given in Section 9.3.2.3.

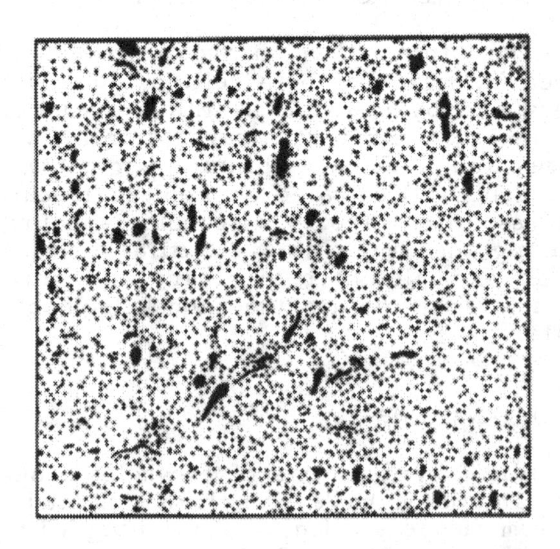

Fig. 9-13 Vessels found in the kidney tissue image (Section 9.2) outlined and ORed with residues of the cell nuclei.

After thresholding at this level, a cellular lowpass filter was used to smooth the boundaries of these vessels. Points within each vessel were output as 1-elements and ORed with the results given in Figure 9-3. An exoskeleton was then made using an algorithm which prevented branches of the exoskeleton from forming in those regions ascertained to be vessels (Figure 9-14). Each branch of the exoskeleton is approximately the perpendicular bisector of the line connecting a pair of nuclei, with the union of all these branches forming the exoskeleton. The polygons defined by the branches of the exoskeleton are directly related to those of Voronoi (1949). Variability in the size of the polygons is related, to some extent, to the variability of cell size in the tissue section whose image was processed. In order to demonstrate this, Figure 9-15 was produced in which are displayed not only the cell nuclei enhanced by marking with expanded residues but also the exoskeleton. The result is descriptive of the organization of the cells within the tissue sample.

9.5 EXTRACTION OF OTHER FEATURES

Reduction to residues, endoskeleton analysis, and exoskeleton analysis form the basis for simple measurements of image context by means of cellular transforms. The execution of these transforms is relatively simple using either the cellular logic machine or the array automaton. This is especially true for those systems where high-level programming languages are employed. In both GLOL and TASIC (Chapter 13) only a few program statements (in some cases a single program statement) are all that is needed to execute these relatively complex operations.

When still more difficult work must be executed, a high-level language contributes compact code which is relatively easy to read and extremely fast in its execution. Section 9.3.1 shows that a GLOL routine of only 40 statements is all that is necessary to carry out traditional chromosome analysis. In some cases, however, the pattern recognition task in hand requires several hundred lines of a high-level language due to the fact that a multiplicity of feature-extraction tasks must be performed. Such is the case in the automated analysis of images of human white blood cells as carried out by the GLOPR cellular logic machines produced by Coulter and installed in the diff-series of robot microscopes for hematology. In this section, examples are given of some of these more powerful feature-extraction sequences as specifically related to blood cell image analysis. Besides simple local features such as the centroid (as extracted by residue analysis) and simple global features such as that produced by the cell nuclei exoskeleton,

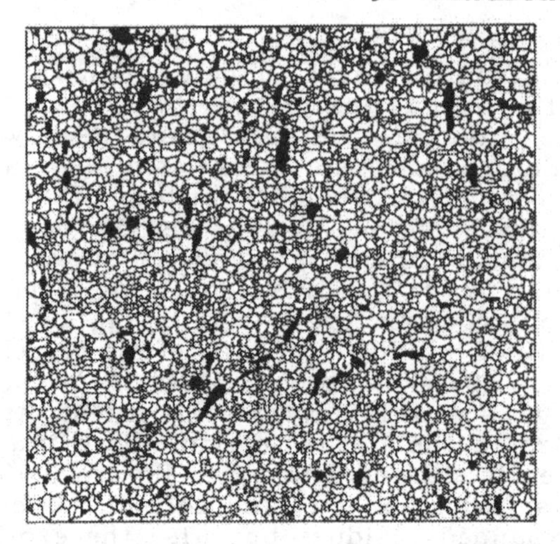

Fig. 9-14 Vessels from the kidney tissue image shown in black with the exoskeleton of the residues of the cell nuclei.

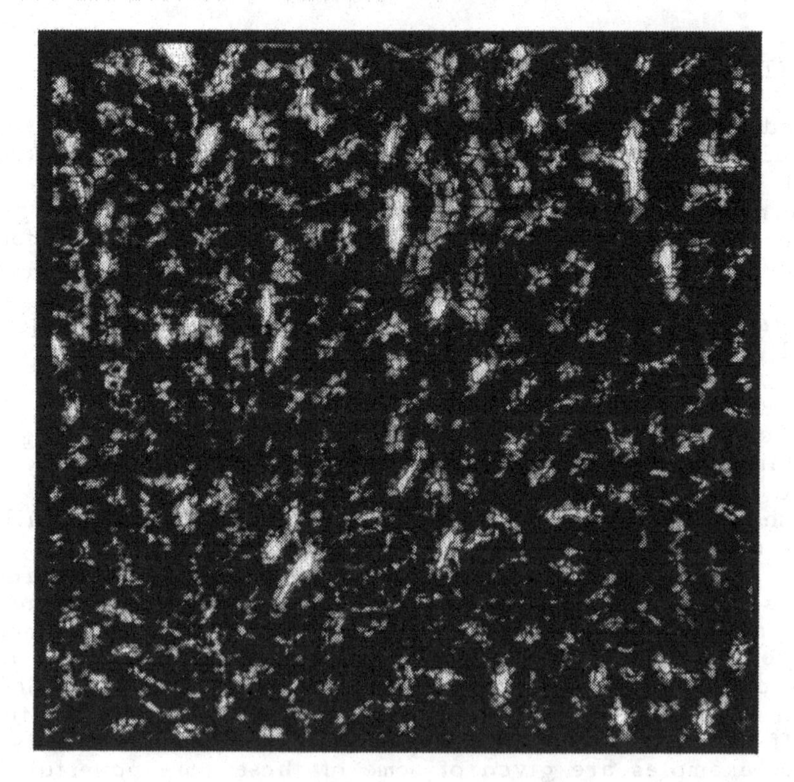

Fig. 9-15 The original graylevel kidney tissue image marked with both the replicated cell nuclei and the exoskeleton of their residues.

more elaborate features may be generated. Although the examples given below are excerpted from the analysis of human white blood cells, each of these examples illustrate routines which have obvious general applicability.

9.5.1 Boundary Filtering

When the probability density function of the image data extracted from an isolated human white blood cell is analyzed, a trimodal graph is ordinarily obtained. The mode which is highest represents image data extracted from the bright background; the next mode, from data taken from the cytoplasm; the lowest mode, data taken from the nucleus. This phenomenon was initially observed and reported by Mendelsohn and coworkers (1969). A *cartoon* of the image may be generated by thresholding the data at the minimum between the background and cytoplasm modes, at the minimum between the cytoplasm and nuclear modes, and at the nuclear mode itself and ORing the results (Figure 9-16). In general, the outlines of both cytoplasm and nucleus are irregular not only due to the basic variability of the optical transmission through the cell along these boundaries but also due to the noise processes in the scanning mechanism employed for extracting this data.

One of the first desires of the image processor is to *idealize* these boundaries by performing some type of *smoothing* operation. This is readily carried out by either the cellular logic machine or the array automaton in that, as described in Chapter 2, many cellular logic operators work by a *boundary propagation*. If a ragged boundary is first propagated inward followed by an outward propagation (an *overture* or *opening* in the Serra sense - see Serra, 1982), then the result is a more uniform boundary as shown in Figure 9-17. The usefulness of this operation is that, after boundary smoothing, the total length of the periphery of an object may be more accurately measured than prior to boundary smoothing. This is because many small non-uniformities in the boundary can lead to large errors in the calculation of boundary length. Figure 9-17 demonstrates how this problem is solved.

9.5.2 Reduced Boundaries

As has been demonstrated in Section 9.2 on the endoskeleton, when one desires to reduce an object to its endoskeleton and, finally, to a residue during size and shape analysis, step-by-step cellular logic transforms may be employed. Either the masks of Levialdi (1972) or the subfield approach of Golay (1969) may be used for this purpose. As an example, Figure

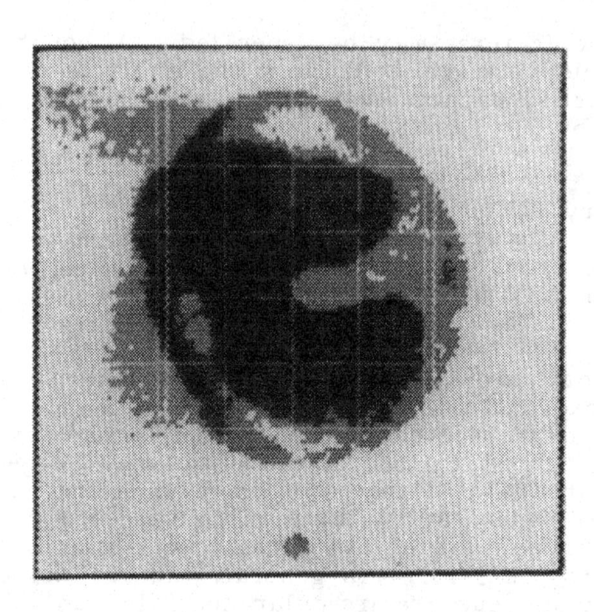

Fig. 9-16 Four-level *cartoon* of a human white blood cell digitized at 0.2 micrometer per picture point in the hexagonal tessellation.

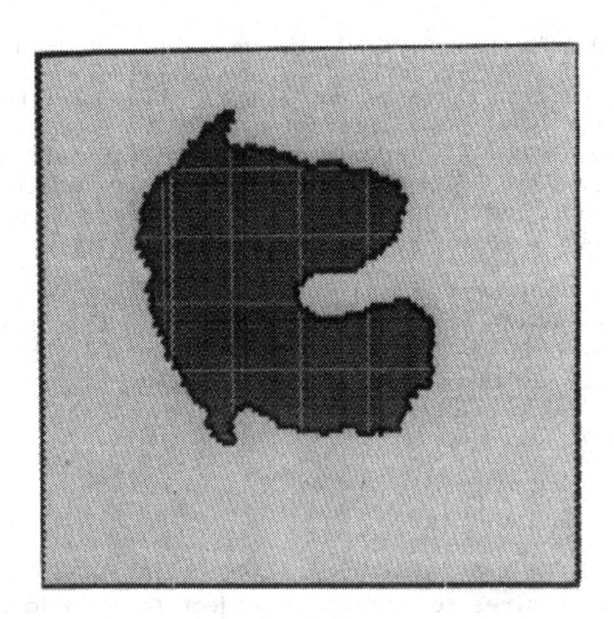

Fig. 9-17 *Idealized* cell outline produced by using a cellular logic low pass filter on the original shown in the above *cartoon*.

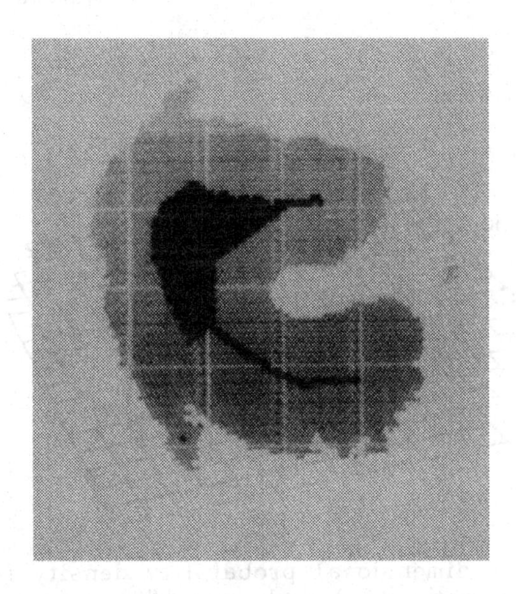

Fig. 9-18 The partial endoskeleton of the image of the nucleus of the human white blood cell being analyzed.

9-18 shows the reduction to a partial endoskeleton of the nucleus given in Figure 9-7. It has been found that reduced perimeter and area measurements, obtained in this step-by-step process, may be more useful in object classification than initial measurements of these quantities. An experiment was performed by one of the authors (Preston, 1971) in which the reduced perimeter and reduced area were measured as a function of the number of operational cycles during the process of endoskeletonization and reduction to residues. A set of several hundred bilevel images of human neutrophil nuclei and of human lymphocyte nuclei were examined and the two-dimensional probability density function of their perimeter/area measurements was plotted (Figure 9-19). No bimodality is observed. However, after partial reduction towards the endoskeleton of all images in this same set of objects, a plot of the same probability density function gave the result shown in Figure 9-20. Here the bimodality is clearly evident. As further cycles in this operation were performed and new probability density functions plotted, the bimodality vanished since each and every object eventually reduced to a single residue. The plots given in Figure 9-20 show the optimum point of separability for images of these two types of nuclei in the perimeter/area measurement domain. It clearly demonstrates the advantage of using reduced perimeter and reduced area for the purpose of shape discrimination.

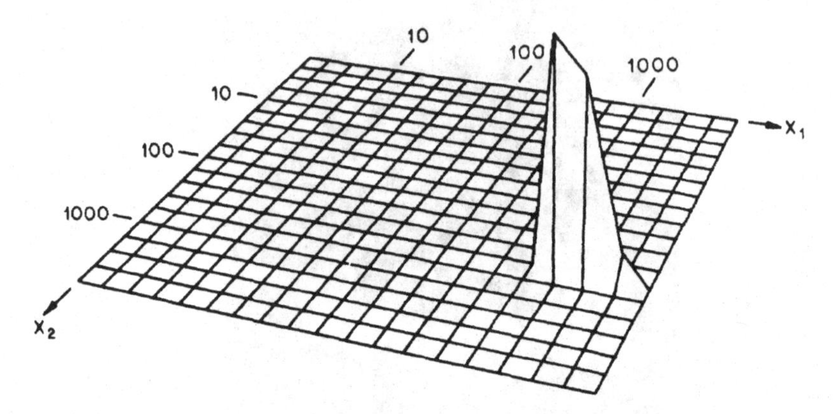

Fig. 9-19 The two-dimensional probability density function for initial nuclear perimeter (X1) and area (X2) measurements for two classes of human white blood cells.

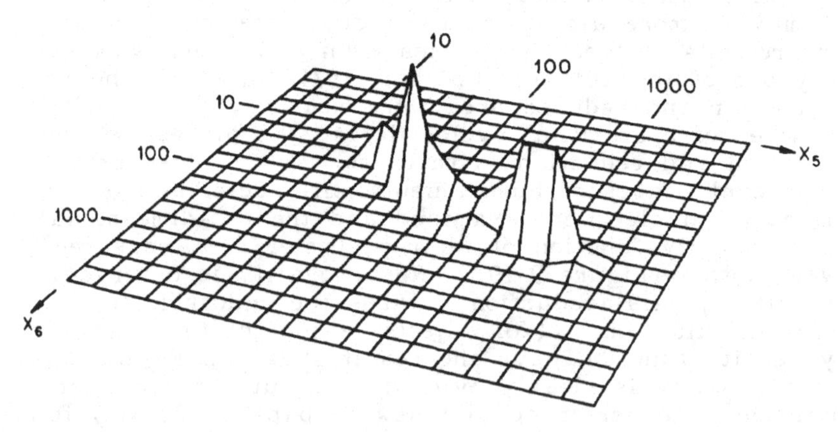

Fig. 9-20 The two-dimensional probability density function for reduced perimeter/area measurements showing bimodality enhancement.

9.5.3 Reduced Convex Hull

Chapter 5 indicates how the cellular logic transform may be employed to form the octagonal convex hull in the Cartesian tessellation and the hexagonal convex hull in the hexagonal tessellation. This technique is also illustrated in Figure 9-7 where it was employed in the initial phases of locating candidate arm tips in chromosone analysis.

Analogous to the reduced perimeter/area measurement given above is the reduced convex hull. This method of feature extraction is useful in performing an inventory of the indentations or concavities in the perimeter of an object. For example, if the convex hull of a bilevel image is formed, an exclusive-or performed with the original, and the result counted, low counts indicates that the object being analyzed has a boundary which is primarily convex. A high count indicates an object whose boundary has concavities (regions which are convex inward). In order to measure the number and size distribution of these concavities, the convex hull may be reduced by propagating its boundaries inward and, as a function of the number of propagation steps, a count may be made of the result obtained from the above exclusive-or. If simultaneously, the result is reduced to residues, a concavity count may be obtained. Figure 9-21 gives an example for the nucleus shown in Figure 9-16 where the major concavity has been located using this method. In this particular case, the convex hull was generated, reduced by 12 cycles. An exclusive-or was then executed with the original. The resultant count of 1-elements gives the area, while reduction to residues would yield unity, thus indicating that only one major concavity is present. This analysis is part of the currently executed white blood cell image processing routing used in the diff-series microscopes.

9.5.4 Partial Exoskeleton

In Section 9.2 it was demonstrated how the exoskeleton is formed from the residues of cell nuclei. This exoskeleton consists of Voronoi polygons. Exoskeletons may also be formed from the nuclei themselves, in which case the branches of the exoskeleton are no longer the perpendicular bisectors of the lines between pairs of nuclei centroids. The polygons so formed are no longer Voronoi polygons. Interestingly, an exoskeleton may be formed from the single nuclear outline shown in Figure 9-16. Recalling that the exoskeleton is the endoskeleton of the background, it is clear that any protrusion of the background, such as that created by an indentation in the nuclear outline, will in many cases produce a branch. This branch will form and then will eventually disappear as the

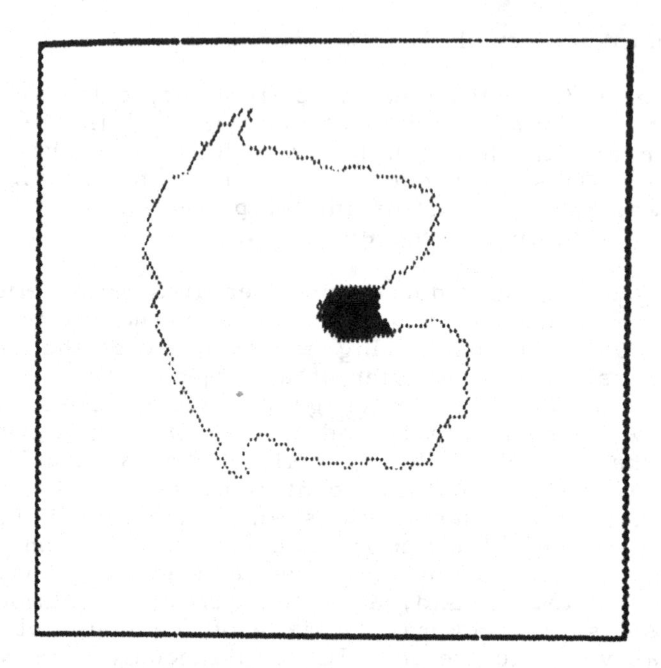

Fig. 9-21 The reduced nuclear convex hull EXORed with the original nuclear image may be used to highlight concavities in the nuclear outline.

exoskeleton of the background goes to completion. Formation of such branches may be used as a measure of the shape of the nuclear outline just as the concavities are measured from the reduced convex hull (Section 9.5.3). Figure 9-22 illustrates the formation of a branch in the *partial exoskeleton*. This measure is also used in the diff-series microscopes as another important feature useful in the differentiation of human white blood cells. Here is is found that a partial exoskeleton forms in the case of both lobulated and indented nuclei.

9.6 GRAYLEVEL FILTERING

The use of cellular logic machines and cellular automata in image filtering operations is described in Chapter 2. In this chapter these filtering methods were illustrated by results obtained by multithresholding, binary cellular logic transforms, followed by arithmetic summation. Chapter 4 describes similar filtering operations done directly on the array of integers obtained from digitizing the image data. Results are comparable since, as pointed out by Justusson (1981), there is a reciprocity relationship between the cellular logic transform applied to a multithresholded image and ranking transforms applied direct-

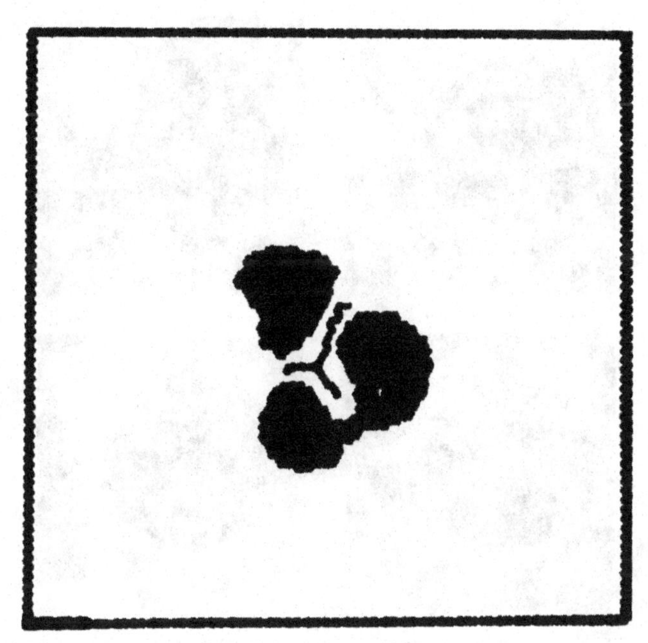

Fig. 9-22 The white blood cell nucleus and its partial exoskeleton formed due to the existence of nuclear indentations.

ly to the original array of integers which represents the image. Which methodology is preferred usually depends on the configuration of the cellular logic machine or cellular automaton employed. Brenner and his coworkers (Lester et al., 1978) have applied these transforms to the images of individual human white blood cells as discussed by Brenner (1981). Figure 9-23 is an illustration of this work which shows the image of a metamyelocyte found in the peripheral blood whose nuclear boundary is badly obscurred by toxic granulation (the fine structure overlaying most of the cell). A boundary found by thresholding at the minimum in the probability density function lying between the nuclear and cytoplasmic mode provides an unsatisfactory picture of the nuclear outline. However, after cellular logic filtering, the boundary may be found readily. These steps are illustrated in the sequence shown in Figure 9-23.

Similar filtering operations for boundary enhancement have also been investigated by one of the authors (Preston, 1983). Figure 9-24 illustrates this and shows an image of human liver tissue in which the boundaries of cell nuclei and of vessels are obscurred by the background structure of the cytoplasm. By using a cellular logic transform for 4 cycles and a value

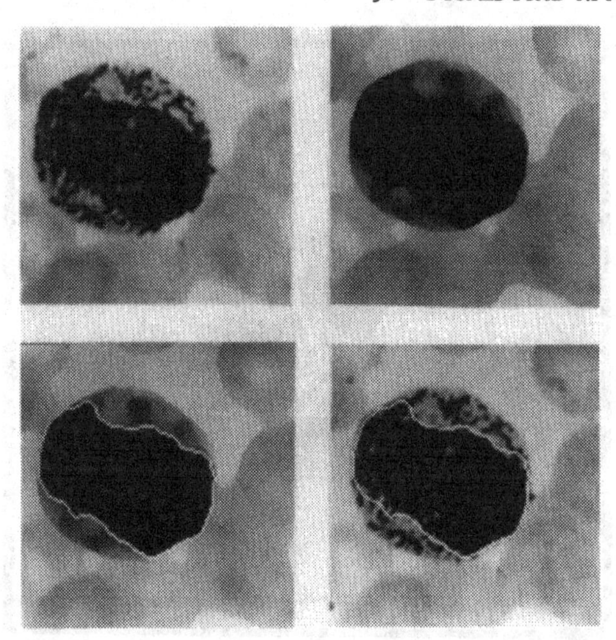

Fig. 9-23 Image of a metamyelocyte showing how cellular logic filtering may be used to find the nuclear boundary. (Courtesy, J. F. Brenner, Tufts University, Boston, Massachusetts.)

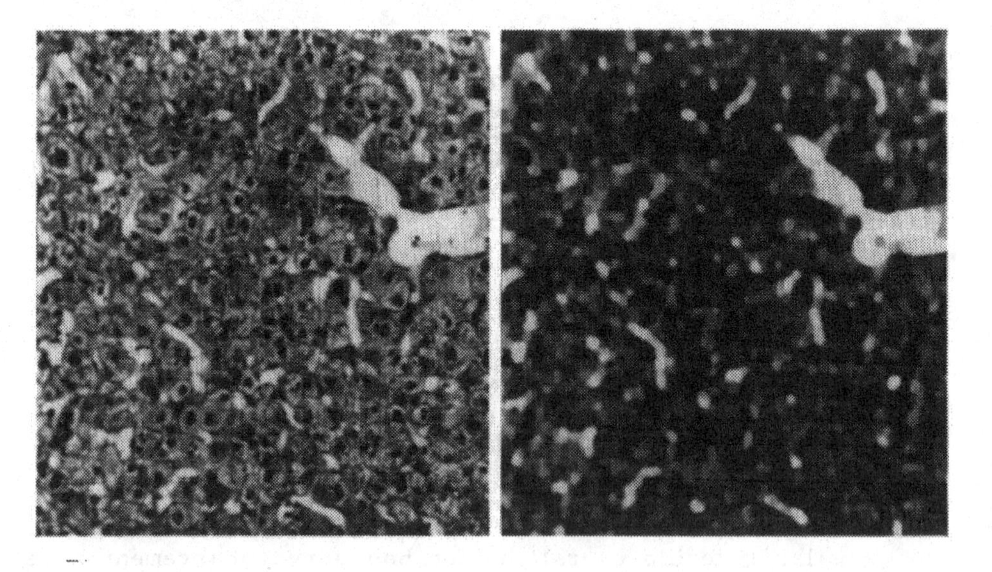

Fig. 9-24 A graylevel image of human liver tissue and the version obtained by cellular logic filtering for the purpose boundary enhancement.

of Ξ equal to 1 followed by a similar number of cycles with a value of Ξ equal to 7, the boundaries of both cell nuclei and vessels are clarified.

9.7 TOMOGRAPHY

As has been shown by many workers (e.g., Bracewell and Riddle, 1967), tomographic projection data and the two-dimensional image from which the projections are obtained form a Fourier transform pair. In practice, the Fourier transform is filtered in the process of tomographic reconstruction to remove high-frequency noise. Thus computed tomographic image reconstruction requires the calculation of the Fourier transform of a two-dimensional function multiplied by a two-dimensional filter function. The Wiener-Khintchine Theorem states that this is equivalent to a convolution in the spatial domain followed by summation. Digital computation of convolution and two-dimensional summation is usually carried out in computed tomography, rather than employing the Fourier transform. When the two-dimensional summation is executed digitally, a problem arises in that interpolation is necessary in finding the final values in the spatial domain. All such operations lend themselves to execution by the cellular automaton. The convolution equation implies that the filtering process may be achieved by adding neighboring values of the projections with various weights to form the final results. These weights may be precalculated and applied identically for all projections. Thus the convolution for all points can be performed at the same time in the cellular automaton with the weight of one projection's ith neighbor having the same weight as any other projection's ith neighbor. In the summation phase, one set of projection data is spread across the whole result field and accumulated in a wholly parallel addition. Typically, two projection values will contribute to the result at a particular picture point in the array so that two addresses and a relative weight are needed per array point, pre-calculated for that particular geometry.

Software for conducting this operation using CLIP4 has been written in CAP4 (Chapter 13) by Clarke (1981). All of the projection data are placed in the CLIP4 array as a series of lines so that the convolution may be achieved by a series of shifts, multiplies, and adds. The convolution in the experiment described here is a binary approximation where most multiplication factors are powers of two and can be executed by moving the binary point. After this, the required arithmetic is performed, a further shift taken, and so on. Interpolation in the summation required for back projection was carried out by effectively taking the nearest neighbor value, i.e., correct placement to within one-half a picture point.

The projection data is spread across the array by means of propagation using a linear template stored in another plane. If the connected points are straight lines, then this process is equivalent to replacing points with lines. In this manner, a set of projection data can be placed in the middle of the CLIP4 array and alternate 1-elements in the set propagated along alternate lines. Complementing the original lines so that they are 010101 instead of 101010 enables the remainder of the 1-elements in the projection data to be propagated. ORing together the two results and repeating the operation for all other bit planes in the projection data completes the back projection for a particular angle. Adding the results into an accumulator is a fast parallel process for summation. This procedure is repeated for all projection angles with the data being set up as appropriate in the middle of the array. The result is the tomographic projection of the original data. However, at this point, the data is spatially distorted so that a geometric transform is required comparable to a *stretching* operation with linear interpolation. This can be executed by shifting sections of the data using a series of mask and shift operations. The final stage of division may be accomplished by moving the binary point.

Examples of these operations are shown in Figure 9-25 where sixty projections are stored in the CLIP4 array and are operated to reconstruct the original cross section. In one case shown, the original data represented a disk containing a hole. The other data set was for two disks of different sizes separated by a distance equal to the diameter of the larger disk. Figure 9-25 shows how the projection data is entered in the array and the result of the reconstruction. The time required for these examples, when run on CLIP4, was approximately one-tenth second. Of this time 5% was necessary for the convolution operation. Back projection and stretching operations required equal amounts of the remaining time.

9.8 THREE-DIMENSIONAL SKELETONIZATION

Hafford and Preston (1984) developed a three-dimensional skeletonization algorithm using the tetradecahedral tessellation and have applied this algorithm to the analysis of three-dimensional neuronal structures. Thirty cross sections of neurons from the preanal ganglion of the soil worm *Caenorhabditis elegans* were manually digitized from electron micrographs and entered into planes 2 through 63 of the 64×64×64 workspace shown in Figure 9-26.

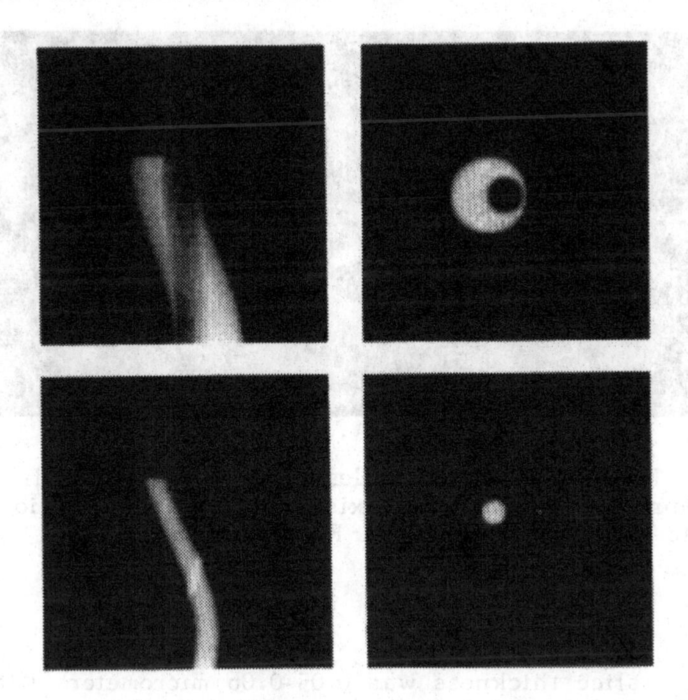

Fig. 9-25 Tomographic projection data as entered in the CLIP4 array with the resultant reconstructions. Execution time is 100ms.

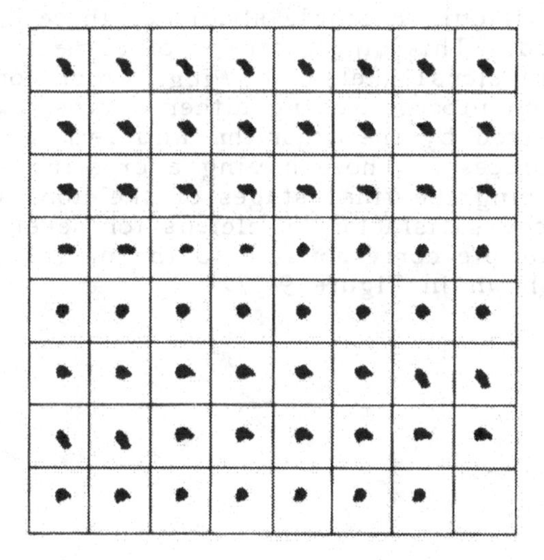

Fig. 9-26 Serial cross sections formed by digitizing nerve cell boundaries, connecting points along the boundaries, and filling each cross section with 1-elements.

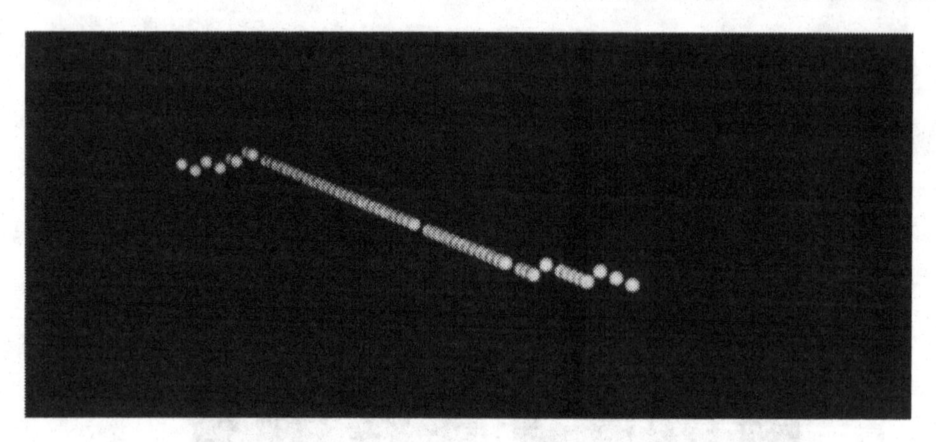

Fig. 9-27 Three-dimensional endoskeleton, wherein individual spheres represent the skeletal axis for each cross section, corresponds to the neuron shown in Figure 9-26.

The average slice thickness was 0.05-0.06 micrometers with the cross sectional diameter being approximately 0.2 micrometers. The three-dimensional cellular logic algorithm described in Chapter 7 was applied so as to cause the surfaces of the solid represented by the cross sections shown in Figure 9-26 to propagate inwards until only a single strand of three-dimensional elements remained. This single strand of elements was defined as the three-dimensional skeleton. *Rings* which formed during the skeletonization process having either a crossing number of 2 or 6 were detected by the algorithm and removed during the skeletonization process. Those having a crossing number of 4 were removed during the final stages of skeletonization. The algorithm produced satisfactory skeletons for several different neurons. An example corresponding to the neuron shown in Figure 9-26 is shown in Figure 9-27.

10. CELLULAR LOGIC MACHINES

10.1 INTRODUCTION

Chapter 1 describes the early work of von Neumann (1951) on cellular automata. This work was not accompanied by reductions to practice in hardware as it was impractical at that time to build machines having the millions of devices required. Only with the development of high-density integrated circuitry in the 1970s has this feat now been accomplished (Chapter 11). Therefore, during the 1950s cellular automata were emulated using the general-purpose computers which were available at that time. The work of Moore (1966) and Kirsch (1957) at the National Bureau of Standards, of Ulam (1962) at the Atomic Energy Commission, and Unger (1959) at Bell Telephone Laboratories are outstanding examples of this work. All of these workers simplified von Neumann's 29-state processing element and concentrated their efforts on studying arrays of 2-state (binary) processing elements. In the 1960s a new trend began with the construction of the first cellular logic machines by one of the authors (Preston, 1961). These and other special-purpose machines emulated the cellular automaton by using a single high-speed processing element to operate sequentially on an array of binary data. With the introduction of the diff3 in the 1970s (Graham, and Norgren, 1980) cellular logic machines were manufactured having several processing elements. Then Sternberg (1981) introduced a pipelined cellular logic machine, called the Cytocomputer, having approximately 100 processing elements. The Cytocomputer was also the first cellular logic machine to include numerical (multi-state) processing elements in addition to binary processing elements, thus making the transition from high-speed special-purpose machines limited to bilevel data to a system which could manipulate multi-state data. This chapter concentrates on the evolution and architecture of the cellular logic machines which have been built over the past two decades, all of which handle bilevel data arrays. Machines of this kind are currently in wide use both for commercial and research purposes in image processing. Despite their limitation to bilevel data, they are also useful in graylevel image processing due to their ability to convert graylevel images to binary images by multiple thresholding and, after performing logical operations at these thresholds, to convert results to graylevel output by arithmetic summation (Chapter 2).

10.2 THE DECADE OF THE 1960s

This section describes early cellular logic machines. All of these were built in the United States during the 1960s. The first was CELLSCAN (Preston, 1961), followed by GLOPR (Preston, 1971), and then by BIP (Gray, 1971). Each machine used a single binary processing element which was "scanned" over the data array. The array for CELLSCAN was initially fixed at 64×300 and later changed to 64×64. For GLOPR the array size was electronically variable. It could be 32×32, 64×64, or 128×128. In BIP, the array size was determined by program control in the host computer and was, therefore, essentially infinite. The major differences between these three machines were both in their speed and flexibility. CELLSCAN could process binary image data at approximately 4,000 picture elements per second; GLOPR, at about 300,000 picture elements per second; BIP, at about 10,000,000 picture elements per second. CELLSCAN was limited to a few specific logical operations on a 3×3 neighborhood. GLOPR was designed to decode in parallel any combination of the Golay primitives (Chapter 2) in the hexagonal neighborhood. BIP could perform general-purpose operations in both the hexagonal and square neighborhood.

10.2.1 CELLSCAN

In the strictest sense, the CELLSCAN system was not a cellular automaton emulator. This was because CELLSCAN used "look-back" circuitry. At the ith processing step for one neighborhood the result of the (i-1)th processing step for the adjacent previous neighborhood was employed. Since CELLSCAN was a neighborhood logic machine which operated upon two-dimensional arrays of binary data, its detailed description is included here because of its significant contribution to progress in this field.

CELLSCAN (Figure 10-1) was used in experiments on the automated recognition of naturally occurring objects from their televised images. The objects selected for automated recognition were the white blood cells of the human peripheral blood stream. The system addressed a microscope-slide-mounted blood smear via a Dage Data-Vision slow-scan television system, using as optics a Leitz Ortholux microscope. The Dage scanner ran at a rate of 60 horizontal lines per second and produced a total of 300 lines at 300 resolution elements per line. Due to data storage limitations in the CELLSCAN computer, a video-rate, one-dimensional reduction circuit was employed to convert each one-line series of 300 binary picture elements to 64 binary picture elements. The reduction algorithm was carried out using a circuit technique developed by Taylor (1954) which reduced negative-going (dark-polarity) video signals by five picture elements and simultaneously augmented positive-going (light-polarity) signals by the same amount. The video signal was then sampled at 64 picture elements per line resulting in the 64×300 binary data field. Preston (1973) gives details.

CELLSCAN was a true Turing machine in that the 19,200 binary values of the picture elements were recorded on an endless magnetic tape at 3,840 picture elements per second. (An ordinary audio tape-deck was used for this purpose with two complementary tracks.) Using the data tracks on the magnetic tape as synchronizing signals, the image stored was serially transferred to memory registers and operated upon using various Boolean and neighborhood logic algorithms. These algorithms included (1) image complementation, (2) reduction without the retention of residues, and (3) reduction with the retention of residues. CELLSCAN was controlled by switches on the control console which (1) selected the algorithm and (2) selected the number of iterations by an amount geometrically variable from 4 to 128. After the specified number of iterations were executed, the computer automatically reverted to a mode of operation wherein the tape continued to be read and the image rewritten without further change. At any point during the cycle the computer could be instructed to count the number of residues in the 64×300 results array. By this method CELLSCAN could generate the residue histogram (Chapter 7). The residue histogram was used in identifying the object being scanned by the microscope. Data processing was carried out at the originally synchronized rate of 3840 picture elements per second for approximately 10 minutes (about 2 million picture element operations) for a typical image recognition task.

A general block diagram of the CELLSCAN system is shown in Figure 10-2. The entire system was a special-purpose design carried out by one of the authors (Preston) in late 1960. The machine itself was build by the Research Engineering Department

Fig. 10-1 Photograph of the complete CELLSCAN cellular logic computer system built in 1961 by the Navigation Computer Corp. for the Perkin-Elmer Corp.

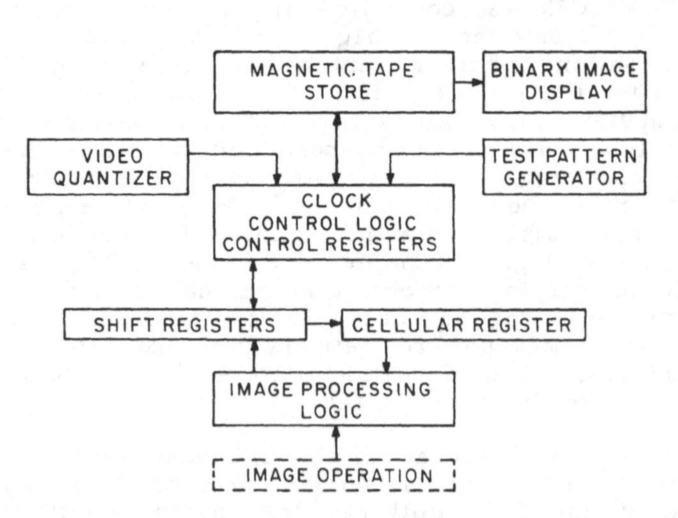

Fig. 10-2 General block diagram of the CELLSCAN computer.

of the Perkin-Elmer Electro-Optical Division under contract to the University of Rochester and funded by the United States Atomic Energy Commission. Fabrication of the digital portion of the machine (including the magnetic tape store) was carried out by the Navigation Computer Corporation and delivered to Perkin-Elmer in mid-1961. The system included both a display of the slow-scan video image from the microscope and a binary display of the result image from the computer as it was written on the endless magnetic tape. (For an example, see Chapter 1.) A test pattern generator was included in CELLSCAN for the purpose of exercising the image processing logic without relying on an image generated from the microscope itself.

Figure 10-3 shows the details of the cellular register and its associated shift registers which held two full lines of binary image data as the image data were moved in and out of the computer. The computer consisted of three control registers, two 60-bit shift registers holding incoming image element values, one 60-bit shift register holding the values of processed elements, and a 9-bit register holding the 3×3 neighborhood data. The bits in the neighborhood register were arranged as shown in Figure 10-3. The X element corresponded to the image ele-

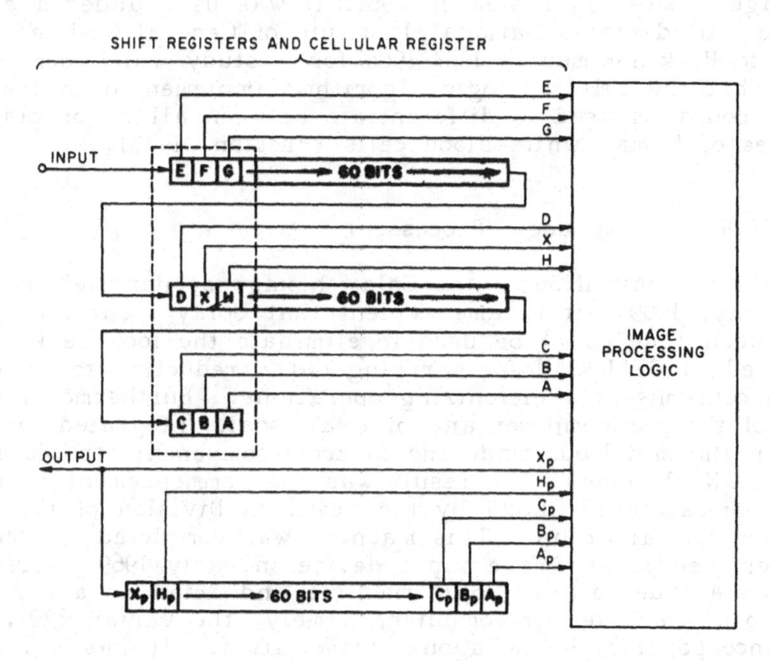

Fig. 10-3 The cellular logic portion of the CELLSCAN computer consisted of incoming and outgoing binary data registers, the cellular register, and the logic required for locating edge, interior, and linking elements.

ment being processed. The elements A,B,...,H were its eight immediate neighbors in the square tessellation. This configuration, was in fact, an exact reduction to practice of the original invention of Golay filed with the United State Patent Office in 1959 (Chapter 1). Finally, the computer included the look-back data, namely, the three-bit register used to hold elements Ap, Bp, Cp, and Hp corresponding to the look-back values of the A, B, C, and H elements.

As the image element data shifted through the registers, the computer processed one image element at a time and wrote the results on the endless magnetic tape. At the same time control registers were used to perform various counting functions. One control register was used to indicate the start and end of each line in the binary data field, another was used to indicate the number of iterations completed, and, when required, a third was used to totalize the number of residues. The functions carried out in the data processing logic are given in detail in a discussion by Izzo and Coles (1962). The circuitry was entirely discrete-component transistor logic.

After the completion of the original CELLSCAN computer and image processing system in 1961, it was used under a grant from the United States National Institute of General Medical Science to Perkin-Elmer in 1964-1966 for a study which demonstrated that the cellular logic algorithms implemented in the machine could be used to differentiate between all major classes of images of human white blood cells (Preston, 1973).

10.2.2 GLOPR (Golay Logic Processor)

With the invention of the Golay hexagonal parallel transform (Golay, 1969), it became evident that Golay's subfield processing technique could be used to eliminate the look-back methods used in CELLSCAN for carrying out a reduction to residues and the concomitant skeletonizing operations. Furthermore, the advent of the minicomputer and of small-scale integrated circuitry in the mid-1960s made the discrete-component structure of CELLSCAN obsolete. The result was the commencement in 1967 of the fabrication of GLOPR by the Research Division of the Perkin-Elmer Optical Group. This machine was completed in 1968 and interfaced to an image input device in early 1969. GLOPR served as a true cellular logic machine and acted as a peripheral to a 32K-byte minicomputer, namely, the Varian 620i. GLOPR incorporated the hexagonal tessellation. It was a vectored machine in that it performed all the instructions required to process an entire data array with Golay's hexagonal parallel transform in response to only two 16-bit microcode instructions from the host computer. The entire system interfaced to a 7-track read-write magnetic tape drive for the purpose of image storage and retrieval. The user interface was via the teletype. Image displays included the display of the binary image being processed, an analog display of the image produced by the input scanner, and a contourograph display for further image-analytic purposes.

A general block diagram of GLOPR is shown in Figure 10-4. The image scanner was an oscillating mirror device instead of the vidicon of CELLSCAN. An RCA 8645 10-stage photomultiplier was used for light detection which, with an illumination level of between 1 and 10 nanowatts per image element, could furnish a signal-to-shot-noise-ratio of about one hundred to one. The horizontal scan rate was 40 hertz with a frame time of three seconds. During each horizontal scan a selected group of 128 elements was transferred to the host computer at approximately thirty thousand image elements per second. Each image element was originally digitized to 8 bits with pairs of values being transferred to the host computer as a 16-bit word. As the scanner retraced, the image data was quantized by the host computer on a line-by-line basis at a level predetermined from the photometric histogram. The result was a 128×128 binary data field stored in 2K bytes of main memory.

A detailed diagram of the GLOPR image processor is shown in Figure 10-5. Five 16-bit registers were used for parallel-serial transfers to the host computer. Two registers were employed to hold the microcode control words. One control word was used to select the Golay primitives for the particular Golay transform to be computed. The second control register determined the selection of subfield order and controlled the length of the internal image data shift registers according to the dimensions of the data field being processed. Two further registers were used to transfer both the input data field and the template data field from the host. The fifth and final register was used to transfer either the computed binary data or the digitized image from the television microscope to the host computer.

The variable-length primary shift registers (which included the cellular register) were shifted at a rate of 500 kilohertz causing the image data to flow sequentially through the cellular register. After each shift of the primary shift registers, the six binary image elements contained in the hexagonal neighborhood were transferred to a special circular shift register. This register shifted at four megahertz and its contents were sampled every 250 nanoseconds so as to address fourteen comparators in parallel. Each of these comparators was matched to one of the fourteen Golay primitives. When subfields were

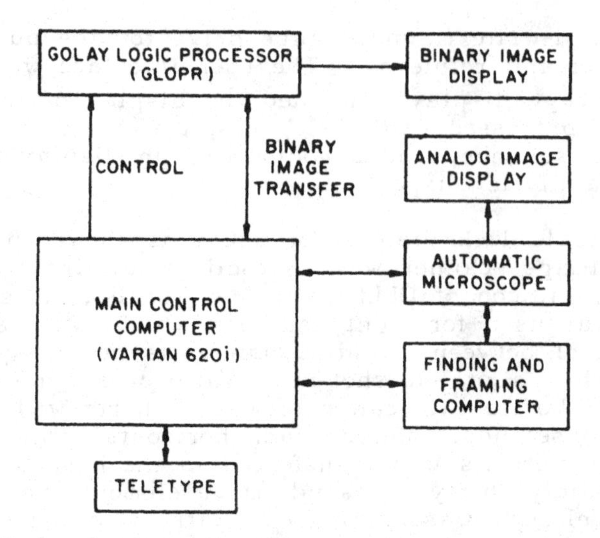

Fig. 10-4 General block diagram of the Golay Logic Processor (GLOPR).

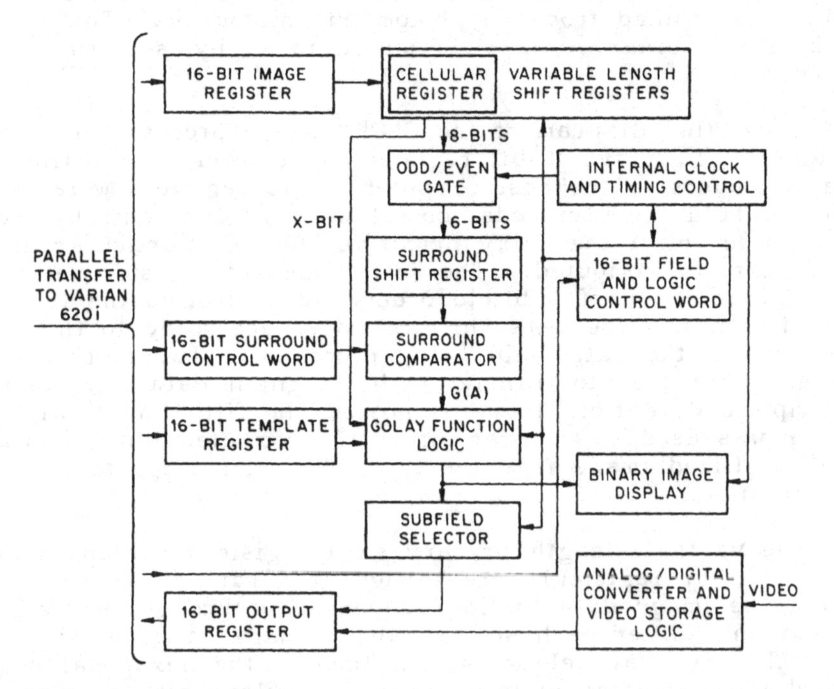

Fig. 10-5 Detailed diagram of the GLOPR computer showing the four 16-bit input registers holding the image and template plus two control words and the 16-bit output register for either digitized video or the result image.

utilized, the subfield logic disabled comparators except when the proper subfield was positioned in the circular register. Both subfields-of-three and subfields-of-seven were provided in the GLOPR configuration.

The basic transform carried out by GLOPR was the Golay hexagonal parallel transform as expressed by the equation

$$D = M[G(A)B+G'(A)C]N1...N2,N3,N4 \quad (10.1)$$

which also appears in Chapter 2 as equation (2.13). Two microseconds were needed to generate the timing sequence required for processing a single picture element. After GLOPR operated upon a group of 16 picture elements, additional time was required for input/out transfers to and from the host computer. These transfers added approximately 50% overhead, resulting in an overall allocation of three microseconds for processing each picture element. Thus 48 milliseconds were required to process the 128×128 data field, 12 milliseconds for the 64×64 data field, and three milliseconds for the 32×32 data field. GLOPR was therefore about one hundred times faster than CELLSCAN.

The primary applications of GLOPR were not only in the automated recognition of human white blood cells on microscope slides but also in reconnaissance and earth resources image processing. The success of GLOPR as an ultra-high-speed image processor led to the first commercial production of a cellular logic machine. This machine is the diff3 which is described in Section 10.3.

10.2.3 BIP (Binary Image Processor)

At approximately the same time that the GLOPR was under construction at Perkin-Elmer, Gray (1971) at Information International Inc. built the first version of a machine called BIP. This early BIP was constructed for use in studies in character recognition and was shortly followed by the production of a full-scale BIP by Information International which was constructed in late 1970. Unfortunately, little information has been published on the architecture of this interesting machine other than one or two pages in the reference to Gray's work given above. The machine is part of an overall system for use in alphanumeric data processing and storage called GRAFIX I and manufactured by Information International. It is understood that a few BIP's were produced with some still operational in the 1980s.

Despite the lack of published material one of the authors (Preston) in correspondence with Stephen Gray at Information

International has obtained considerable additional material on BIP. This material is contained in an unpublished internal report (Gray, 1972). The following comments are summarized from this report.

BIP and the GRAFIX I system (Figures 10-6 and 10-7) constitute a far more elaborate image processing complex than any other system previously constructed. This system forms the connecting link between systems of the 1960s and those of the 1970s. (In fact, although design started as early as 1966 with preliminary fabrication in 1968, the version of BIP described below was not operational until early 1971.) BIP is controlled by a major mainframe, the 36-bit Digital Equipment Corp. PDP-10. Interactive use under a time-sharing operating system is available for up to four users. BIP, besides having cellular logic hardware, also incorporates a numerical comparator and nine parallel Boolean comparators for template matching. The below discussion concentrates on its cellular logic features with brief reference to other special features of interest.

As with GLOPR, BIP operates serially and may take its input from the binary neighborhood of one data field, from the binary central element of the same data field and, simultaneously, from a template data field. The template size is limited to 36×36 while the data field may be extended to 36×511. For BIP this represents a string of approximately fifteen characters in a typical character recognition operating environment. It should be noted here that BIP was designed for commercial automatic character recognition wherein the symbols to be recognized are in a stylized machine-printed font with their location in memory assumed accurate to within about one picture element of the template.

The processing rate of which BIP is capable is faster than the input/output rate of the host computer. Thus BIP's performance is host memory-transfer-rate limited. Within this limitation BIP operates at one picture point operation each 100ns. Thus it is 30 times faster than GLOPR, although four times slower than the diff3 GLOPR (Section 10.3.1). Without this limitation, the operational rate of BIP would be identical to that of the diff3 GLOPR, namely, 25ns per picture point operation.

The simplest operation carried out by BIP is to do nine simultaneous Boolean logic comparisons against the template and all eight versions of the template obtained by shifting by one picture element in the horizontal, vertical, and ± 45° directions. The Boolean logic comparison is the same for all nine directions and results are separately counted for each picture element and stored in nine correlation counters. The condition specified may be selected from any one of the maps given for two binary input

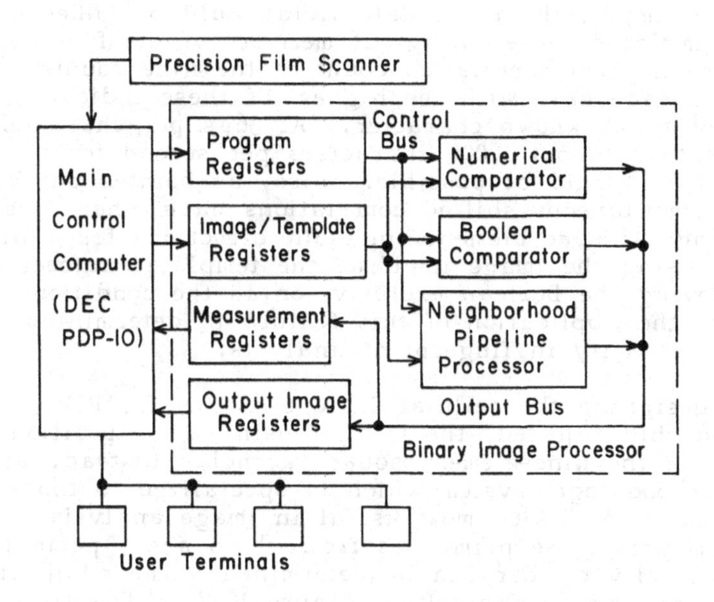

Fig. 10-6 The GRAPHIX system of Information International Inc. includes a Digital Equipment Corp. PDP-10 as host with the BIP as a high-speed peripheral.

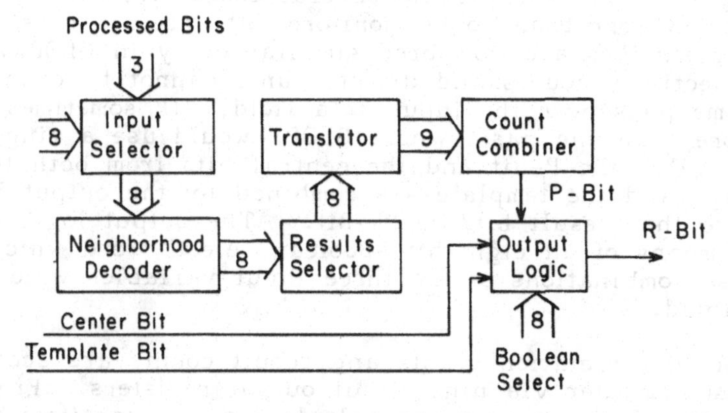

Fig. 10-7 Block diagram showing the flow of image data and measurements through the BIP.

variables by Karnaugh (1953). When used in character recognition, for example, the input data field would be ANDed separately with templates representing all members of the font being read. The highest correlation count of the nine counts would be used in each case with the highest of these indicating the identity of the unknown character. At 36μs per character, a reading rate of about 1,000 characters per second for an all-upper-case font should be possible. Also, as pointed out by Gray, the nine simultaneous shifted correlations have other uses, such as counting all edge elements in eight directions for a binary image by using the image itself as the template (autocorrelation) and employing the Boolean exclusive-or as the condition. Gray illustrates the application of this feature in determining local edge directionality in fingerprint analysis.

In designing the cellular logic circuitry for BIP, Gray considered, but rejected, the idea of using a 512-position look-up table for the nine-element square kernel. Instead, BIP uses a neighborhood logic system which is specialized to those operations which were deemed most useful in image analysis. Gray's design computes those primitives from which area, perimeter, and connectivity number can be determined. The neighborhood logic is arranged as a pipeline (Figure 10-7). The first stage selects either the eight neighborhood inputs, as with GLOPR, or previously computed look-back data, as with CELLSCAN. The selected values are then either operated upon by a neighborhood decoder (for determining connectivity) or sent directly to a translator for counting. The translator produces a one-out-of-nine result in nine separate output channels. These outputs are separately gated in the count combiner by a nine-bit control word and then ORed to produce the "processed bit" or "P-Bit" on a single wire. This permits several connectivity or neighborhood count conditions to be monitored at once, just as all 14 Golay primitives are monitored simultaneously in GLOPR. But, since connectivity counts and direct counts cannot be counted in the same pass over the input data field, BIP sometimes requires more than one pass, where GLOPR would use a single pass. Finally, the P-Bit and the central bits from both the input image and the template are combined by the output logic to generate the "result bit" or "R-Bit." The output logic is gated by means of an eight-bit "Boolean select" word which flags those combinations of the three input variables which are to be enabled.

Data on correlation counts and result counts are accessible to the host computer via nine 36-bit output registers. From this data the host computer can calculate such quantities as total area, the number of connected regions, the average size of connected regions, etc.

Using the real-time integer comparator, BIP also has image histogramming capabilities and can be used both for thresholding (to produce binary images) and for such operations as shading correction, wherein an integer template is employed for reference. Also available are such geometric transforms as image scaling, rotation, and reflection. Finally, image mosaics much larger than 36×511 can be processed by modifying the host computer control program. Control of all BIP operations is via six 36-bit words of control code stored in BIP by the host computer. The first word provides control information to the cellular logic circuitry, the next word indicates one of about 30 operational codes, while the remaining four provide addresses and control information for geometric transforms.

10.3 THE DECADE OF THE 1970s

In the decade of the 1960s (Section 10.2) the usefulness of the cellular logic machine for emulating the cellular automaton in many applications of image analysis was demonstrated (especially in automated medical microscopy). Because of the large effort expended in medical microscopy every year (several billion dollars in the United States alone), commercialization of the cellular logic machine for this use followed the feasibility demonstrations of the 1960s. The result was the diff3 robot microscope designed at Perkin-Elmer and in production in 1977 by Coulter Biomedical Research, Inc. Simultaneously, the European computer science and engineering community (in particular that of Sweden) through visits to industry in the United States (in particular to Information International and Perkin-Elmer) became aware of the power of the cellular logic machine. This led to the construction of PICAP-I by the University of Linkoping (Kruse, 1973) while, at the same time, the Bureau of Mines in France began a program leading to the AT4 which was commercialized by Leitz as TAS. Somewhat later, the University of Delft in Holland produced the DIP-1. The United States Department of Defense commenced a program under the auspices of the United States Air Force leading to the Cytocomputer built at the Environmental Research Institute of Michigan (Sternberg, 1981). All these machines (with the exception of the diff3) were considerable departures from their predecessors (CELLSCAN, GLOPR, and BIP) and differed markedly from each other. All of them carried out cellular logic operations, but also most carried out numerical calculations, not only over the 9-element square neighborhood and the 6-element hexagonal neighborhood, but also much larger neighborhoods for performing high-speed numerical convolution operations. The major features of most of these machines are discussed below with the text outlining their overall data processing features, and concentrating primarily on their cellular logic capabilities.

10.3.1 The diff3 Analyzer

The diff3 analyzer is a robot - a complete clinical blood smear handling system which operates entirely without human intervention, except for the transferral of microscope slides from the mechanical and chemical preparation units to the microscope cassette. A technologist is always in attendance, not only to transfer slides from the sample-preparation units to the automated microscope, but also to visually review those white blood cells whose images the analyzer states are abnormal. (See Chapter 9 for more details.) The design study for this machine was commenced by the Perkin-Elmer Instrument Group in 1972 and production was initiated in 1977. At that time, the diff3 product line was sold to Coulter Electronics, which founded the Coulter Biomedical Research Corp. where a research, development, and manufacturing facility for the diff3 product line was established. By 1980 a new version of the system, the diff3-50, was in production and in 1982 production was started on a multi-microprocessor-controlled unit called the diff4 (Graham, 1983). In all of these versions of the Coulter robot microscope, the GLOPR cellular logic machine is the same.

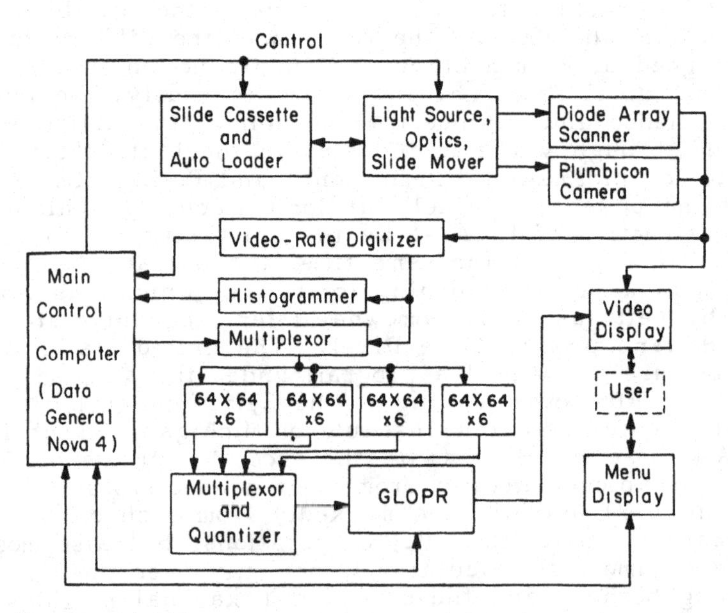

Fig. 10-8 The diff3 robot microscope system includes the GLOPR for image analysis, eight hardware image memories, a video-rate histogrammer and comparator all under control of a Data General Nova 4.

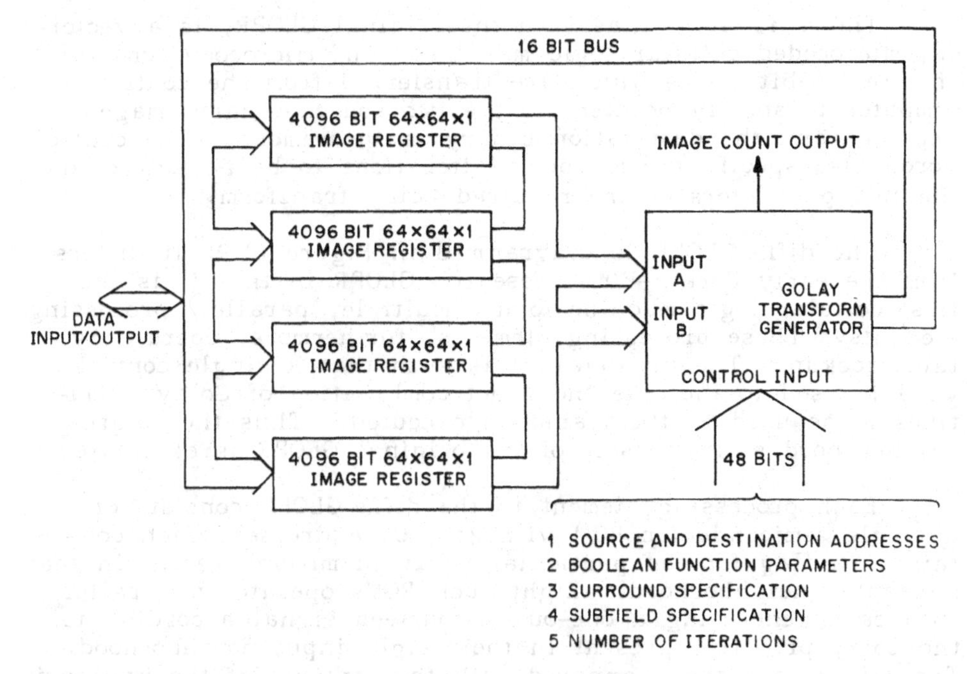

Fig. 10-9 The diff3 GLOPR is controlled by three microcode words which specify the source and destination buffers, the Golay transform, the number of iterations, and subfields.

Figures 10-8 and 10-9 shows the components of the diff3 analyzer - an enormously complex system described thoroughly by Graham and Norgren (1980). After auto-loading each microscope slide, the diff3 must automatically establish focus and then initiate a search for white blood cells. Once located, the image of each white blood cell (recorded with both red and green illumination) is transferred via a video-rate digitizer and multiplexor to one of four 64x64 six-bit graylevel memories. A video-rate histogrammer, whose output is available to the control computer (a Data General Nova4), is used to select one or more thresholds for the purpose of generating binary images to be stored and processed in the diff3 GLOPR. A special-purpose comparator and memory multiplexor is used for thresholding.

The user interacts with the system via an alphanumeric menu display where either an individual item or menu page is selected by cursor control. At the same time, the user may observe the video display of both the television image, the contents of any one of the graylevel digital memories, or any of the GLOPR binary memories. (Some examples are shown in Chapter 9).

The diff3 GLOPR, as with the original GLOPR, is a vectored, microcoded cellular logic machine. The microcode consists of three 16-bit words which are transferred from the control computer to specify pointers to the two source binary image memories and the destination binary image memory. The control words also specify the number of iterations to be performed and the full parameters of the required Golay transform.

The diff3 GLOPR is diagrammed in Figure 10-9. It differs from the early Perkin-Elmer research GLOPR in that it is the first cellular logic machine to use multiple, parallel, processing elements. These processing elements, furthermore, operate by table lookup. The arrangement is such that a single control word is used to indicate the exact combination of Golay primitives to be used in the transform executed. Thus the rotating neighborhood shift-register of the original GLOPR is eliminated.

Each processing element in the diff3 GLOPR consists of a single semiconductor ROM with $2^6 = 64$ addresses, each containing a flag for the particular Golay primitive present in that hexagonal neighborhood. Eight such ROMs operate in parallel with each generating a óne-out-of-fourteen signal according to the Golay primitive present in their eight input neighborhood. These flags are then compared with the contents of the appropriate microcode control register along with the present values of the picture elements being processed in order to determine the output values of the picture elements. In addition, eight picture elements from a template image may be employed in the comparison. In this fashion, all eight picture elements are processed during one clock cycle (200ns), with a picture element operation time of 25ns. Thus, the diff3 GLOPR is about one hundred times faster than the original GLOPR of Perkin-Elmer and approximately ten thousand times faster than CELLSCAN.

Figure 10-10 shows the logic diagram of one of the eight processing elements of the diff3 GLOPR. The ROM is addressed by the neighborhood from image A and its output is delivered via a set of gates specifying the Golay primitive to the combinational logic circuitry whose other inputs consist of the center element from the image A and the center element from the template B. The outputs of eight processing elements are placed on a bus which connects to the destination binary image memory. As the eight outputs are generated, a set of eight counters is used to sum the binary 1s appearing on each of the eight wires of the bus to the destination image. At the end of a complete Golay transform cycle, the contents of these counters are added to produce a final count. This count, which is the only measurement generated by the diff3 GLOPR, is then transferred to the control computer.

The advantage of the diff3 GLOPR architecture is that the microcode used to specify the Golay primitives for a particular transform consists of a single 16-bit word. This avoids the alternative of storing a complete lookup table consisting of $2^7 = 128$ words of microcode. Microcode for any Golay transform is loaded many times faster using this architecture than with an architecture comprised of multiple lookup tables. The disadvantage is, of course, that certain specialized functions, i.e., those which allow image translation, orientation-dependent measurements, etc. cannot be executed. (See Chapter 2.) The total microcode provided by the control computer consists of three 16-bit words specifying the following five functions:

1. The destination of the two source and one destination image registers.

2. The Boolean Golay function specification supporting all possible combinations of data from the two source registers.

3. The 14 Golay primitives.

4. The number of iterations for which the transforms is to be performed.

5. The subfield and the subfield sequence specification.

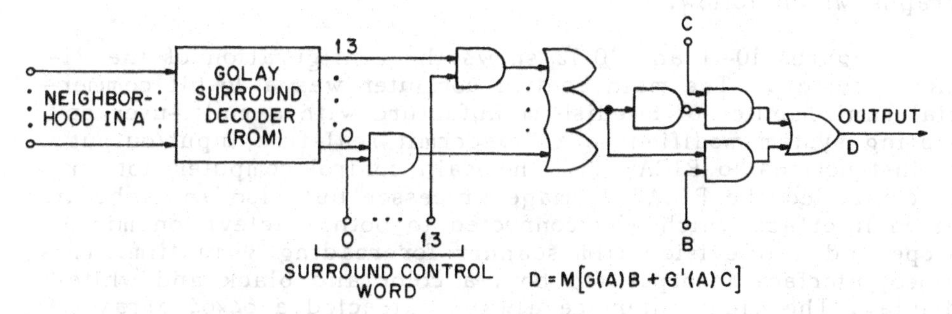

Fig. 10-10 Logic diagram of one of the eight processing elements of the diff3 GLOPR. The ROM decodes all of the Golay primitives in parallel while the combinational logic handles the center element values of both the primary image and the template.

One other interesting feature of the diff3 GLOPR is that a selected binary image memory may be used as a template for the video-rate histogrammer. This permits real-time histogram analysis of a particular region in the image (such as the nucleus of a blood cell) while simultaneously removing from the histogram calculation all graylevel picture element values not corresponding to that region.

10.3.2 PICAP (Picture Array Processor)

In the early 1970s the Department of Electrical Engineering, University of Linkoping (Sweden), proposed the construction of a parallel picture processing machine initially called PPM and, later, PICAP (Kruse, 1973). The design was finished in 1972 and construction commenced. The first PICAP system (PICAP I) combined features of GLOPR (cellular logic) and of BIP (high-speed two-dimensional template matching and numerical convolution) in an overall system for neighborhood image processing. PICAP I was a cellular automaton emulator in that, among other features, it included processing elements which sequentially processed binary image data using a 3×3 kernel.

As with BIP, PICAP I was a multi-user, special-purpose image processing system. It was also a major departure from the systems of the 1960s in that it could not only compute numerical convolutions over a 3×3 kernel but could also employ what were called "condition templates" in this kernel. It was also the first to employ big disc storage technology for image data. This led to a far greater range of neighborhood operations than were available in CELLSCAN, GLOPR, or BIP. Although this section concentrates on the cellular logic capabilities of PICAP I, some other features are outlined in the paragraphs which follow.

Figures 10-11 and 10-12 shows the configuration of the PICAP I system. The main control computer was a 16-bit commercial minicomputer of Swedish manufacture with a multi-user operating system modified so as to permit real-time input/output of instructions to PICAP I. The main control computer not only controlled the PICAP I image processor but also an elaborate video interface which was connected to both a television microscope and a television film scanner for reading 35mm film. This video interface incorporated both a color and black and white display. The video interface always extracted a 64×64 array from the incoming television data and digitized the picture elements in this window as 4096 four-bit words. By means of oversampling, the video interface could cause these words to be selected over a specified region in the entire video frame or, at

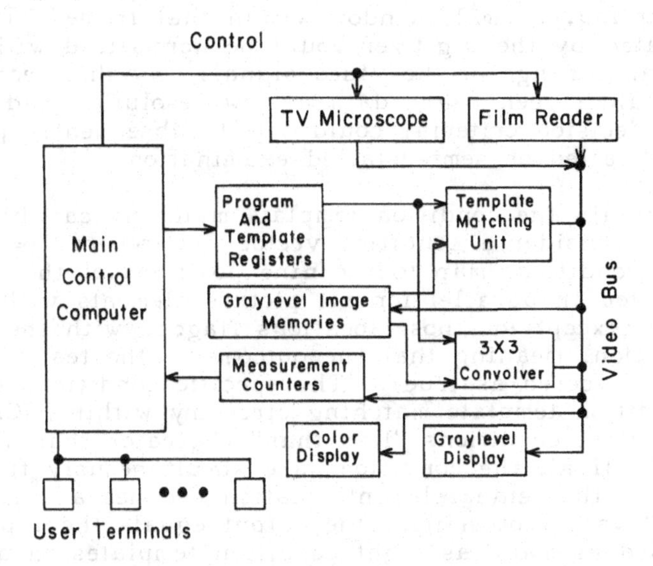

Fig. 10-11 In the PICAP I system a video bus and interface introduced all image data. The main image processing units were the template matcher, with which cellular logic could be performed, and the 3×3 convolver.

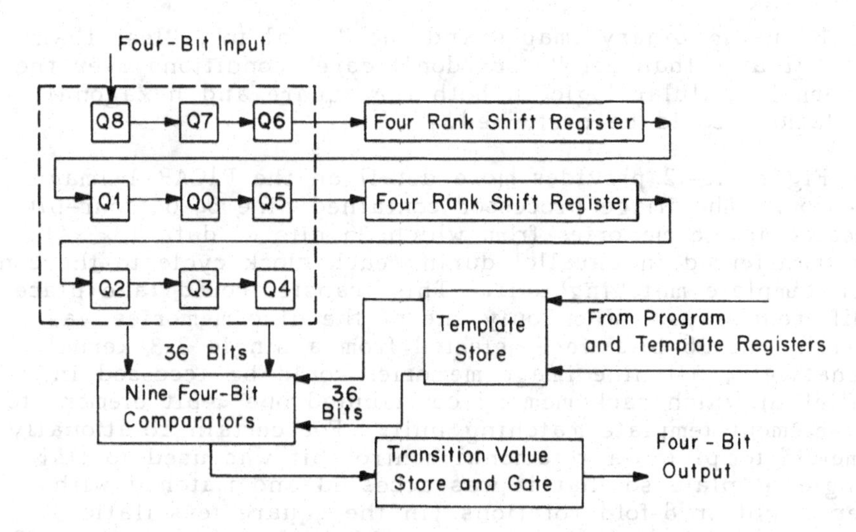

Fig. 10-12 The template matching unit of PICAP I consisted of nine numerical comparators operating in parallel on data stored in the 3×3 cellular register.

normal sampling, a small window within that frame. The four bits generated by the digitizer could be normalized with respect to the dynamic range of the video signal. By this technique the user could screen image data at low resolution and, based on certain decision criteria, could select subsequently particular areas for detailed or semi-detailed examination.

To explain the condition template matching capabilities of PICAP I, consider a condition vector C(I) where I = 1,...,9. These nine conditions map to the nine positions of the 3×3 kernel and are tested in parallel for all picture elements within the 3×3 window (except for those locations flagged with the "don't care" condition, meaning that the outcome of the tests at those locations are scored as true). The specific conditions examined by the condition template matching circuitry within PICAP I were the logical conditions "less than", "greater than", and "equal to." If all the conditions are simultaneously true, then the output at the center element location becomes a specified "transition" value; otherwise, the output equals the input. PICAP I used as many as eight condition templates in one pass over the input array. However, in the case where more than one condition template was stored, these templates were examined sequentially (in a priority list). When the first all-true condition match was obtained, its specified transition state became the state of the output picture element. If no true matches occurred, then, as before, the picture element value was unchanged.

By using binary images and the "equal to", "less than one", "greater than zero", or "don't care" conditions over the 3×3 kernel, cellular logic in both the square and hexagonal tessellations could be performed.

Figure 10-12 provides more detail on the PICAP I image processor. The image processor contained nine 64×64 four-bit dedicated image memories from which 36 bits of data (3×3×4) were transferred in parallel during each clock cycle to the condition template matching unit. This transfer could take place in different ways. When only one of the nine memories was accessed, the 36 bits were selected from a single 3×3 kernel. Alternatively, all nine image memories could be accessed in parallel in which each memory contributed one 4-bit element to the 9-element template matching unit. For certain rotationally symmetric templates a rotational control bit was used to flag a single template so that it was accessed and matched with either 4-fold or 8-fold rotations (in the square tessellation). Thus the condition template store could emulate as many as 32 or 64 condition templates, respectively.

PICAP I required 1.2 microseconds for a single template-matching operation per picture element. Thus, for operations that could be carried out in a single match per picture element, the speed of PICAP I was slightly faster than the original GLOPR built by Perkin-Elmer. However, in more complicated operations, requiring multiple templates, PICAP I was slower. Instructions for PICAP I were stored in the main control computer and passed one instruction at a time to the PICAP I image processor. Each instruction was a variable-length sequence of 16-bit words. The first word was always the same and included the operational code and the addresses of the source and destination images. Following this were at least five words containing the condition template and the transition output. When the final word of these five was appropriately flagged, this indicated an end-of-instruction condition, otherwise, more templates followed. Thus PICAP I was a vectored machine (as with GLOPR) which carried out a full cycle over 4096 picture elements in response to one set of instruction words.

The output of the PICAP I image processor consisted of the contents of the measurement registers. There were 32 of these registers which were allocated to count all matches for each of the eight templates (eight registers) and to totalize the number of picture elements where matches occur (one register). There were also 16 registers used as counters to generate a histogram of the 4-bit output image data. Three additional registers were used to hold the measurements of the maximum value, minimum value, and mean value of the picture elements. Four additional registers were used to store the minimum and maximum values in both the vertical and horizontal directions where non-zero picture elements were found.

In summary, the PICAP I system went far beyond earlier cellular automaton simulators in that it was capable of performing almost any binary or numerical cellular operation with an extraordinary amount of flexibility. (Perhaps one disadvantage was programming complexity.) Applications over a wide range (from medical microscopy to finger print analysis) were investigated by the faculty and students at the University of Linkoping using PICAP I. Success with this system led to the production of PICAP II in the 1980s as described in Section 10.4.

10.3.3 Delft Image Processor (DIP-1)

Fabrication of the Delft Image Processor (DIP-1) commenced with a design study by the Pattern Recognition Group of the Applied Physics Department in conjunction with the Laboratory for Computer Architecture of the Electrical Engineering Department at the University of Delft in 1976. The system has been fully operational since early 1979. Prior to 1976 the Pattern Recognition Group had assembled a digital television system including various video input devices and a 256×256 8-bit-per-element video frame memory. This was linked to a Hewlett Packard 2100 minicomputer in the Applied Physics Department having both disc and magnetic tape storage facilities. This in turn, was networked to the IBM 370/158 in the University of Delft central computing facility.

The DIP-1 was introduced into the existing computer system by the end of the 1970s. Included with the DIP-1 hardware was a Hewlett Packard 1000 minicomputer with 448K bytes of main memory which acted as control computer (Figure 10-13). As can be seen from Figure 10-13, there are three input devices, namely a Plumbicon camera, a flying spot scanner, and a television microscope. Unlike PICAP 1, there is no centrally controlled video interface system but, rather, the main control computer acts as the primary receiver of image data. This data may then be transferred to 20M byte disc for reformating and delivery, as required, to the DIP-1. Displays are directly as-

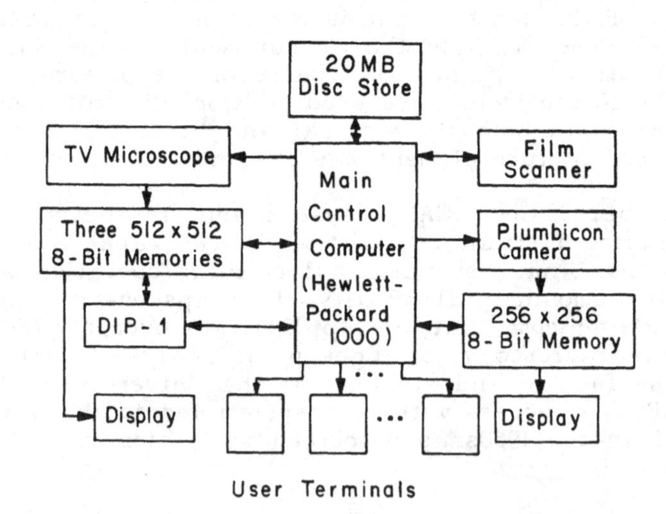

Fig. 10-13 The DIP-1 system has three separate television scanners. The 512×512×8 memories associated with one of these serve as the primary store for the image processor which, in turn, is controlled by the HP 1000 host.

sociated with the Plumbicon camera system as well as with the television microscope but not the flying spot scanner. All of the local memories for the television input devices operate at full video rates when acquiring data. This rate is readjusted for input and output to both the main control computer and the DIP-1. Finally, memory associated with the Plumbicon camera acts as the working image memory for the DIP-1.

The DIP-1 differs considerably from both GLOPR, BIP, and the PICAP I. In particular the DIP-1 contains a writeable control store and microprogram sequencer to control its various data processing modules. Thus DIP-1 is dynamically microprogrammable and operates under the control of microprograms which specify the sequence of signals needed to control the DIP-1 modules in performing image processing operations. The control store is addressed by a microprogram sequencer in order to determine which microinstruction should be performed next. This permits parallel processing within the DIP-1. All operating modules produce condition flags stored in flag registers for routing the sequence of microinstructions to be executed.

The operating modules of DIP-1, unlike those of GLOPR, BIP, and PICAP I, are traditional arithmetic logic units, multipliers, and numerical conversion tables. In addition to this there are cellular logic tables. The organization of DIP-1 is distinct from that of the dedicated cellular logic processor of GLOPR, the dedicated Boolean comparator of BIP, or the dedicated condition template matching unit of PICAP I. Not only may the image data within the DIP-1 be handled as 8-bit picture element values, but also floating point representations are available (12-bit mantissa; 6-bit exponent). Also 12-bit integers or logical fields may be specified. Convolutions may be executed over a region as large as 16×16.

This section will not attempt to deal with all of the features of the DIP-1, but rather will concentrate on its cellular logic capabilities. The two arithmetic logic units (ALUs) of DIP-1 are functionally identical ALUs. Each may operate either in floating-point mode using 18-bit inputs or in integer mode using 12-bit integers or 12-bit logical fields. An ALU operating in the integer mode may not only add and subtract but also may increment, decrement, or carry out Boolean operations. The single multiplier in floating-point mode uses 18-bit normalized floating-point numbers as operands. The multiplier may also use 12-bit integer numbers and cperate in an integer mode with the product lying in the range -2048 to +2047. The way in which these units are arranged is shown on the following page in Figure 10-14.

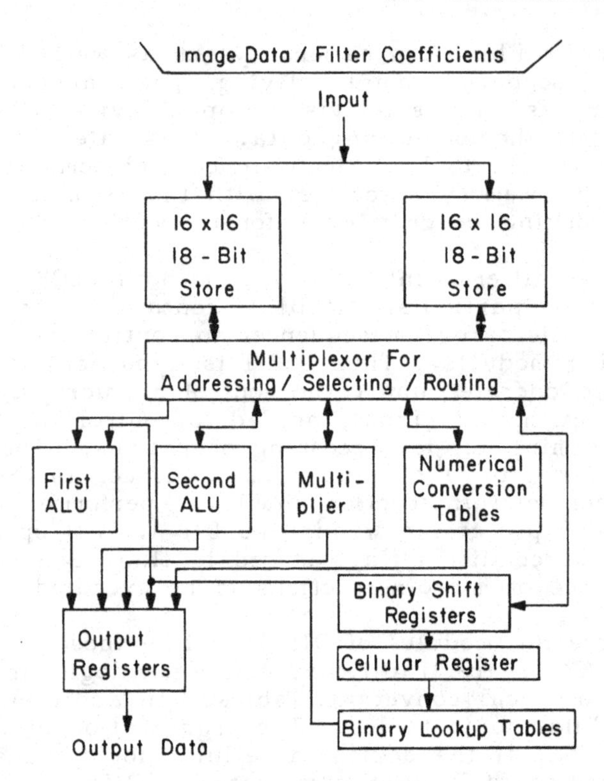

Fig. 10-14 The DIP-1 uses two 18×16×16 stores for 16×16 convolutions employing ALUs and the multiplier. Also binary shift registers and a cellular register exist for cellular logic transforms.

The neighborhood data buffers each have memory associated with them which can contain up to 256 eighteen-bit numbers. For large area convolutions (16×16) these memories are used to store filter coefficients. They may also be used as row or column buffers for temporary storage of intermediate results from the data-manipulation modules. When performing in this fashion they have both auto-increment and auto-decrement features using both absolute and relative addressing.

The conversion tables are used to perform integer or floating-point transformations by table lookup in one microinstruction cycle. Each of these conversion tables consists of 4096 words of eighteen bits each. These tables may be loaded by the host computer. (They are also directly loadable from within the data loop.) One purpose of these tables is to execute real-time graylevel point operations.

Of particular concern for this chapter is the binary lookup table module which has eight 512-entry tables in which all configurations of the 3×3 binary neighborhood are included. The cellular logic module is organized to incorporate two 256-bit binary shift registers and two 3×3 cellular registers which are used to store incoming data and template (as in GLOPR). There is another 256-bit shift register and a partial "look-back" cellular register (exactly as in CELLSCAN). Electronically variable selector switches route data from the cellular portion of these registers to address a specific one of the 512-position lookup tables. By providing eight tables, selectable under program control, the user no longer needs to reload frequently used tables, but only flags one of them with a control word. These tables have outputs which are masked (with the mask being supplied by the microcode). Their outputs are, in turn, delivered to OR-gates which provide outputs to the result registers.

By using the "look-back" feature, recursive cellular logic transforms may be performed. With look-back disabled, a traditional cellular automaton simulator generating non-recursive results is implemented. (This feature is similar to BIP.) By extending the shift registers, it is possible to add an exterior boundary to the image which is either set to binary 0 or binary 1. Further flexibility is provided in that the masked output of the lookup tables may be used as input to one of the ALUs. By this method, setting the proper bit in the ALU makes it possible to remember previous results for an entire image processing cycle and to use these results to stop the next image processing cycle if no changes have occurred.

The host computer is used to write the various conversion tables, the celluar logic tables, the neighborhood data buffers, and to provide the control words for the control store. The DIP-1 is arranged so that it can be started and stopped by the host computer and can be caused to execute microprograms on a single-step basis for diagnostic purposes. Debugging facilities are also provided by permitting the host computer to interrogate specific registers and gather intermediate results on microprogram activity.

The DIP-1 has been used by the University of Delft in various image processing tasks ranging from medical microscopy to industrial inspection. Timing of the DIP-1 as a cellular logic machine has been analyzed and reported by Gerritsen (1982). If the cellular logic lookup table is not resident in the DIP-1, then 40ms is required to transfer it from disc to the host computer (all cellular logic lookup tables are pre-computed) and another 40ms to transfer it to the DIP-1. Another 100ms of overhead is required by the applications program resident in the host computer. For a 256×256 image the time required to fetch data from the digital image memory, operate upon that data, and replace it in the digital image memory is 97ms. Thus the entire operational time (277ms) is dominated by systems overhead. Using this overall figure, the DIP-1 requires 4.2 microseconds per picture point operation but, if the cellular logic lookup table is already loaded and further iterations using the same table are performed, then the picture point operation time for the subsequent cycles is 1.5 microseconds. Thus the DIP-1 is comparable in speed to the original GLOPR and PICAP I, but is a few hundred times slower than either the diff3, GLOPR, or BIP.

10.3.4 The Cytocomputer

In the mid 1970s the Environmental Research Institute of Michigan (ERIM) proposed to the United States Air Force a new cellular logic architecture. The machine which was then implemented was called the "Cytocomputer." The first public description of this machine was presented by Sternberg (1978) and later by Lougheed and McCubbrey (1980). (See also the Sternberg patent issued in the mid-1970s.) The general cytocomputer architecture is a pipeline of CELLSCAN-like processing elements each of which contains a 3×3 cellular register with intervening (N-3)-length shift-registers. Each processing element in this pipeline performs cellular logic by means of a full 512-position lookup table. The output from one processing element is delivered to the input of the next shift register. The contents of each lookup table are, in general, different from those of the others. Therefore, most of the lookup tables are transmitted separately to the cytocomputer by the host computer. The host computer also delivers the image to be processed as a bit stream synchronized with the flow of data through the cytocomputer system itself. The cytocomputer built by ERIM has a total of eighty-eight pipelined processing elements. An individual stage of the pipeline is shown in Figure 10-15.

The cytocomputer architecture is advantageous when executing a known algorithm repetitively. In this case the lookup tables are fixed and are loaded at the onset and left in position from there on. This is particularly useful for Air Force applications where a continuously scanning reconnaissance system interrogates the terrain scan-line-by-scan-line, thresholds the resultant graylevel information into bilevel format, and submits the data picture-element-by-picture-element to the cytocomputer system. Such a system would use a permanently stored algorithm for the detection of all objects of interest in the terrain. (Several cytocomputer pipelines could also be used - one for each type of object of interest.)

Disadvantages of such a system appear in a research en-

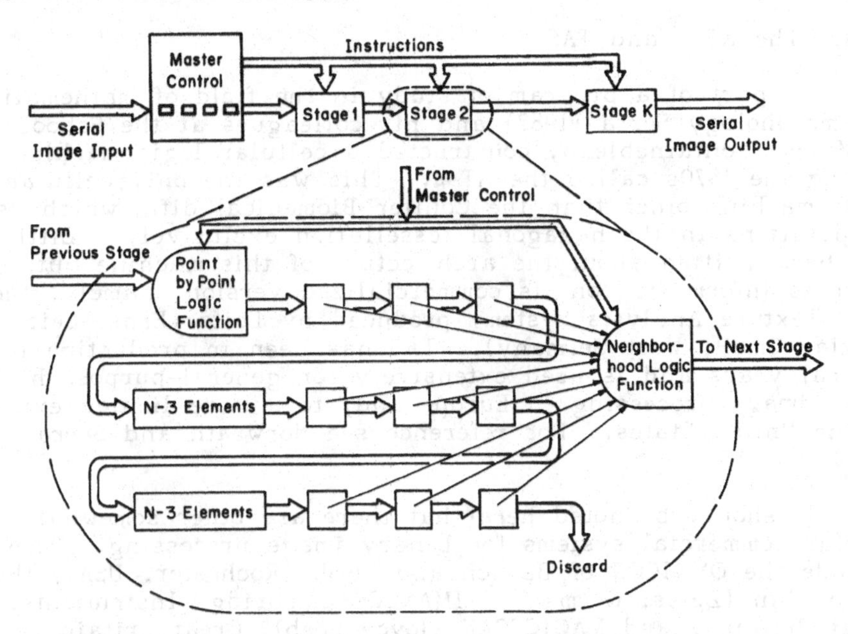

Fig. 10-15 Block diagram of the cellular logic portion of a single cellular logic stage of a cytocomputer.

vironment where algorithms are frequently modified. There is significant overhead in the lookup table loading procedure. Furthermore, the fixed value of shift register length within each processing element prevents variable-size images from being processed readily. Another disadvantage is that, if short algorithms are to be handled, only a few pipeline stages are used after which the image data must travel down the rest of the pipeline before any results are obtained. Processing elements at the end of the pipeline perform no function other than to pass data from processing element to processing element.

In the original cytocomputer built for the Air Force by ERIM the operating time for one step of the pipeline is 650ns. Thus, with an eighty-eight step algorithm, the effective picture point operation time is 8ns. This makes this cytocomputer system approximately ten times faster than BIP and three times faster than the Coulter Biomedical Research diff3 GLOPR. It is also considerably faster than all other existing cellular logic machines.

The main application of the cytocomputer system has been in military reconnaissance target detection. However, Sternberg (1983) has reported several biomedical applications of this system in the automatic location of cell nuclei in limb-bud tissue and in the analysis of gel electrophoretograms.

10.3.5 The AT-4 and TAS

As part of a program of study in the field of mathematical morphology Serra (1982) and his colleagues at the School of Mines, Fontainebleau, constructed a cellular logic machine during the 1970s called the AT-4. This was the only cellular logic machine other than the Coulter Biomedical diff3 which used computations in the hexagonal tessellation exclusively. Little has been written about the architecture of this machine but there is information on its commercialized version, namely, the TAS (Texture Analysis System) produced by Leitz (Ernst Leitz Wetzlar GmbH, West Germany). TAS has been in production for several years and is used extensively for general-purpose high-speed image processing in Europe and, to a more limited extent, in the United States. For reference see Norwrath and Serra (1979).

It should be noted here that there are other somewhat similar commercial systems for binary image processing. These include the OMNICON of Bausch and Lomb (Rochester, USA), the Video Plan (Zeiss, Germany), IMANCO (Cambridge Instruments, Great Britain), and MAGICSCAN (Joyce Loebl, Great Britain). These latter machines all have some cellular logic capability, i.e., can perform augmentation and reduction of binary images and, in some cases, skeletonization. These machines are also characterized by the ability to perform many other numerical functions. Frequently these are not carried out in special-purpose hardware, but rather by a host minicomputer. The cellular logic architecture of all of these image processors would take too long to describe and, for purposes of brevity, the TAS will be taken as a case-in-point.

The TAS system is shown in Figure 10-16 as described by Norwrath and Serra (1979). A television camera and thresholder are used to generate binary images which are then stored in one of eight 256×256 image memories. The system is controlled by a minicomputer (the Digital Equipment Corporation LSI-11/2) and the user controls the system via a CRT terminal. Two image processing units are available for use in carrying out two-dimensional logical transforms. One of these units is hardwired while the other is programmable. The general mode of operation is for the user to set up a "test pattern" or "template" in the hexagonal tessellation which would correspond, for example, to one of the Golay primitives in all of its orientations. The image then is delivered to one of the two-dimensional processing units and the required cellular logic operation is executed. A programming language has been devised for TAS which is used in Chapter 14 as an illustrative case-in-point for commercial cellular logic machines. The TAS user may execute several operations in parallel on one or more of the eight binary images. These operations may include not only cellular

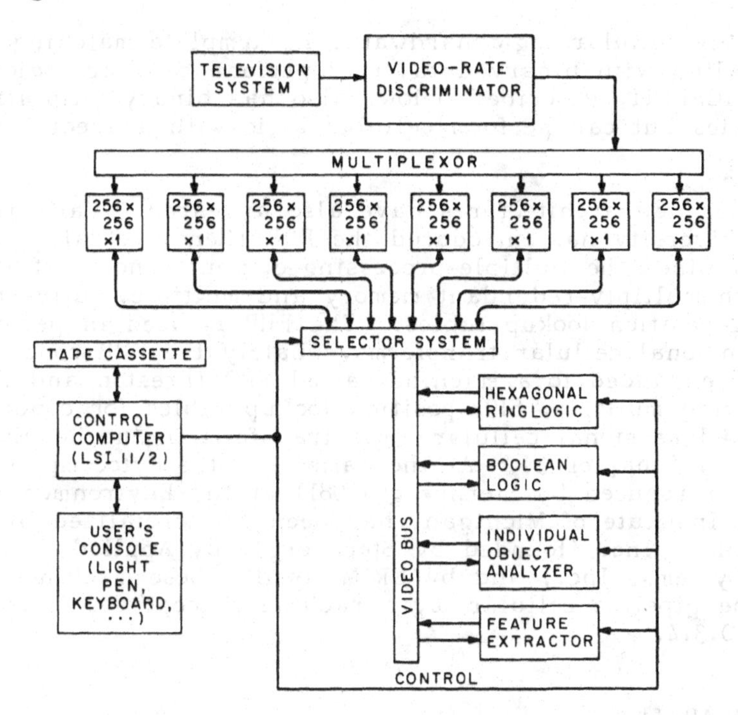

Fig. 10-16 Block diagram of the TAS hexagonal tessellation cellular logic machine manufactured by Ernst Leitz Wetzler GmbH.

logic computations but also the simultaneous Boolean combination of either inputs or outputs. This particular feature provides speed and efficiency somewhat beyond that available in the Coulter Biomedical diff3. Otherwise the two systems are very much the same.

10.4 PROGRESS SINCE 1980

The 1980s have seen the continued use of the now-commercialized pure cellular logic machines such as the diff3 of Coulter Biomedical, in the field of hematology and, for general-purpose applications, such machines as the TAS of Leitz. Many hundreds of these machines are in use world-wide. In general, however, the current trend is away from these purely logical machines. Lately the preference is to embed the cellular logic machine in more general-purpose image processing architectures. Typical of this trend is the DIP-1, which combines cellular logic hardware in a structure which also includes general-purpose arithmetic logic units, multipliers, and numerical conversion tables. Another example is PICAP I where, although there is no identifiable cellular logic hardware, its template matching unit, when dealing with binary data, may perform cellular logic transforms. PICAP II, described below, also has binary templating capabilities but can perform cellular logic with a special processor.

Other new architectures have also appeared. Carnegie-Mellon University has introduced the PHP (Herron et al., 1982) which combines the multiple-processing-element concept of the diff3 with multiply-redundant memory and multiple, fully-structured 512-position lookup tables. The PHP is used in performing two-dimensional cellular transforms. Lately the PHP concept has been expanded to a machine called TRO (Preston and Ragusa, 1983) having multiple, 8192-position lookup tables for conducting the three-dimensional cellular logic transform in the hexahedral tessellation (Chapter 3). At the same time the cytocomputer concept, introduced by Sternberg (1981) at the Environmental Research Institute of Michigan, has been commercialized by Machine Vision, Inc. (founded by Sternberg), by Applied Intelligence Systems, Inc., and by ERIM itself. These machines all follow the pipeline cellular logic machine concept described in Section 10.3.4.

10.4.1 PICAP II

As described by Kruse et al. (1982), the individuals at the Department of Electrical Engineering, Linkoping University, who had produced PICAP I were planning the next generation machine by the late 1970s. PICAP I was limited both by its slow execution rate, by its small (64×64) image memory size, and by narrow dynamic range (16 gray levels) in its nine graylevel image memories. By the end of 1979 the complete architecture of the next generation machine had been formalized. This machine, called PICAP II, was planned as an "instruction-set processor" as defined by Reddy (1980). An instruction-set processor is a multi-processor system which employs several processing units which are specific to certain computational functions. In effect, PICAP I was a dual-instruction-set processor in that it had separate modules for both video interfacing and for template matching. PICAP II is, instead, comprised of seven processors which are capable of simultaneous operations. Furthermore the operating speed of PICAP II is considerably enhanced over PICAP I in that it is controlled by a host computer whose bus rate is 40 megabytes per second. (See Figure 10-17.) By this token the speed of PICAP II is approximately 20 times that of PICAP I. Finally, the image memory of PICAP II has been expanded by a factor of 20 and contains four megabytes of storage in the form of sixteen 512×512 eight-bit graylevel image stores.

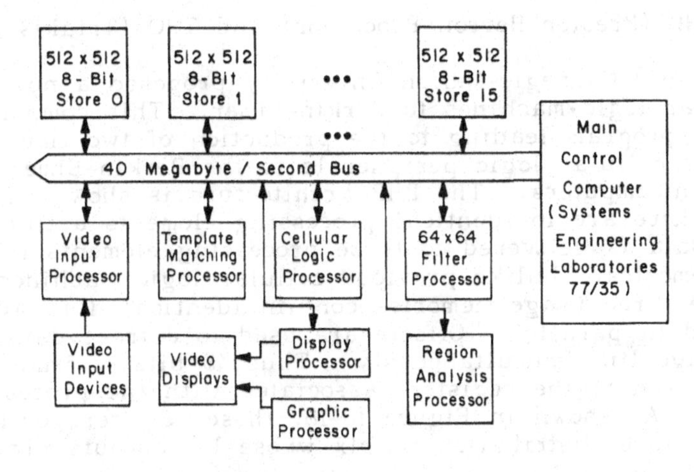

Fig. 10-17 The PICAP II system of the University of Linkoping (Sweden) is an instruction-set processor with seven specialized image processing peripherals operating in parallel. These share a common 40 megabyte per second bus with the host computer.

Only the cellular logic processor is specific to the discussion of this chapter. This unit discards the complexity of template matching for executing cellular logic transforms in favor of an architecture which is specific to image data packed as bit streams. In this manner it is very similar to the special-purpose cellular logic processor of DIP-1. The user employs this particular processor rather than the template matching processor for cellular logic operations using binary image data.

The other processors in the PICAP II system are the template matching processor (described in Section 10.3.2), the filter processor for carrying out numerical convolution operators in neighborhoods as large as 64×64, and the region measurement processor which is capable of both region labeling, direction-code encoding of region boundaries, and the generation of other region measurements, such as area, perimeter, etc. Finally, the video display is controlled by three processors for (1) display, (2) graphics, and (3) video interfacing. The display processor provides the usual lookup tables for three-color image display and also has dynamic addressing for the purpose of scale expansion (zoom) and translation (scroll). The graphics processor is used for generating both symbols and vector-graphics for image overlay purposes. The video interface processor is similar to that of PICAP I in that it controls both the region from which video data is to be extracted from an input scanner as well as the instantaneous dynamic range of the image digitizer.

10.4.2 PHP (Preston-Herron Processor) and TRO (Triakis Operator)

In 1979 Carnegie-Mellon University proposed a new concept in cellular logic machines to Perkin-Elmer. This company then funded a program leading to the production of two machines designed as cellular logic peripherals to the Perkin-Elmer 3200-series minicomputers. The PHP architecture is shown in Figure 10-18. There are 16 identical processing elements acting in parallel. Data is delivered to these processing elements from three image memories. Unlike previous cellular logic machines, however, the three image memories contain identical data and are addressed in parallel. Offsets are used to extract data from three image lines simultaneously. Thus 48 bits of image data are delivered to the registers associated with the processing elements. As shown in Figure 10-18, these registers deliver 54 inputs to a distribution matrix whose 144 outputs provide

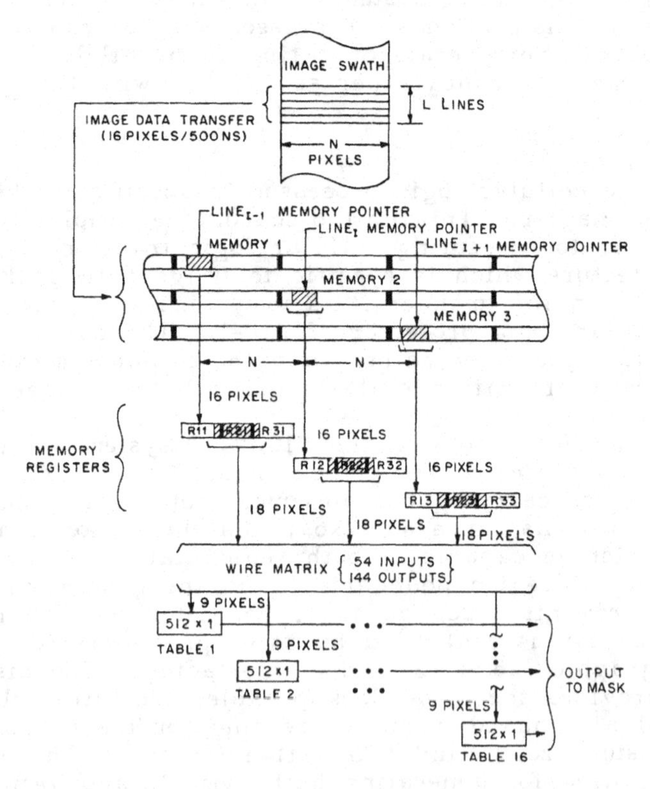

Fig. 10-18 The Carnegie-Mellon PHP (built for Perkin-Elmer in a joint program with the University of Pittsburgh) introduced the concept of triply-redundant memory to achieve high speed cellular logic. The redundant memories (above) provide 48 pixels simultaneously to registers connected to the sixteen 512-position lookup tables (below).

the 16 nine-bit addresses to all 16 lookup tables. The outputs of the lookup tables then are gated to the PHP output bus which transmits them to the host computer via a subfield mask.

The Perkin-Elmer host computer is used to load the appropriate lookup table (which is identically copied in parallel to all processing elements), to carry out the thresholding operations necessary to convert graylevel images to binary images, and to load and unload the image data from the PHP as a thirty-two thousand bit image slice. The three identical image slice memories are each 2048×16. These memories are then read and processed by table lookup. The overall image dimensions are determined not by the PHP but are under software control in the host computer. For example, a square image as small as 16×16 up to an image slice as large as 3×10912 can be processed by the PHP. Any intermediate image size can be handled as long as the number of lines in the image are selected modulo 1 and the number of pixels per line modulo 16.

The host computer is capable of loading and unloading at the rate of five megabytes per second (40 million picture elements per second). Thus an image slice consisting of thirty-two thousand 1-bit elements could be processed in 1.6ms, i.e., in one input cycle plus one output cycle. This implies that 12.4ms would be required to process a 512×512 image. However, in its present configuration, the host computer of the PHP must transfer all binary images from disc through main memory to the PHP and, after processing, return them via the same route. This adds significant overhead and slows the entire system to one twentieth of its potential. A total computation time of 250ms is now needed for a 512×512 image. It has been estimated (Herron et al., 1982) that, by eliminating disc transfers, this time could be reduced by one half and that, by changing the driver software so as not to release the channel to the PHP after processing every image slice, the 512×512 image could be transformed in approximately 50ms (200ns picture point operation).

There are currently two PHPs systems now in operation: one at the Wooster Heights Facility of Perkin-Elmer and one at the Biomedical Image Processing Unit of the University of Pittsburgh. The success of these machines inspired the construction in 1982 of a three-dimensional cellular logic processor having identical architecture but more memory and larger lookup tables. This machine is the TRO (TRiakis Operator) which is patterned on the analysis developed in Chapter 3. TRO has seven identical memories each of which can contain eight planes of a 64×64×64 workspace. As with the PHP, the TRO runs 16 processing elements in parallel. Each processing element is addressed from the 13 appropriate neighborhood values gated from the memory registers. Thus, each lookup table has 8192 positions. These tables are loaded in parallel from the host

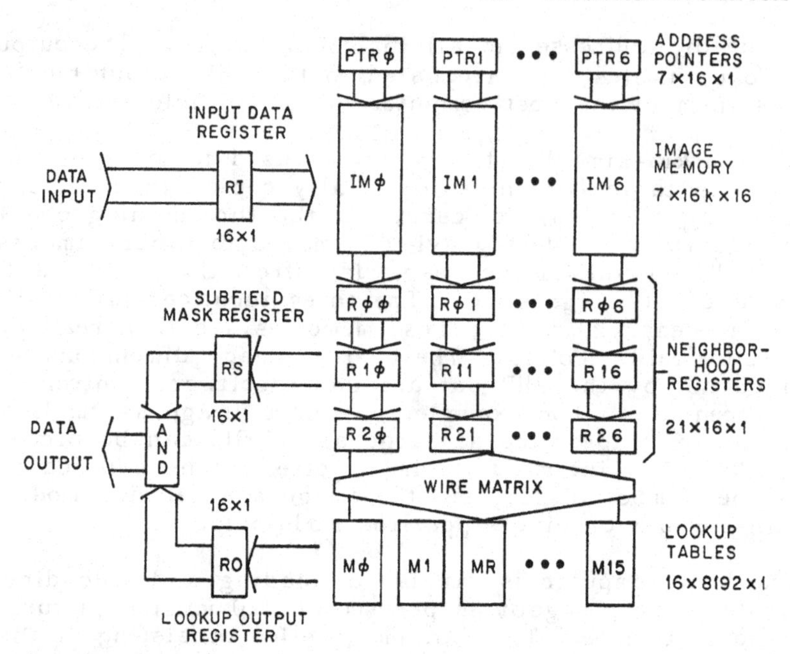

Fig. 10-19 The TRO computer is the three-dimensional equivalent of the PHP with seven-fold memory redundancy furnishing 112 bits in parallel each instruction cycle to the processing element registers.

computer, the seven redundant memories are filled simultaneously with eight 64×64 planes, the information is processed, returned to the host computer, and a new 64×64×8 slice is extracted from the workspace. The total time required for the entire workspace is approximately 200ms (exactly comparable to the PHP and for the same reasons). TRO is diagrammed in Figure 10-19.

10.4.3 Retrospective

The 1980s have witnessed the successful commercialization of the cellular logic machine both in the form of Coulter Biomedical diff3 and the Leitz TAS. Earlier machines such as the Perkin-Elmer CELLSCAN and research GLOPR have been retired. The only machine remaining from the 1960s is the Information International BIP of which a few are still in operation.

Research on cellular logic machines now appears to be divided into three areas: (1) pure parallel cellular logic machines such as the PHP and TRO, (2) pure cellular logic machines of a pipelined architecture such as the various cytocomputer systems, (3) and cellular logic machines which are em-

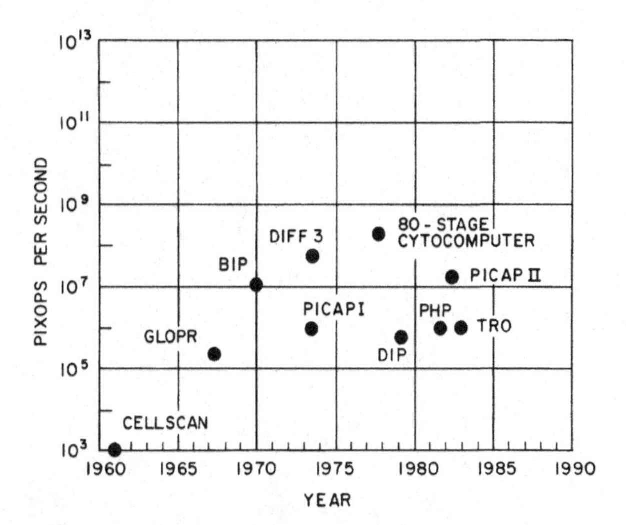

Fig. 10-20 The rate of increase in machine speed, measured in pixops/sec (picture point operations per second) for the cellular logic machines discussed in this chapter is given above for the interval 1960-1983.

bedded in larger systems whose other modules are either instruction-set oriented or of a more general-purpose computational nature. Examples of the latter are, of course, PICAP and DIP.

As shown in Figure 10-20, the rapid increase in speed of cellular logic machines from 300ms per picture point operation (CELLSCAN, 1961) to 20ns (diff3, PICAP II) has ceased and there appears to be considerable concentration on improvements in cost performance. (The PHP costs only $6000.) The full-array cellular automata now in operation (Chapter 11) outclass the cellular logic machines in speed but not necessarily in processing flexibility. The cellular logic machine will continue to have the advantage over the full-array processor in that the size of the image data which may be handled is far larger. This should be kept in mind when considering that some imaging systems (weather satellites) are now producing images as large as 15000×15000. Therefore it is unlikely that a full-array automaton will ever be practical in handling all image processing tasks. Rather, a merger of the two technologies may occur as witness the CLIP7 cellular automaton described in Chapter 11.

11. ARRAY AUTOMATA

11.1 INTRODUCTION

The idea that two-dimensional data arrays could be processed by matching arrays of computers was first introduced by von Neumann (1951). It was not until the paper by Unger (1958) that it was seriously proposed to construct such an array. Unger proposed using simple one-bit computers (processing elements) and, although Unger's work inspired many subsequent proposals (e.g., McCormick, 1963), it took some fifteen years for technology to develop to the point where array construction was an economic proposition.

At the time of writing, twenty-five years after the publication of Unger's paper, three array automata, each having several thousand processing elements, are in operation: CLIP4, DAP, and MPP. It is the purpose of this chapter to describe and compare these array automata and to suggest in what ways automata of this type might be expected to develop in the future.

11.2 CLIP (Cellular Logic Image Processor)

The CLIP series of array automata has been evolved by the Image Processing Group of the Department of Physics and Astronomy in University College London (UCL) under the supervision of one of the authors (Duff). It is interesting to observe that it was not originally the intention of Duff and his co-workers to construct a general purpose cellular automaton. Their research plan, drawn up in the early 1960s, had been to investigate the possibility of loosely modeling the function of the human retina and visual cortex by a layered construction of matrices of logic elements, the first of which would include an array of photosensors and the last an array of light emitters. Between these extremes, logic layers would exist to perform operations on the image data so that a transformed image would appear at the output within milliseconds of an image being projected onto the input sensors. The machine dubbed UCPR1 (Duff, 1967) embodied these principles and was first demonstrated in 1967. UCPR1 made use of fixed local neighborhood operators. Specifically, these operators located all ends and junctions in line images. The work of Levialdi et al. (1968) showed that propagation between processing elements could be turned to advantage in image analysis, relating naturally to the concept of connectivity between image elements (Chapter 5). Duff (1969) then generalized the use of propagation by proposing a square array of simple one-bit processing elements, interconnected by diodes, so that propagation could take place simultaneously in the four principal array directions. This led to the fabrication of the so-called Diode Array which was assembled in 1969 (Duff, 1971). Here electro-mechanical relays were used to configure internal connections in each of the 25 processing elements of a 5×5 processor array. This array, although technologically crude and so small as to be virtually useless as a general-purpose image processor, laid the foundations for the immediately following series of CLIP machines. The Diode Array, however, was actually capable of implementing a wide variety of basic transformations on binary images. These included edge detection, region labeling, augmentation and reduction, etc.

11.2.1 CLIP1 Through CLIP4

In order to see whether two-dimensionally connected arrays of logic gates would perform correctly and not exhibit unexpected oscillations, due to the differing signal paths between various extremes of the array, CLIP1 was constructed in 1971 (Duff, 1971). This was a 10×10 array which used off-the-shelf integrated circuits and implemented certain of the Diode Array functions. The satisfactory behavior of CLIP1 encouraged the immediate construction of the next array in the series, (CLIP2), which was completed in 1972 (Duff, 1974).

CLIP2 could be programmed by push-buttons to perform sequences of up to 32 instructions, but it soon became apparent that more generality was needed in each processing element. Many desirable and even necessary operations were not possible in this system. For example, image data could not be shifted across the array nor could any operation be performed which defined the relative positions of the components of an image. In short, the array had no *sense of direction*. By elaborating the processing element to a full Boolean processor with gated inputs from each of the eight array neighbors, CLIP3 was then formulated to eliminate these difficulties (Duff, 1975). CLIP3 was a 16x12 array constructed in 1973 and, although still too small to be effective as an image processor, it was completely and generally programmable, i.e., it could in principle be programmed to perform all operations on two-dimensional data arrays. It therefore constituted a general purpose computer optimized for image processing and image analysis.

As with the earlier members of the CLIP series, CLIP3 was still too small an array to be used sensibly as an image processor. However, by scanning the CLIP3 array over 96x96 images, incorporating suitable edge data stores around the processing array to provide for propagation, and also using local neighborhood extension from the 16x12 data segments, greylevel picture processing on a 96x96 by six-bit image was achieved in 1974. At the same time, a custom-designed large-scale integrated circuit development program was initiated, leading to the specification of the CLIP4 processor. CLIP4 closely resembled CLIP3, incorporating all of its major features other than the edge interconnection threshold gate; the addition of this feature would have required an uneconomical number of gates for only a relatively small contribution to computational power. Some additional features were included, such as a few gates, external to the double Boolean processor, which provided a single-operation facility for full binary addition. Unfortunately, integrated circuit custom-design services were only barely available in Great Britain during this period and it was not until early 1980 that a full 96x96 CLIP4 array could be put into operation (Figure 11.1). Since then, this particular cellular automaton, comprised of 9216 processing elements, has been in continuous use and has provided the central image processing capability for the UCL Image Processing Group in a wide range of application projects and also in more fundamental studies of image processing algorithms.

11.2.2 The Architecture of CLIP4

The structure of the CLIP4 processing element (Figure 11.2) is deliberately and almost classicially simple. In order to maximize the ratio of computing power to cost, all unimportant functions are eliminated, thus permitting the incorporation of eight processors, with 32 bits of local image memory per processor, in a single integrated circuit package. For all non-arithmetic operations, the processing element consists of two identical Boolean processors, each of which has the same two inputs, P and A. Both processors can generate any of the sixteen possible Boolean functions of their two inputs. They

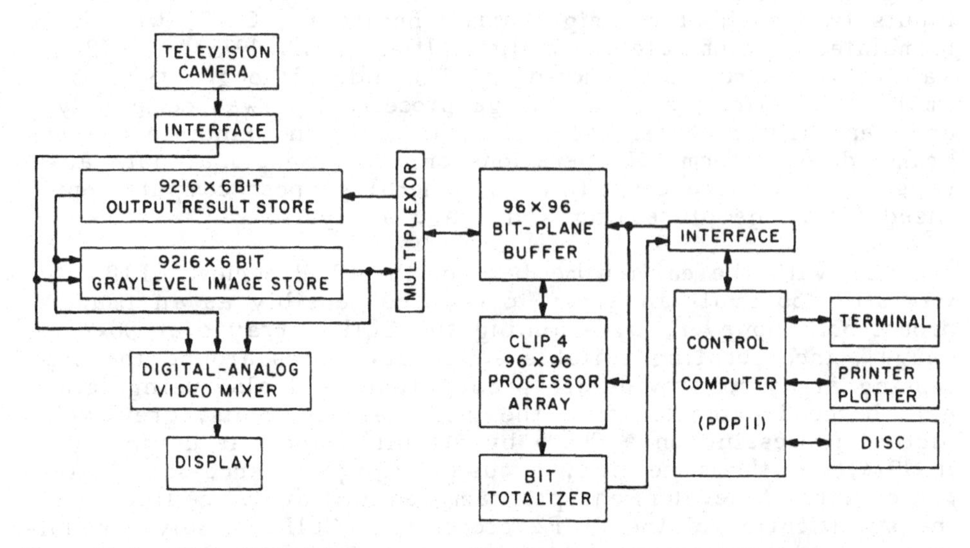

Fig. 11-1 The CLIP4 system at University College London incorporates a television camera, a storage register for a full 96×96 graylevel field, the array of 9216 processing elements, and a control minicomputer.

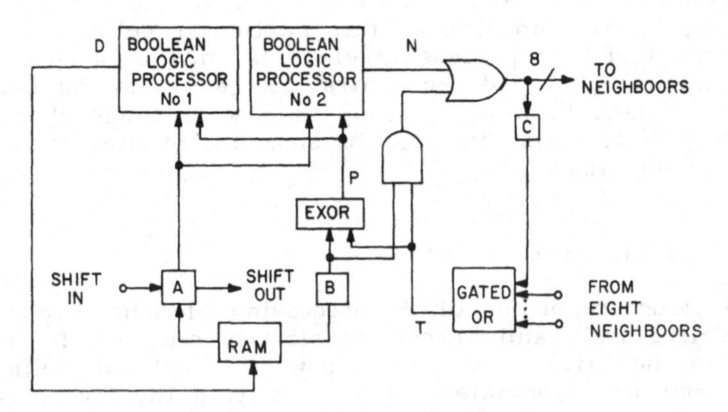

Fig. 11-2 The processing element of the CLIP4 automaton has gated parallel inputs from and outputs to eight adjacent processing elements, carry save, and dual Boolean processors.

are controlled by two sets of four control lines. One of the Boolean processors has an output D which can be clocked into any one of the 32 memory locations associated with the processing element (D0...D31). The other Boolean processor produces an independent output N which is immediately transmitted to the eight neighboring processing elements in the array as well as being stored in the one-bit buffer C for use in subsequent operations. Each processing element also receives eight inputs from its eight neighbors and these are individually gated into an OR-gate along with the previous value of C to produce an output T. The choice of inputs used to produce T is the same for every processing element in the array. T is EXORed with D as described below to generate B.

The inputs to the two identical Boolean processors are A, a single bit of image data, and P, which is the EXOR of B, another single bit of image data, and the value T described above. Both A and B are held in one-bit buffers which can be loaded from selected image memory locations.

11.2.3 CLIP Operations

A CLIP operation can be decomposed into the following parts:

- **LOAD A** A single bit of image data is taken from a specified memory location and loaded into the one-bit A buffer.
- **LOAD B** A single bit of image data is taken from a specified memory location and loaded into the one-bit B buffer.
- **SET** The Boolean functions for both Boolean processors are specified as well as the neighborhood interconnection gates to be enabled. Interconnection inputs to processing elements at the edge of the array are all set at 1 or 0 as required. The array connectivity may be chosen as square or hexagonal. The additional arithmetic gates described below are either enabled or disabled.
- **PROCESS and STORE** The controls specified by the set operation are switched from zero to their required values. The selected interconections are enabled and the array

> edge conditions established. When propagation has stabilized so that the outputs of the Boolean processors are constant at all points in the array, the output D is clocked into the selected D location and, if desired, N is stored in the one-bit buffer C.

The use of the processing element to provide arithmetic functions has been described in principle in Chapter 4. If A and B are the corresponding bits of two bit-stacks and if the chosen Boolean functions are

$$\begin{aligned} D &= A.EXOR.P \\ N &= A.AND.P \end{aligned} \tag{11.1}$$

then N will be the carry and D the sum of A and B. If the carry is now stored in the one-bit buffer C, B and T (which is equal to C) are ORed together to produce P, and the next most significant bits in the addition are loaded into A and B, the same functions will give

$$\begin{aligned} N &= A.AND.(B.EXOR.C) \\ D &= A.EXOR.B.EXOR.C \end{aligned} \tag{11.2}$$

Finally, the N output is ORed with (B.AND.C) giving the full adder functions

$$\begin{aligned} CARRY &= (A.AND.(B.EXOR.C)).OR.(B.AND.C) \\ SUM &= A.EXOR.B.EXOR.C \end{aligned} \tag{11.3}$$

Further details on programming CLIP4 in the CAP4 programming language are given in Chapter 13.

Besides the arithmetic operations, there are three other main classes of functions which can be performed by CLIP4. *Point operations* may be executed between images or for single images. Such operations include image duplication (copying), binary complementation, and all of the Boolean functions. *Cellular operations* may perform Boolean functions on the square or hexagonal neighborhood and *propagation operations* may either label contiguous regions or regions connected to edge elements.

Other ancillary operations involve input/output, controller register functions, array testing, and bit counting. Input and output frame buffers are provided for the 96×96 pixel pictures as shown in Figure 11.1. The input buffer is loaded from a television camera, via a six-bit analog-to-digital converter. Transfers between the array and the frame buffers are effected on single bit-planes, column-by-column, using 96 independent row shift registers which form part of the image processing circuitry, i.e., each bit of the shift register is actually a buffer, suitably reconnected for input/output (Figure 11.2). The controller presents sequences of instructions to the array and also performs register operations to control looping and image memory indexing. Every D output is also taken to an OR gate for the entire array which can be used to detect empty arrays. Finally, a tree of parallel adders can be used to sum the one-elements in a bit plane, taking approximately 96 clock cycles while the bit plane to be counted is shifted across the array so as to address the adding tree.

11.3 DAP (Distributed Array Processor)

The design of a pilot version of DAP using a 32×32 array of processing elements was started in October 1974. The pilot hardware was completed in the Spring of 1976 (Flanders, et al., 1977). A 64×64 commercial version appeared in 1980 (Hunt, 1981). Unlike CLIP4, DAP was not designed with image processing in mind, being intended for numerical and logical operations on two-dimensional arrays of data for the purpose of solving problems in meteorology, finite element analysis, operations research, etc. The designers were apparently not aware of the work of Unger (1958) and it is interesting to see how DAP, although developed independently, bears a great resemblance to Unger's original concept.

11.3.1 General System

The memory of DAP consists of a 4K-bit unit for each of the 4096 processing elements (16Mb total) which is simultaneously addressable by both the host computer and the input/output selectors of the processing elements. The host is the International Computers Ltd. (ICL) model 2900 and the shared memory is, in fact, a standard ICL 2MB memory module (Figure 11.3). Another novel feature is that 32-bit ICL main control unit (MCU) instructions, for use in operating DAP, are stored in the same memory module. However, these instructions must be transferred

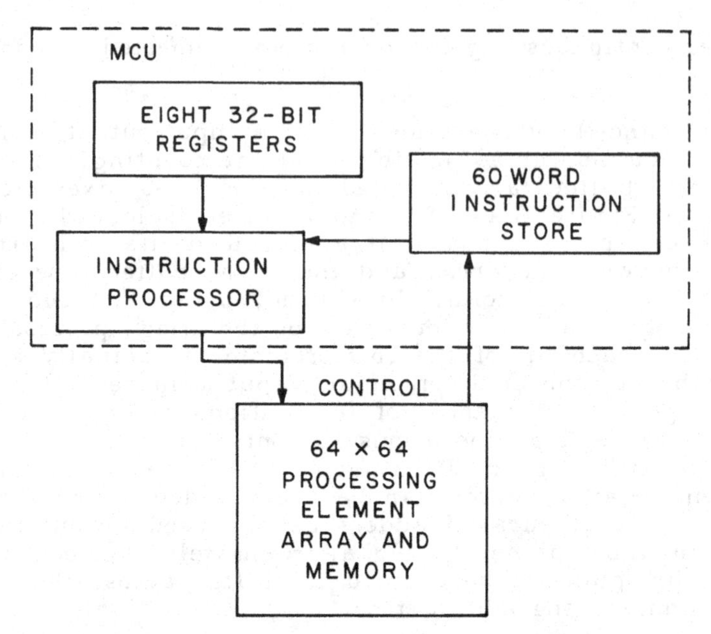

Fig. 11-3 The DAP system memory shares a 16Mb memory module with the ICL 2900 host.

from memory to the MCU for execution since there is no mechanism for their interpretation by the processing elements themselves. For matrix operations by DAP data are stored in *vertical format* where the numerical value of each matrix element is stored within one 4K-bit module (one processing element's memory). On the other hand, vector operations use *horizontal format* where the value of a vector element is spread over one row or one column of processing elements.

Also contained in the ICL 2900 host are eight registers which connect to the processing elements via the row and column network. There is a bit in each register which has one-to-one correspondence with each row or column. The MCU controls the contents of these registers and their transfer to the appropriate processing elements.

11.3.2 The DAP Processing Element

The processing element of DAP is shown in Figure 11.4. The single-bit full adder takes its inputs from Q, a single-bit accumulator, from C, a single-bit carry register, and from an input mutiplexor P which receives the carry outputs from four neighboring processors. This same multiplexor can also receive

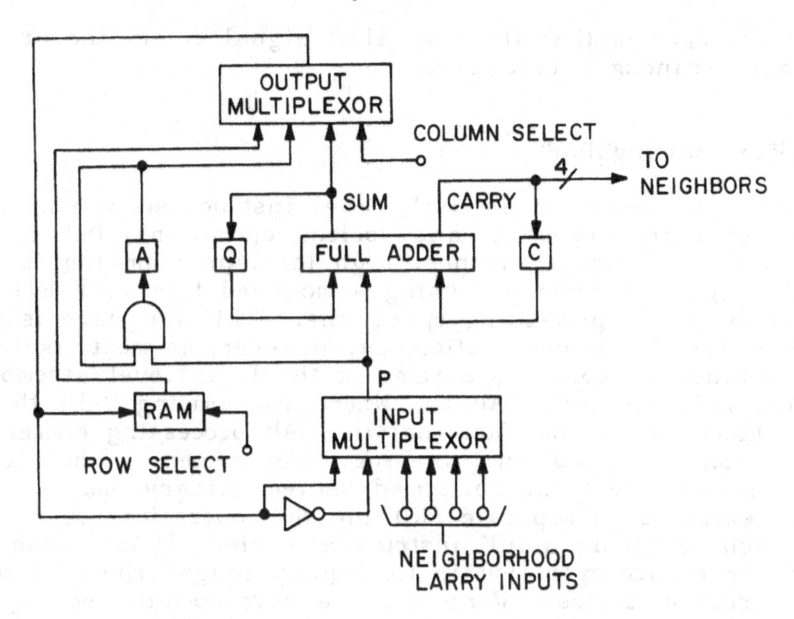

Fig. 11-4 In the DAP processing element neighborhood inputs and outputs are multiplexed to and from a full adder having 1-bit sum and carry storage.

a signal from the local output multiplexor (either direct or inverted). Besides selecting Q as an input, the same input may be set to 0 or 1. Also either Q, C or 0 may be selected at the "carry" input. The output from the full adder returns the sum to the accumulator and the carry to the C register. The carry output also propagates to the input multiplexors of the four neighbors.

As mentioned above, there are both input and output multiplexors in the DAP processing element. These multiplexors make it possible for all paths within the processing element to be one bit wide. The input multiplexor, as well as going to the full adder, is also connected to an AND gate whose other input is the output of the A register. The contents of the A register serve to indicate the *activity* status of the processing element and via additional gates (not shown in Figure 11.4) can control the storage of data from this processing element in memory. The output multiplexor selects its output from one of four inputs: (1) the random access memory, (2) the sum output of the full adder, (3) a column select signal broadcast to the array, or (4) the state of the A register. Finally, the output from this multiplexor returns to the input multiplexor (both direct and inverted), goes to memory, and to the main control unit. Note that the row select signal enters the processor via the random access memory.

11.3.3 Programming DAP

Using sequences of assembly level instructions which support all standard arithmetic and Boolean operations, DAP is programmed in a bit-serial manner. High level programming is also possible. This is performed using a modified form of FORTRAN adapted for array processing operations. This language is called DAP-FORTRAN. Highest efficiency, however, is most likely to be obtained by coding programs in the lower level assembly language which is called APAL. When programming DAP, the user is hampered by the fact that the DAP processing element does not contain a Boolean processor. For example, when logical functions are to be performed between binary images, they must be executed as separate AND and NOT operations operations, each requiring a full instruction cycle. Thus a single augment or reduce operation in the square neighborhood needs nine instruction cycles. Worse still, a more complicated logical neighborhood operation, such as the Golay transform would require many more instruction cycles.

11.4 MPP (Massively Parallel Processor)

In 1971 one of the authors (Preston, unpublished Perkin-Elmer Engineering Report, 1971) proposed to the United States National Aeronautical and Space Administration (NASA) at the Goddard Space Flight Center the GPP (Golay Parallel Processor) which was to be a 128×128 array automaton. The processing elements (subcontracted to Texas Instruments) were to duplicate the full logical neighborhood decoder of the Perkin-Elmer research GLOPR (Chapter 10) and thus would perform the Golay transform in a fully parallel array automaton. Each processing element would have memory for one bit each from both the two source images and the one destination image with the other bit contained in the other elements of the array. NASA Goddard set aside the idea of the GPP while pursuing an optical data processor, called the TSE computer (Schaeffer, 1979), until 1977 when TSE was abandoned as impractical and work on a 128×128 automaton was revived. Discussions with the Goodyear Aerospace Corp. (which had built TSE) led to the definition of a more general purpose processing element that would be capable of performing integer arithmetic and would have processing elements with far more memory than GPP. The result was the 128×128 MPP which was delivered by Goodyear to NASA Goddard in 1983. This system is described by Batcher (1980, 1982) and Potter (1982, 1983).

In general the features of MPP are once again similar to those of earlier systems, from Unger's machine to DAP. A general block diagram is shown in Figure 11-5. The MPP consists of the array of processing elements with its internal memory, an external memory called the "staging memory," the array control unit (ACU), and a control minicomputer which was initially a DEC PDP11, later augmented by a DEC VAX11/780.

The principal components of the processing element of MPP are shown in Fig. 11-6. The array comprises 128×132 elements which are arranged in two-row-four-column sub-arrays on custom VLSI HCMOS chips. Eight processing elements are integrated per chip, but each processor requires two additional 1K×4 memory chips so as to supply each processor with 1K bits of storage. A block of 128 rows by 4 columns can be by-passed by gates incorporated in the chips, so that a faulty circuit can be eliminated with no disturbance to the array operation. The array normally operates with an arbitrary block by-passed unless a fault develops, so that the logical size of the array is always 128×128. All connections to and from by-passed processing elements are automatically routed to the subsequent neighbors so that the user of the array is not aware of the by-passed elements. This is perhaps an expensive solution to fault correction, involving some 3% more processors in addition to the by-pass gates, and can only be justified under conditions where manual single chip replacement is operationally undesirable.

Arithmetic operations are performed in the full adder, which has three one-bit inputs, buffered by the A, P and C one-bit registers. The one-bit sum output is stored in the one-bit B register and the output carry is returned to the one-bit register C. In order to avoid the extra instruction cycles required to return the partial results of integer sums and multiplications to local storage, a shift register of variable length (2,6,10,14,18,22,26 or 30 bits) is provided. This shift register may be used in conjunction with the full adder as desired.

Logical operations are performed in a logical unit associated with the one-bit register P. The data bus D interconnects all the one-bit registers in the processing element and interfaces to the random access memory. P can be loaded from D or from any one of its 4-connected neighbors. The logical unit associated with P may then be used to form one of the 16 possible Boolean functions of the current state of P with the current state of D, and the result is stored in P.

At any stage of the arithmetic, logical, or routing operations, data can be transferred via D to the 1024-bit random access memory associated with each processor. As in the DAP, all operations can be made conditional upon the state of a one-

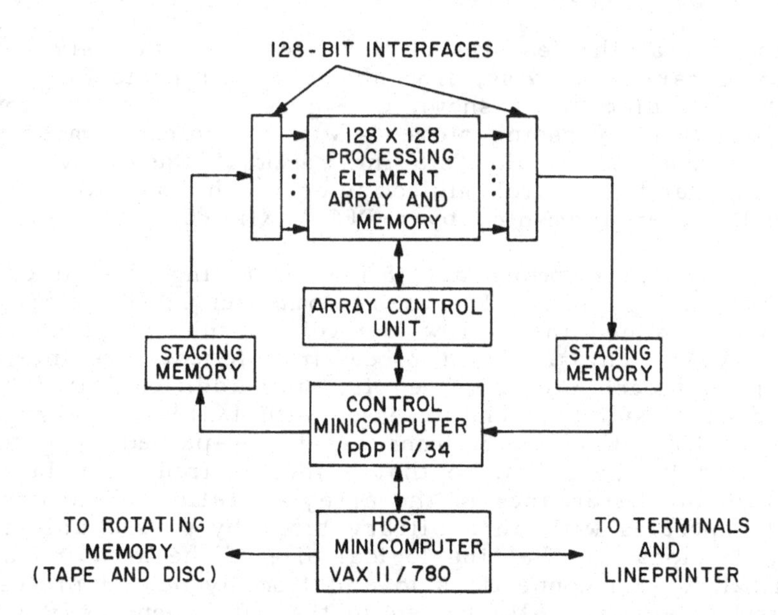

Fig. 11-5 The MPP is controlled via a hierachy of computers: A 32-bit minicomputer, a 16-bit minicomputer, and the ACU (Array Control Unit).

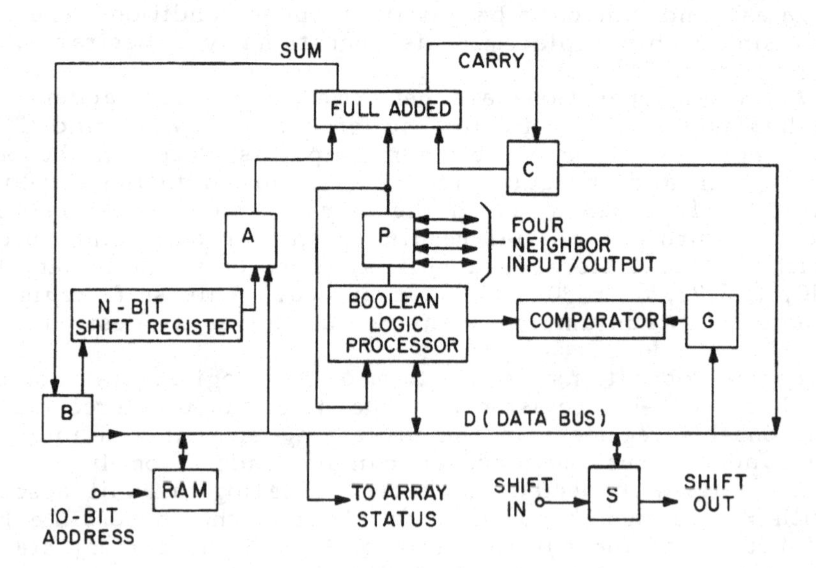

Fig. 11-6 The processing element of the MPP has parallel I/O to four neighbors and a 1-bit data bus connecting to registers for both the Boolean processor, full-adder, and variable-length shift register.

bit mask or activity register G. Finally, an overall OR-gate connects to all D busses to detect the empty array condition.

Data is input to or output from the array by shifting columns of one-bit data through row shift registers formed by the one-bit registers S in each processor. This may be done independently without interrupting the action of the processing elements. Transfers to D are then made in a single parallel operation during one instruction cycle once the full array of data is in the appropriate S locations. Array operations in MPP are based on a 100ns instruction cycle.

In addition to the control of the array of processing elements, other operations are carried out in the ACU. This unit has three sections which are involved in (1) processing element control, (2) I/O control, and (3) primary control. The processing element control section is responsible for array operations such as arithmetic, logic and routing; the primary control performs all scalar arithmetic for addressing and other purposes; the I/O control manages the flow of data in and out of the array. The staging memories are table-controlled systems used in re-order and re-formatting the data as the input/output flow occurs. Serial-to-parallel and parallel-to-serial conversion is possible as well as windowing and mosaicing. The control minicomputer acts as a program and data management unit (running under a standard real-time multiprogramming system) and is responsible for coordinating the programs for operating the three sections of the ACU.

MPP may be programmed at assembly level. Now a higher level language (Parallel Pascal) has also been developed by Reeves and Bruneu (1980). No doubt many other proposals will follow as MPP comes into regular use.

11.5 COMPARING THE ARRAYS

The three array automata considered in this chapter are very similar in the way in which their processing elements are conceived. Each has a single-bit structure offering in some form or other both a full adder and logic capability; each communicates with at least its four immediate array neighbors; each has local storage associated with the processor; all the systems operate in a full parallel mode with each processor receiving the same instruction. The differences between the systems to some extent reflect the various motives of the designers.

CLIP4 was intended primarily for research on images and its cost was minimized so as to make it available for research use. The structured nature of image data can best be investigated by using multiply-connected processors, so CLIP4 was designed to have parallel access to both hexagonal and square connectivities. On the other hand, an emphasis on binary picture processing, coupled with the desire for cost effectiveness, led to a minimal 32 bits of on-chip storage per processor, implemented using a commercially available NMOS custom process. The interconection OR gate also reflects the image processing emphasis, as does the double Boolean processor and its propagation facility. The parts of the system external to the processing element array itself are also image oriented. In particular, a 96×96 sub-image from a standard television camera is digitized and input to the array during a single television cycle, so that real-time image analysis is a feasible proposition.

In contrast, DAP was not intended for image processing but rather was designed to manipulate multiple dimensionality data, where parallelism could be used to advantage. Here the emphasis was on an extension of a large mainframe computer with the idea that the array of processing elements could be treated as *active memory* integral to the mainframe memory itself. Interprocessor communication may be less rich than in CLIP4 but host-to-array communication has been optimized. Off-chip memory has been generously provided so that floating-point and double-length computations are comfortably handled. Conversely, real-time image processing is not a feasible proposition since no good data channels to peripherals (such as TV input devices) are provided.

The third array automaton MPP to some extent tries to combine the two philosophies represented in CLIP4 and DAP. However, it pays a severe penalty in not providing parallel inputs from neighboring processing elements. It also can be argued (Gerritsen, 1981) that another fault in the design is that the contents of its shift register cannot be used to address its memory, thus making table-driven computations infeasible. Image data transfer rates and arithmetic capability surpass those available in CLIP4 but not as much as one would surmise from a simple comparison of instruction execution times.

Figure 11-7 plots the operating speed (in picture point operations per second) of CLIP4, DAP, and MPP. This figure is based on executing a simple function, such as a Boolean operation between two image arrays. More complex operations, e.g., skeletonization, would lead to a considerably different graph. See Figure 10-20 for a comparison with cellular logic machines.

11.6 FUTURE POSSIBILITIES

Although CLIP4 is the only member of the CLIP family in operation at the time of writing, three more versions are

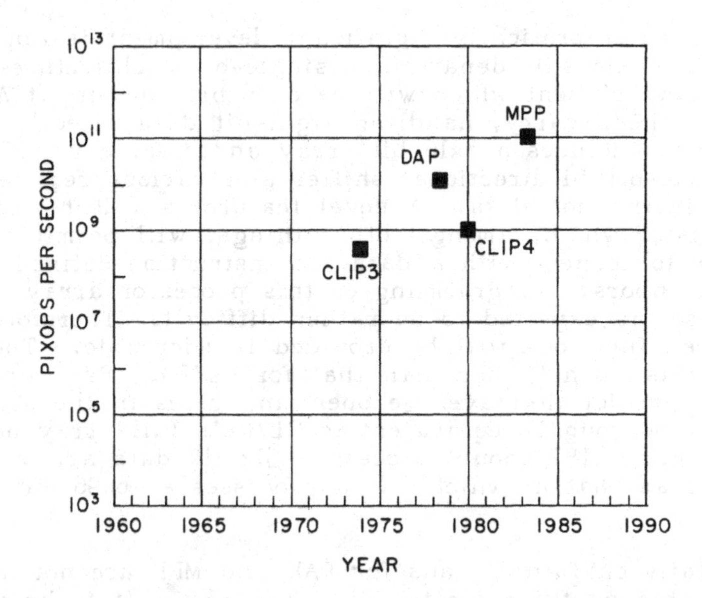

Fig. 11-7 Machine speed in pixops/sec (picture point operations per second) for the array automata described in this chapter.

near completion. Two of these (CLIP4R-1 and CLIP4R-2) differ mainly in their input/output facilities. The third (CLIP4S) is arranged as a 512×4 array of processing elements which is scanned over an image of 512×512 eight-bit integers. The controller is designed so that the user sees the system as a full 512×512 array. CLIP4S is arranged to run the same assembly language programs as the other CLIP4 machines. A modular version of CLIP4 is now being manufactured by Omicron Electronics Ltd., so as to permit any array size to be built up based on a 32-column, 4-row module, with 32 bits of array memory per processor. Furthermore, additional off-chip memory is offered interspersed with the array boards as required. The only penalty paid is that this additional memory exhibits a slightly longer access time.

The CLIP series is continuing with the introduction of improved processor designs. In particular, CLIP5 has been proposed by Fountain (1982) as a production-engineered version of CLIP4, incorporating sixteen processors in each integrated circuit and using external local storage (which is therefore not limited by the chip dimensions). Other features include better on-chip control and more efficient data paths.

A more theoretically significant development is appearing in CLIP7. This will depart from single-bit architectures with a processing element which will be a 16-bit dual-input ALU with look-ahead carry, handling eight-bit data directly. The circuit also includes a 4x16 bit-array addressable register block, a 16-bit bi-directional shifter and various registers to store instruction data. A novel feature is a 16-bit condition code register which, amongst other things, will permit each processor to connect with a data- or instruction-defined subset of its neighbors. Programming of this processor array by the casual user is expected to be rather difficult. Therefore commonly used functions will be provided in microcode. The speed for CLIP7 is much higher than that for CLIP4. Preliminary estimates predict that average operating rates in the scanned mode will be roughly equivalent to CLIP4's full array performance, i.e., CLIP7 should process a 512×512 data array in the same time as that at which CLIP4 processes a 96×96 data array.

Details on future plans for DAP and MPP are not as readily available as those for the CLIP program. It is known that ICL is planning to produce a cheaper, slower-speed version of DAP as a peripheral to minicomputers such as the ICL PERQ. These new DAP systems should play a useful role in image analysis as they should be far less expensive than the present DAP, which can only be procured as part of a million-dollar mainframe. This will permit their purchase by research laboratories, rather than as computing service units, which has been the fate of the original systems. Finally, as regards MPP, it has been mentioned (Batcher, 1980) that a smaller satellite-borne version is under design. No further information is available at the time of writing.

12. PATTERNS OF GROWTH

12.1 INTRODUCTION

No book on cellular automata would be complete without a chapter on *patterns of growth*, especially the most popular generator of these patterns, namely, John Horton Conway's cellular automata game "Life" (see Gardner, 1971). Long before the invention of Conway's Life, Moore (1968) at the United States Bureau of Standards and Ulam (1962) at the Los Alamos Scientific Laboratory were analyzing growth patterns using digital computers. Moore and Ulam used the digital computer to simulate the action of a cellular automaton consisting of an array of processing elements far simpler than the 29-state processing elements of von Neumann (1951). They wrote computer programs to simulate an array of two-state processing elements exhibiting either d1-connectedness (Ulam) or d2-connectedness (Moore). (See equations 6.1 and 6.2.)

12.1.1 Shell Custering

Moore, a member of the Metallurgy Department of the National Bureau of Standards, was primarily concerned with using the digital computer in the automatic recognition of the various grain structures observed in micrograms of metallurgical specimens. In conducting his studies he designed a computing language called STRIP (Standard Taped Routines for Image Processing) which ran on the National Bureau of Standards SEAC digital computer to which, in the year 1956, he had attached a drum scanner for the purpose of digitizing micrograms. Moore's simulated cellular automaton consisted of an array of 168×168 processing elements. He discovered a set of transition rules which produced an effect that he called "shell custering" or, simply, "custering" that caused "growth rings" to be produced depending upon the configuration of the initial two-state (binary) input pattern. The transition rules for custering are that all 1-elements in the input pattern change to 0-elements while all 0-elements adjacent to 1-elements in the input pattern are changed to 1-elements. If the original pattern consisted of a collection of contigious regions of 1-elements, then the first cycle of custering causes 1-elements to appear only at the exterior edges of these regions. Henceforth growth rings propagate both inward and outward from these edges until a stable, oscillating pattern is achieved with a period of two cycles.

An example of shell custering taken from the later work of Preston (1971) is shown in Figure 12-1. This shows both the binary input pattern and one state of the final stable oscillating output pattern. This result was obtained using the original GLOPR and a single Golay transform, namely,

$$A = M[G(A)A']1\text{-}13,N1, \qquad (12.1)$$

with N1 sufficiently large (greater than 64 for the GLOPR 128× 128 array) so that growth continues until the final oscillating output pattern is achieved. Growth, of course, would continue infinitely for an infinite automaton.

12.1.2 Further Transition Rules

As pointed out by Gardner (1971) other transition rules developed by Fredkin at the Massachusetts Institute of Technology using d1-connectedness produce the results illustrated in the upper portion of Figure 12-2. The rule here is that any element with an even number of 1-elements in its neighborhood becomes a 0-element while any element with an odd number of 1-elements in its neighborhood becomes a 1-element. These transition rules are of interest because they cause self-reproduction of certain input patterns (such as the tromino shown in Figure 12-2). It is not difficult to show that four replicas of the input pattern are produced at cycle four; 16, at cycle 16; 64, at cycle 64; etc. Thus growth is continuous (within the bounds of the cellular automaton simulated) and the output never stabilizes.

Similarly, Ulam's transition rule (see lower portion of Figure 12-2) causes growth which continues to infinity. The transition rules here are that, once generated, a 1-element never

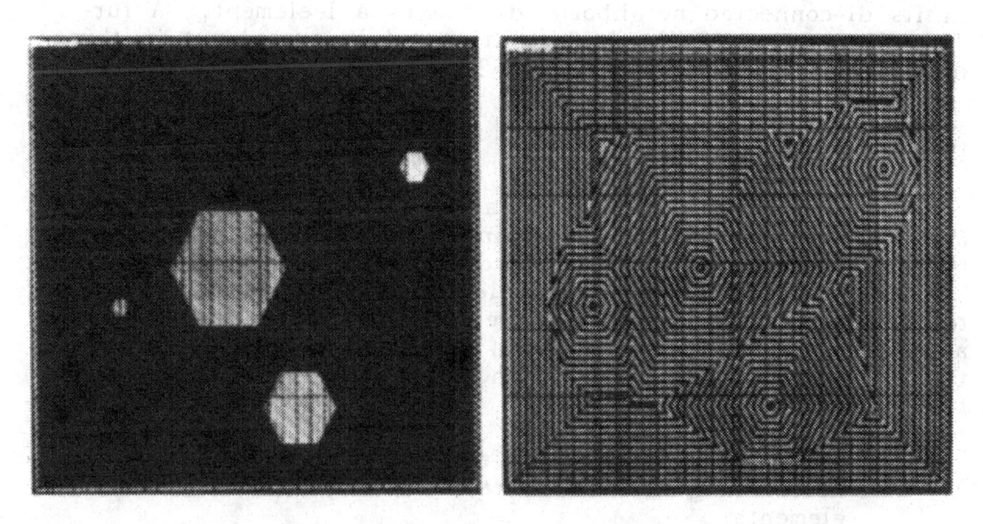

Fig. 12-1 Custering in the hexagonal tessellation is illustrated. The initial pattern is on the left and the result of custering this pattern to oscillatory stability is on the right.

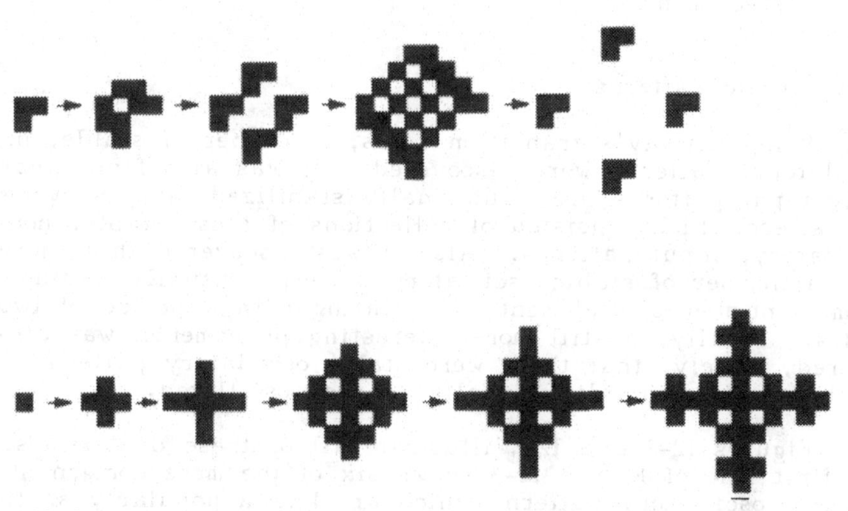

Fig. 12-2 The transition rules of Fredkin lead to a self-reproducing pattern of growth (upper line) while those of Ulam lead to infinite growth which is in general not self-reproducing.

becomes a 0-element while a 0-element with a single 1-element in its d1-connected neighborhood becomes a 1-element. A further illustration of Ulam's transition rules carried out in three-dimensions are given in Chapter 3.

12.2 CONWAY'S LIFE

After the initial work on patterns of growth initiated by Moore in the late 1950s and continued by Fredkin and Ulam in the early 1960s, Conway (see Gardner, 1971) became interested in finding a set of transition rules which, rather than leading to continuous growth, as with the transition rules of Fredkin and Ulam, would lead to bounded growth with eventual stability even for an infinite automaton. Conway's final choice for transition rules were

1. A 0-element becomes a 1-element if its d2-connected neighborhood contains exactly three 1-elements.

2. If a 1-element has in its d2-connected neighborhood either two or three 1-elements, then it remains as a 1-element.

3. If the d2-connected neighborhood of a 1-element contains fewer than two 1-elements or greater than three 1-elements, it becomes a 0-element.

12.2.1 Stable Patterns

Using Conway's transition rules, a number of stable, non-oscillatory, patterns were discovered. It was also found that many input patterns grew but finally stabilized into an overall pattern which consisted of collections of these stable, non-oscillatory, input patterns. Also it was discovered that there were a number of stable oscillatory patterns, usually having a small number of 1-elements, oscillating with a period of two cycles. Finally, a still more interesting phenomenon was discovered, namely, that there were stable oscillatory patterns which translated unidirectionally as they oscillated.

Figures 12-3 and 12-4 illustrate all of these discoveries. The first line of Figure 12-3 shows six of the more common stable, non-oscillatory patterns which are known popularly as the "block", "tub", "beehive", "boat", "pond", and "loaf," respectively. The block and the tub are tetrominos, i.e., they consist of exactly four 1-elements. The boat is the only pentomino

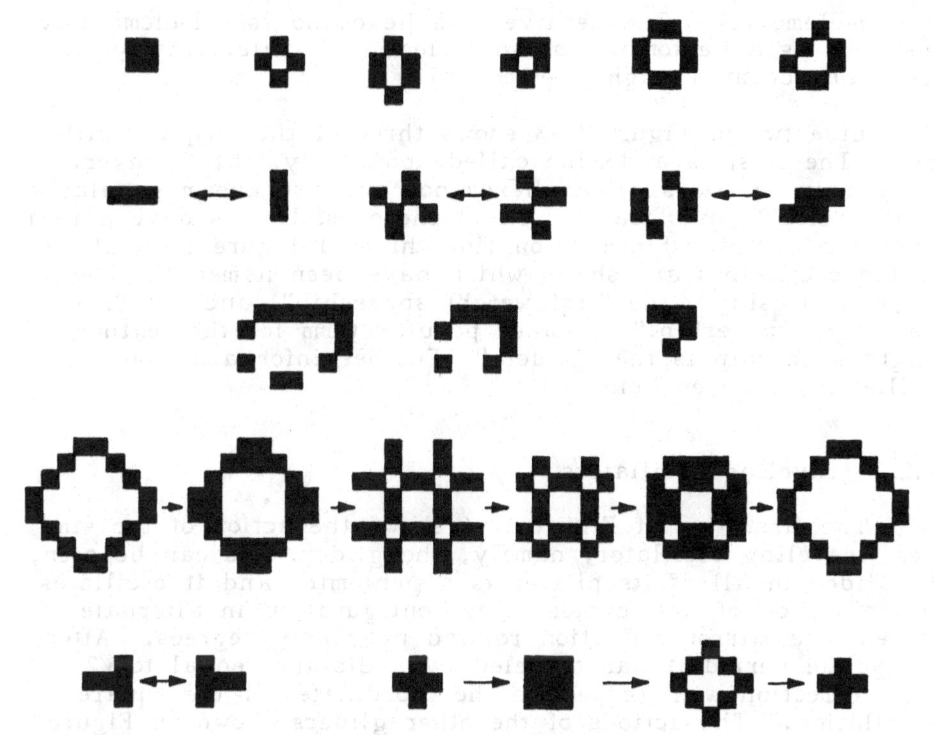

Fig. 12-3 Stable, oscillatory, and translatory patterns generated using the transition rules of both Conway and (bottom line) those for Three-Four Life. (See text for details.)

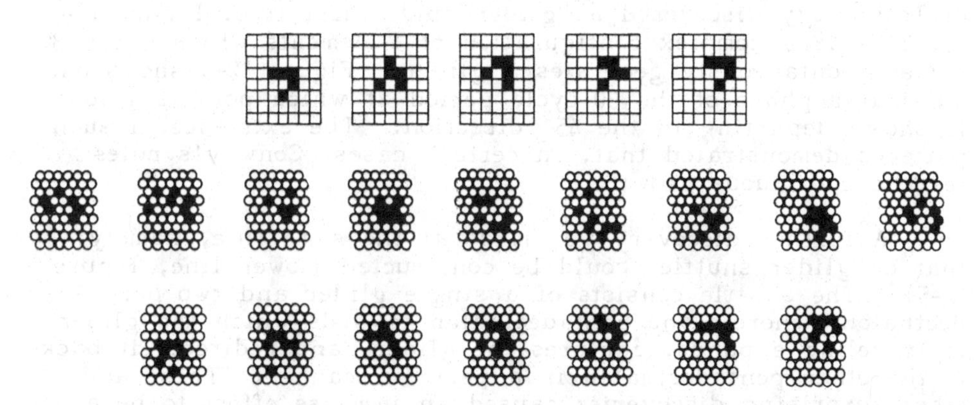

Fig. 12-4 The featherweight spaceship (or "glider") generated using the rules of Conway (upper line), Golay's glider (middle line), and the glider of Preston (bottom line).

(five 1-elements). The beehive is a hexomino (six 1-elements). The loaf, is a heptomino (seven 1-elements) while, finally, the pond, an octomino (eight 1-elements).

Line two of Figure 12-3 shows three of the simple oscillators. The first is a domino called, popularly, the "blinker." The remaining two are both hexominos and are known popularly as the "clock" and "toad." All of these oscillators have a period of two cycles. Finally, on line three of Figure 12-3, three moving oscillators are shown which have been named the "heavyweight spaceship", the "lightweight spaceship", and the "featherweight spaceship." Another popular term for the featherweight spaceship is the "glider." Further information on these oscillators is given below.

12.2.2 Traveling Oscillators

The first line of Figure 12-4 shows the action of the simplest traveling oscillator, namely, the glider. As can be seen, the glider in all of its phases is a pentomino and it oscillates with a period of four cycles. Its configuration in alternate phases is a mirror reflection rotated by ninety degrees. After a complete period it has traveled by a distance equal to $\sqrt{2}$ in a 45° direction with respect to the coordinates of the square tessellation. The actions of the other gliders shown in Figure 12-4 in the hexagonal tessellation are described in Section 12.3.

Initially it was assumed by Conway that his transition rules always led to eventual stability. However, in late 1971, within a few months of his initial invention, workers at the Artificial Intelligence Laboratory of the Massachusetts Institute of Technology discovered a "glider gun" (first line, Figure 12-5). This is a complex configuration of 1-elements which every 30 cycles spontaneously generates a glider. Figure 12-4 shows one particular phase of the 30-cycle period at which point a glider is shown departing in the 45° direction. The existence of such patterns demonstrated that, in certain cases, Conway's rules lead to continuous growth.

A further discovery was made at the same time, namely, that a "glider shuttle" could be constructed (lower line, Figure 12-5). The shuttle consists of a single glider and two pentadecthalons wherein the pentadecthalon towards which the glider is traveling expands, captures the glider, and redirects it back to the other pentadecthalon in a periodic manner. These and other surprising discoveries caused an immense effort to be expended upon Conway's Life in the early 1970s at dozens of university and industrial laboratories. These results, unfortunately, are mostly unpublished. The next section describes some of the more outstanding discoveries.

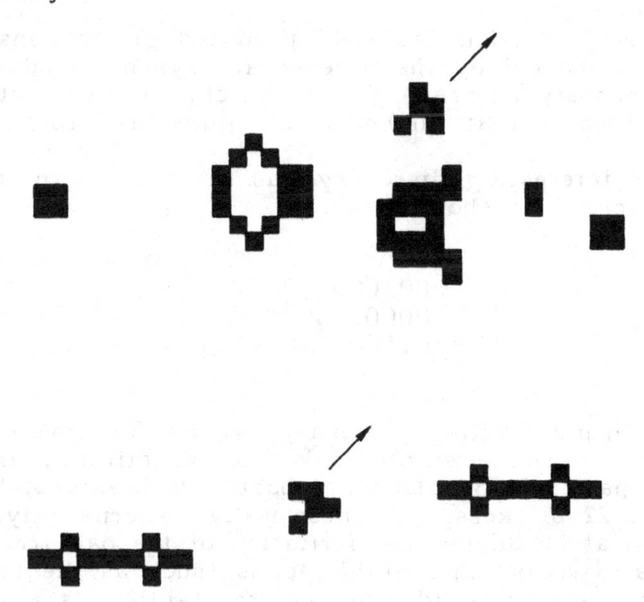

Fig. 12-5 Conway's transition rules applied to the glider gun (upper line) lead to the continuous production of a chain of gliders while two pentadecthalons form a glider shuttle (lower line).

12.2.3 Still More Complex Patterns

By the end of 1971 the result of applying Conway's transition rules to all n-ominos with values of n ranging from three through seven (approximately 8000 patterns in all!) had been tabulated. Results using other starting patterns having many more 1-elements had also been investigated. New oscillators had been discovered. One example is shown on the fourth line of Figure 12-3 which is an oscillator having a period of five cycles with a starting pattern consisting of a hexadecamino (sixteen 1-elements). For this particular oscillator, the number of 1-elements per cycle over the five cycle period goes as 16-24-24-16-24.

The Artificial Intelligence Group at the Massachusetts Institute of Technology not only discovered the glider gun but also found a configuration of thirteen gliders which, upon colliding formed a glider gun. This gun then proceeded to generate endlessly more gliders. Next it was discovered that eight glider guns could be positioned so that the entire ensemble oscillated with a period of 300 cycles producing a middleweight spaceship each period. Finally, a starting pattern was found called a "breeder" which endlessly produced glider guns. All glider guns generated by the breeder are synchronized so that they simultaneously generate gliders which, in turn, interact with the breeder so that further glider guns are produced.

Another interesting discovery was the starting pattern called the "acorn" as shown below

001000000
000010000
011001110

This rather simple starting pattern grows for 5206 cycles at which time it becomes a stable "oak tree" containing, among other stable patterns, 34 blocks, 8 boats, 24 beehives, 5 loaves, 2 ponds, and 22 blinkers. Of these many patterns only the blinkers oscillate. During the formation of the oak tree, the acorn generates many other stable forms (such as the toad oscillator) but these are eradicated before stability is reached.

12.3 OTHER TRANSITION RULES

During the past decade most of the research on patterns of growth has utilized Conway's transition rules. Hundreds of workers have contributed to this effort, primarily during the early 1970s. Sporadically, however, there have been other interesting efforts which are reported in this section.

12.3.1 Three-Four Life

A set of transition rules called "Three-Four Life" was briefly explored in 1972. The transition rules are

1. A 0-element which has in its d2-connected neighborhood exactly three or four 1-elements becomes a 1-element.

2. A 1-element which has in its d2-connected neighborhood exactly three or four 1-elements remains as a 1-element.

3. In all other cases the output is a 0-element.

Using these transition rules it is found that the only non-oscillatory stable form is the block. Oscillators have been discovered having various periods including periods two, three, four, six, ten, and twelve cycles. Two examples are shown in the bottom line of Figure 12-3. The first is a tetromino with a period of two cycles, while the other is initially a pentomino having a number of 1-elements varying 5-9-8 over its three-cycle period.

12.3.2 Golay's Hexagonal Life

In the hexagonal tessellation a sequence of Golay transforms (Chapter 2) may be used to express a series of three transition rules similar to those of Conway. They are

$$B = M[G(A)A']S1\ldots,,$$

$$C = M[G(A)A]S2\ldots,,$$

$$D = M[G'(A)A]S3\ldots,, \qquad (12.2)$$

where the set S1... refers to the set of Golay surrounds which must be present if a 0-element is to become a 1-element; S2..., the set of surrounds for which a 1-element is to be retained as a 1-element; S3..., the set of surrounds for which a 1-element is to become a 0-element. After the arrays B, C, and D are generated, then the algorithm is carried to completion by or-ing the arrays B, C, and D to obtain the outcome.

Golay investigated a simplified version of these transition rules which is expressed simply by a single statement, namely,

$$A=M[G(A)]2/11,N, \qquad (12.3)$$

which states that, when either a 1-element or a 0-element has in its hexagonally connected neighborhood two 1-elements exhibiting either Golay surround number 2 or Golay surround number 11, then the outcome is a 1-element. In all other cases the outcome is a 0-element. Some results obtained from the use of this transform are shown in Figures 12-6 and 12-7. First, it is is found that there are no non-oscillatory stable patterns under this transform. The simplest oscillatory stable pattern is the domino as shown in the first line of Figure 12-6. This pattern behaves much as Conway's blinker. The second line of Figure 12-6 shows a twin domino which oscillates with a period of six cycles with the number of elements present varying according to 4-6-6-4-6-6. The next line in Figure 12-6 shows the result of applying this transform to the equilateral tromino which

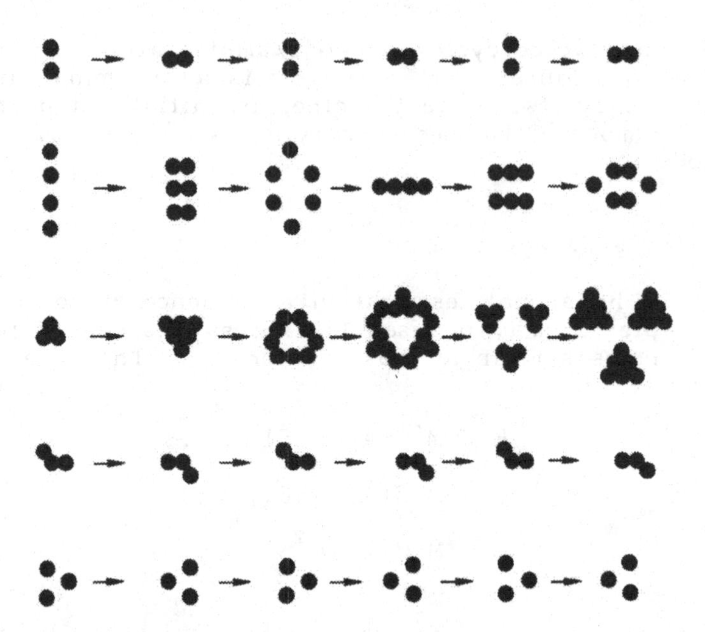

Fig. 12-6 In the hexagonal tessellation Golay's transition rules lead to the growth of the equilateral tromino (middle line) or to various types of oscillators.

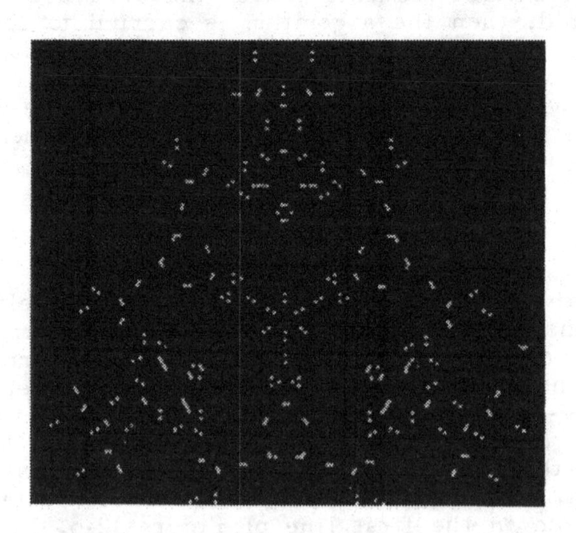

Fig. 12-7 An oscillatory stable pattern (above) is produced by applying Golay's transition rules to the equilateral tromino. This configuration oscillates with a period of six cycles and is shown on the 321st cycle from the original pattern.

starts out by triply replicating mirror images of itself after four cycles and then continues for 317 additional cycles before reaching oscillatory stability. The diagonally connected tromino (fourth line Figure 12-6) oscillates with a period of two cycles and has been called a "bird." Finally (last line Figure 12-6), another oscillatory tromino is Golay's "blinker" which also oscillates with a period of two cycles.

The final outcome of applying Golay's transition rules to the equilateral tromino is as shown in Figure 12-7. This figure shows the outcome at cycle number 321 at which point there are 288 1-elements comprising 105 dominos, nine twin dominos, six blinkers and six birds. Note that three of the twin dominos exist at one phase; the other three, at another phase. This stable oscillatory pattern, therefore, cycles with a period of six. However, the total count of 1-elements cycles with a period of three, namely, 288-294-300-288-294-300, etc.

12.3.3 Preston's Hexagonal Transition Rules

Preston (1971) investigated another set of transition rules which was found to give birth to gliders in the hexagonal tessellation. One of the most interesting of these rules may be expressed by the following set of Golay transforms

$$B = M[G(A)A']2/11,,$$
$$C = M[G(A)A]2/11/12,, \qquad (12.4)$$

with the final outcome generated by or-ing the contents of arrays B and C. When this set of Golay transforms is applied to a starting pattern consisting of a filled hexagon whose sides are alternately of length four and three, six gliders are generated at the 116th cycle (see Figure 12-8). These gliders propagate outwards along six primary directions of the hexagonal tessellation and escape towards infinity. Then, at the 406th cycle, six more gliders are generated which do not escape but are immediately encompassed by the expanding structure of the central configuration of 1-elements. At cycle number 470 six more gliders appear and escape to infinity as did the original six.

As of 1971 only Preston's glider was known in the hexagonal tessellation. The action of this glider is shown in the last line of Figure 12-4. As can be seen, it oscillates with a period of six cycles and its configuration every third cycle is a mirror image rotated by sixty degrees. During the six-

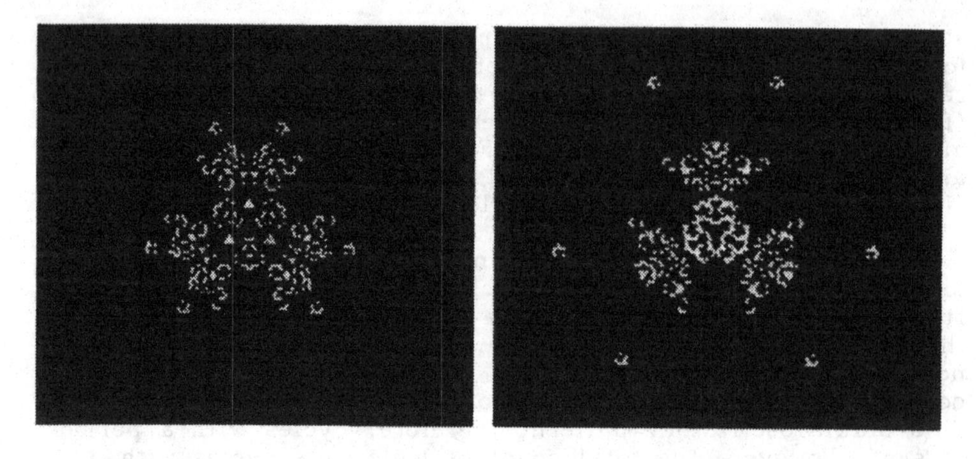

Fig. 12-8 Preston's transition rules lead to the generation of six gliders on the 116th cycle (left) which propagate outwards along the primary hexagonal axis. The image on the right shows the results at the 148th cycle.

cycle period the glider propagates in one of the hexagonal directions by a distance equal to two units.

In 1983 Preston (unpublished) discovered a Golay glider, i.e., a glider which propagated in a stable oscillatory fashion under Golay's hexagonal transition rules. This glider is shown in the middle line of Figure 12-4. Its period is eight cycles and, as with the Preston glider, on alternate cycles, a mirror image is generated with 60° rotation. At the end of its four-cycle period the initial Golay glider has traveled along one of the principal hexagonal directions by a distance equal to two units. Thus, the Golay glider is 1.5 times faster than the Preston glider.

12.3.4 Hexagonal Custering

Hexagonal custering may be performed using the Golay transform

$$A = M[G(A)A']1\text{-}13,N, \qquad (12.5)$$

and is illustrated in Figure 12-1. Another interesting form of custering is that performed in subfields as discovered by Preston (unpublished) in 1969 and shown in Figure 12-9. The re-

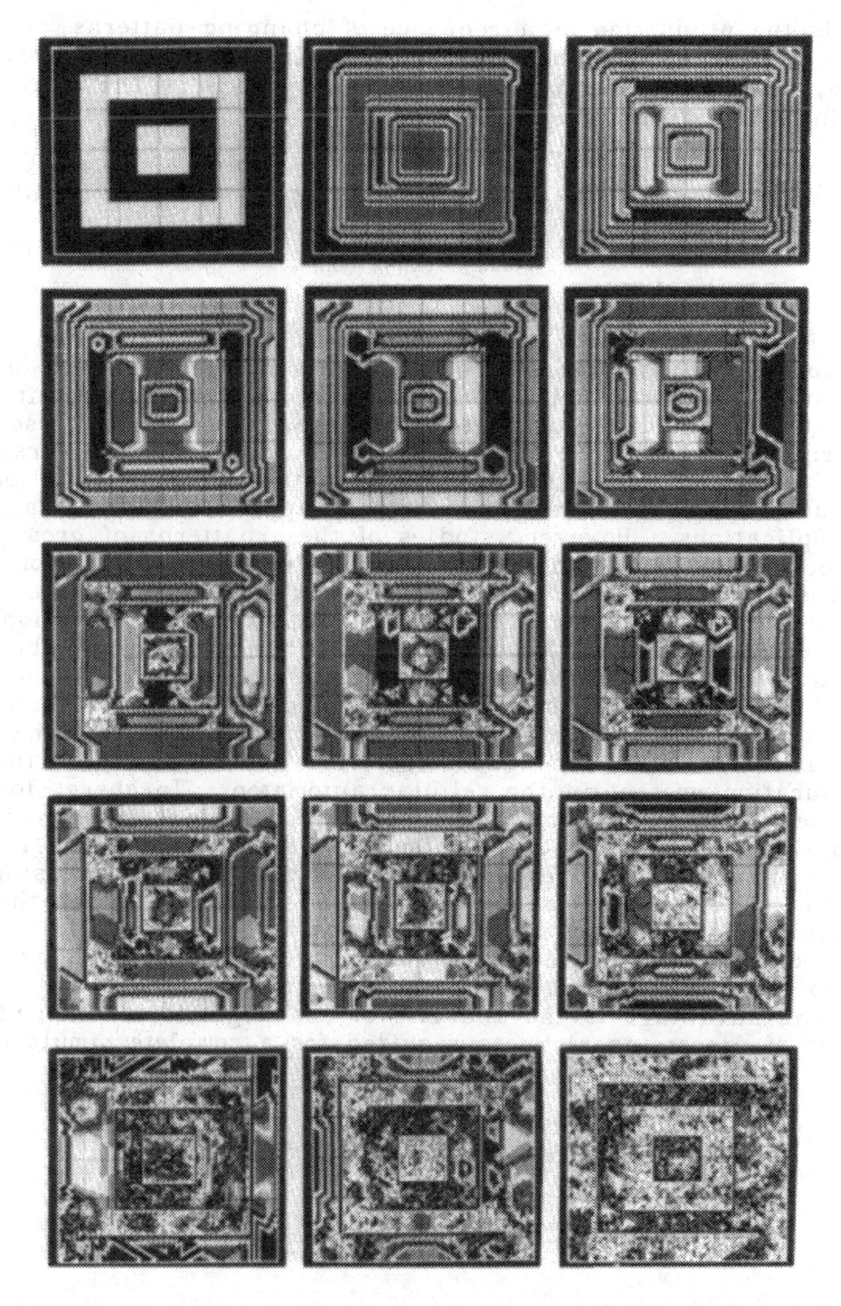

Fig. 12-9 Custering in the hexagonal tessellation using subfields leads to an endless succession of changing, non-repetitive patterns starting with the original pattern (upper left corner) shown above.

sult is the production of a sequence of changing patterns, which, as far as is known today, do not reach oscillating stability, although repetitions must eventually occur. The corresponding Golay transform is

$$A = M[G(A)A']1\text{-}13,N,3 \qquad (12.6)$$

12.3.5 Generalization

Investigations of patterns of growth have captured the imagination of many workers not only because of the challenge of generating transition rules but because of the fascination with the patterns which are generated. In the modern world these patterns may readily be displayed on the computer cathode-ray-tube display in "real time." Other than the mental stimulation generated by these investigations, there appear to be no practical applications. However, studies of these patterns of growth may be related to the theory of games. In the general theory of games it can be said that most board games, such as chess and checkers are actually cellular automata games. Investigations of these games differ only from the games described above by reason of their significantly increased complexity. This complexity arises not only because of the fact that the processing element may take on many states but also due to the fact that the transitions rules are adaptive, i.e., vary according to the configuration present in the cellular automaton. In chess, for example, besides the empty state, each processing element has six active states (pawn, rook, bishop, knight, queen, king). Transition rules are dependent upon the state of the processing element. Furthermore, transition rules are dependent upon the overall configuration of all of the processing elements. In these board games the cellular automaton is limited in its extent to the total number of positions available on the board. Thus one advantage to the analysis of board games is the finite number of processing elements required for a complete simulation.

13. PROGRAMMING METHODS

13.1 INTRODUCTION

The earliest electronic digital computers were programmed by means of sense switches, program boards, and other "hard-wired" techniques. Typical were ENIAC (Electronic Numerical Integrator And Computer) in the United States and COLOSSUS in Great Britain. ENIAC was built during the interval 1943-1945 at the University of Pennsylvania. COLOSSUS was entirely operational by the year 1943. The next great advance in computers was the *stored-program* machine originated by von Neumann in his historical *First Draft Of A Report On The EDVAC* published in March of 1945 (Goldstine, 1972). According to Lavington (1980), the world's first operational stored-program computer was the University of Manchester's Mark I. On June 21, 1948, this machine ran its first program (a factoring program requiring approximately an hour's time). The Manchester Mark I was a 32-bit machine having a 32-line program store and a repertoire of only seven instructions. Programming was done entirely in machine language which was laboriously keyed into the computer in binary form. Machine-language programming was typical of the 1940's and it was only in the 1950's that assembly-language programming and later high-level language programming (e.g., FORTRAN) was introduced. Most of the cellular logic machines described in Chapter 10 and the array automata discussed in Chapter 11 use high-level language instructions, although an assembly-like language (CAP4) is often used for programming CLIP4.

The purpose of this chapter is to illustrate the use of a variety of languages in programming both cellular logic machines and array automata. The sections which follow deliberately contrast the machine-language programming of CLIP4, the FORTRAN-like programming of DAP, the high-level special-purpose programming of both the Perkin-Elmer and the Coulter Electronics GLOPR, and the general-purpose, multi-instruction language employed by the Leitz Texture Analyzer System (TAS).

13.2 EVOLUTION OF PROGRAMMING LANGUAGES

At the time of writing, dozens of languages have been developed for programming both cellular logic machines and array automata. Several surveys of such languages (with comparisons to many other languages in the field) have been written by one of the authors (e.g., Preston, 1983). Examples are DEFPRO and PPL, used by the University of Linkoping in programming PICAP; TAL, developed by the University of Leiden for use as an alternate to TASIC in programming the Leitz TAS; C3PL, used at the Environmental Research Institute of Michigan for programming the cytocomputer systems; and SUPRPIC, written at Carnegie-Mellon University for programming the PHP.

Some of these languages have evolved as self-contained systems including an operating system or monitor for controlling the computer environment, an editor for preparing program code and subroutines, a parser for either interpreting or compiling the program statements, and some sort of file management system. Other languages are themselves coded in a high-level language and are compiled as either major programs or libraries of subroutines in this language. For example, SUPRPIC is coded primarily in FORTRAN. Still others utilize the more modern command substitution systems available in large multi-tasking or time-sharing computer environments both to call and parameterize previously compiled subroutines without the necessity of recompilation. The development cf these languages, unfortunately, has taken place and still continues to take place in an uncoordinated fashion. There is little communication between the various groups of language designers working in the field. No coordinating society or journal exists to act as a vehicle for publication.

The utility of a particular language is hard to measure. One should, of course, consider the following factors

1. The ease with which the neophyte may be familiarized with the language.

2. The facility with which the expert can both code a new program and read and understand those previously coded.

3. The efficiency with which the language employs all of the resources of the computer system with which it is associated.

4. The ease by which old programs may be modified, combined, and re-established for performing new tasks.

5. The ways in which the language may be modified for future growth in its command repertoire.

In the pages which follow, we do not pretend to cover all the aspects of programming languages but, rather, attempt to furnish sufficient breadth in the treatment so that the reader becomes familiar with both the structure of programming methods available and how they are employed in performing simple tasks.

13.3 CAP4

The programming language for the CLIP4 computer is CAP4. The reader should refer to the description of this machine in Chapter 11 where the major CLIP operations, employing the LOAD and SET commands, are explained. The reader should also review the function of the inputs and outputs of the CLIP4 Boolean processors, namely D, which is a logical variable whose value is dependent upon the inputs A and P, as well as N, whose value depends on a different function of A and P.

CAP4 was the original programming language for the CLIP4 and is still in frequent use. It should be mentioned that a new high-level language, IPC, not presented in this chapter, has been added and documentation may be found in Wood (1983). The CLIP4 is interfaced to a Digital Equipment Corp. PDP11/34, a conventional minicomputer, using the UNIX operating system. Most of the generally used image processing functions have been coded in CAP4 as a subroutine library accessible under UNIX via calls from a CAP4 main program. (Alternatively, IPC, using the language C, has been adapted for image data types and this language treats CAP4 subroutines as functions.) Because of its power and simplicity, IPC provides an excellent development language. Most programs run faster if written completely in CAP4. Use of IPC permits sharing of the computational burden between CLIP4 and the PDP11. This greatly facilitates the use of the image memory disk and other standard computer peripherals.

Returning to the subject of CAP4, the operations of the CLIP4 Boolean processors can be summarized by the equations

$$N = A.F_1.B.F_2.T$$

$$D = A.F_3.B.F_4.T$$

where $$T = .OR.(N_1, N_2, \ldots, N_8, C) \qquad (13.1)$$

and where the functions performed (F_1, F_2...) depend on the values enabling the Boolean processor control lines.

The major instruction performed by the CLIP4 array is the SET instruction (Chapter 11). This instruction determines the operation to be executed whereas the other three primary instructions (LOAD, PROCESS, STORE) are concerned only with data transfers. In CAP4, the format of the SET function is

$$\text{SET } F_D,\ [d_r,\ B\ C\ R]\ F_n,\ ES \qquad (13.2)$$

where F_D and F_N are the Boolean functions for the D and N outputs, respectively,
d_r is a list of interconnection directions,
B enables the output of the B buffer,
C enables the output of the C buffer,
R enables the arithmetic gates,
E uses 1 for the edge connection condition (0 is the default condition),
S selects square connectivity (H is used for hexagonal).

Five main classes of operation may be performed. They are

Point Operations
e.g., SET A.P
which performs a Boolean AND of the contents of the A and B buffers;

Local Operations
e.g., SET A+P, [2 4 6 8] A, E
which selects elements which are 1-elements or which have 0-elements in the N, E, S, or W directions, or which are on the array edge;

Propagation Operations
e.g., SET A.P, [1-8] A.P, E
which selects connected sets of 1-elements which touch the array edge;

Labeled Propagation Operations
e.g., SET A.P, [1-8, B] A.P
which select connected sets of 1-elements in A which intersect the 1-elements in B;

Arithmetic Operations
e.g., SET A&P, [B R C] A.P
which adds the contents of A to the contents of B using the carry in C and reloads C with the resulting carry.

Since there are 16 possibilities for the output of each Boolean processor and two values for each of the B, C, R, E, S/H variables, as well as for the eight elements of the direction list, the SET function can take ($16 \times 16 \times 2^{13}$) forms which is just over two million. Many of these forms have no practical significance, e.g., variations of the direction list when neither R is enabled nor when either of the Boolean functions contain P. Others have still to be evaluated. Even so, a very large number of distinct and useful operations are possible.

To provide an example of programming in CAP4, the below series of commands carries out a simple skeletonizing function using the masks of Levialdi, as discussed in Chapter 7.

LOOP:	SET A	Copy image I_1 into the ref-
	LDA I_1	erence image location (I_0).
	PST I_0	
	SET 0, [1 2 8] A, P, S	Determine whether a 1-ele-
	LDA I_1	ment is present at position
	LDB I_2	1, 2, or 8.
	PST I_2	
	SET 0, [4 6] -A, P*A, S	Determine whether a 0-ele-
	LDA I_1	ment is present at position
	LDB I_2	4 or 6.
	PST I_2	

...Repeat last two SET-PST cycles for seven other masks...

```
SET A@P          Compare result (I1) with
LDA I1           original (I0).
LDB I0
PST DUMMY
BNZA LOOP        Branch to LOOP if not
                 zero.
```

The maximum time required to skeletonize over the CLIP4 96×96 array using the above code is about 5ms (100μs per complete skeletonizing cycle).

13.4 HIGH-LEVEL LANGUAGES

Low-level languages, such as CAP4, are so machine specific as to make codification of their command structures impossible. However, a study of *high-level* languages for cellular logic machines and array automata, in particular those used for image processing, has disclosed definite groupings of command categories and sub-categories (Preston, 1980). The major command categories found were: (1) utilities, (2) image display, (3) arithmetic, (4) geometric manipulation, (5) image transform, (6) image measurement, and (7) decision theoretic. These categories and their common sub-categories are given in Table 13-1. In the following sections, this categorization is employed both for the purpose of studying particular programming languages and for elucidating their special features.

The languages which have been selected for study (in addition to CAP4 described above) are the FORTRAN-like language DAP-FORTRAN of International Computers Ltd., Great Britain (Hunt, 1981), the GLOPR Operating Language (GLOL) developed by the Perkin-Elmer Corporation and later by Coulter Electronics Inc. and, finally, the TASIC language which drives the Texture Analyzer System (TAS) developed by Leitz G.m.b.H. of West Germany (Nawrath and Serra, 1979). Further specifics on each of these languages and the programming methods associated with them are illustrated in the sections which follow.

13.4.1 DAP-FORTRAN

The DAP (Distributed Array Processor) is described in Chap-

Table 13-1 Categorization of Image Processing Commands

UTILITIES	GEOMETRIC MANIPULATION
Storage Allocation	Scaling/Rotation
Control	Rectification
Formaters	Mosaicing/Registration
I/O Commands	Map Projection
Test Pattern Generators	Griding/Masking
Help Files	
DISPLAY/GRAPHICS	IMAGE TRANSFORMS
Video Screen	Noise Removal
Hard Copy	Fourier Analysis/Special Transforms
Interactive Graphics	Power Spectrum
	Filtering
	Cellular Logic
ARITHMETIC OPERATIONS	IMAGE MEASUREMENT
Point	Histogramming
Line (Vector)	Statistical
Complex Number	Numerical/Geometric
Boolean	
DECISION THEORETIC	
Feature Select (Train)	
Classify (Unsupervised)	
Classify (Supervised)	
Evaluate Results	

ter 11 and, with the MPP (Massively Parallel Processor), it is one of the two fastest array processors now in existence. DAP is programmable in a special language called DAP-FORTRAN. DAP-FORTRAN is suitable for the development of algorithms involving Boolean or numerical processing and can perform certain simple operations, e.g., AND, OR, EXOR, in a single cycle time. A lower-level assembly language, APAL, is available which permits the construction of macros. These may be assembled so as to form a specialized programming language tailored to a particular application. In this case, the user program would consist of a number of assembly language routines, each being mainly a sequence of macro calls, plus a number of DAP-FORTRAN sub-routines.

DAP-FORTRAN supports matrices and vectors as basic elements as well as scalars. It provides a close match between the capabilities of the computer and most applications requirements, especially for numerical problems. DO-loop facilities are available which take full advantage of the logical capabilities of the processing elements of the DAP array although the assembly language programmer can improve the performance of the hardware (with respect to its manipulation by DAP-FORTRAN). DAP-FORTRAN and APAL can be mixed at the subroutine level so that, under ordinary circumstances, only particularly critical routines must be converted entirely into assembly language. Secondly, the comprehensive macro facilities of APAL enable the user to define operations which are tailored to a particular application and, in effect, the user may set up his own high-level language. DAP-FORTRAN supports the standard repertoire of FORTRAN commands. In addition, the programmer may use any or all of the commands given below in a FORTRAN-like environment.

13.4.1.1 *Utilities*

VECTOR - Designate the name of a linear array.
MATRIX - Designate the name of the two-dimensional array.
COL - Generate a column test pattern.
ALT - Generate a bar test pattern.
ROW - Generate a row test pattern.
LMAT - Generate a matrix test pattern.
EL - Generate a general test pattern.

13.4.1.2 *Arithmetic Operators*

VC - Transfer the elements of a vector to storage.
REV - Re-index the elements of a vector.
SUM - Sum the elements of two vectors.
ALL - Compute the intersection of two vectors.
ANY - Compute the union of two vectors.
MAXV - Calculate the maximum element in a vector.
MINV - Calculate the minimum value of the elements of a vector.
MAXP - Locate the maximum element position within a vector.
MINP - Locate the minimum element position within a vector.
FRST - Determine the first vector of a set of vectors.
MERGE - Merge vectors.
SHRP - Right shift of a single point.
SHLP - Left shift of a single point.
SHRC - Right shift of an entire matrix.
SHLC - Left shift of an entire matrix.
MAT - Store the elements of a matrix.

MATC - Store the elements of a single column of a matrix.
MATR - Store the elements of a single row of a matrix.
REVC - Re-index the columns of a matrix.
REVC - Re-index the rows of a matrix.
MERGE - Merge two equivalent matrices.
SUM - Perform matrix addition.
COLN - Extract a selected column from a matrix.
ROWN - Extract a selected row from a matrix.
SUMC - Add two columns of two matrices.
SUMR - Add two rows of two matrices.
ALL - Intersect two matrices.
ANY - Perform the union of two matrices.
FRST - Determine the first matrix from a set of matrices.
MAXV - Find the maximum value of a matrix.
MINV - Find the minimum value of a matrix.
MAXP - Locate the position of the maximum point in a matrix.
MINP - Locate the position of the minimum point in a matrix.
ANDRWS - Perform the Boolean AND of rows of a matrix.
ANDCOLS - Perform the Boolean AND of columns of a matrix.
ORROWS - Perform the Boolean OR of rows of a matrix.
ORCOLS - Perform the Boolean OR of columns of a matrix.
SHNP - Shift point(s) of the matrix in the upward direction.
SHNC - Shift column(s) of the matrix in the upward direction.
SHSP - Shift point(s) of the matrix in the downward direction.
SHSC - Shift column(s) of the matrix in the downward direction.
SHEP - Shift point(s) of the matrix in the right direction.
SHEC - Shift column(s) of the matrix in the right direction.
SHWP - Shift point(s) of the matrix in the left direction.
SHWC - Shift column(s) of the matrix in the left direction.

13.4.1.3 *Example of Use*

There are no commands in DAP-FORTRAN for image display,

image transformation, image measurement, geometric manipulation, or decision theory. The user must employ either the macro facilities of the language or use FORTRAN in order to code subroutines for these purposes.

To provide an example of the use of DAP-FORTRAN, the below sequence of commands carries out the skeletonizing procedure for a binary image stored in the DAP 64×64 processing-element array. The program consists of a main subroutine which calls the subroutine THIN which, in turn, uses the external logical matrix function SHP. THIN tests for edges in specific quadrants of the 3×3 kernel used to explore the binary image. Subfields are not used so that the technique is similar to that employed in programming CLIP4 using the masks of Levialdi. The kernel parameter DIRN is the quadrant index within the 3×3.

```
      SUBROUTINE MAIN
      COMMON/PAT/INPUT
      LOGICAL INPUT( , )
      LOGICAL A( , ),B( , ),C( , )
      A=INPUT
  100 CONTINUE
      C=A
      DO 200 I=1,4
      CALL THIN(A,B,I)
      IF (ALL(B.LEQ.A)) GOTO 200
      A=B
  200 CONTINUE
      IF(ANY(A.LNEQ.C) GOTO 100
  300 CONTINUE
      RETURN
      END


      SUBROUTINE THIN(IN,OUT,DIRN)
      LOGICAL IN( , ),OUT( , )
      INTEGER DIRN
      EXTERNAL LOGICAL MATRIX FUNCTION SHP
      LOGICAL OP( , ),TEMP1( , ),TEMP2( , )
      LOGICAL FILL( , ),ANY( , ),(MANY( , )
      LOGICAL QUAD( , ,4),Q1( , ),Q2( , ),Q3( , ),Q4( , )
      EQUIVALENCE (QUAD( , ,1),Q1)
      EQUIVALENCE (QUAD( , ,2),Q2
      EQUIVALENCE (QUAD( , ,3),Q3)
      EQUIVALENCE (QUAD( , ,4),Q4)
      INTEGER MYDIRN,QNO,INO
      ANY =.FALSE.
      MANY=.FALSE.
      FILL=.TRUE.
```

```
      OP=SHP(DIRN,IN)
      MYDIRN=DIRN
      DO 100 QNO=1,4
      FILL=FILL.AND.OP
      TEMP1=.NOT.OP
      TEMP2=.FALSE.
      DO 200 INO=1,2
      MYDIRN=MYDIRN+1
      OP=SHP(MYDIRN,OP)
      TEMP2=TEMP2.OR.OP
      MANY(ANY.AND.OP)=.TRUE.
      ANY(.OP.)=.TRUE.
  200 CONTINUE
      MYDIRN=MYDIRN-1
      QUAD( , ,QNO)=TEMP1.AND.TEMP2
  100 CONTINUE
      Q3=Q3.OR.Q4
      OUT=IN.AND.(.NOT.(Q1.AND..NOT.(Q2.OR.Q3)
     *.OR.(Q2.AND.TEMP1.AND..NOT.(Q1.OR.Q3))).OR.FILL)
      RETURN
      END

      LOGICAL MATRIX FUNCTION SHP(DIRN,OP)
      INTEGER DIRN
      LOGICAL OP( , )
      GOTO(10,11,12,13),DIRN
   10 SHP=SHNP(OP)
      RETURN
   11 SHP=SHEP(OP)
      RETURN
   12 SHP=SHSP(OP)
      RETURN
   13 SHP=SHWP(OP)
      RETURN
      END
```

Although this 68 line program is shorter than the 93 line CAP4 program for skeletonizing given in Section 13.3, it is probably more difficult to read and understand. Execution of this program over the DAP 64×64 array for a worst case requiring 32 steps has been estimated to take 3ms (Preston, 1981). Thus, although the cycle time of the DAP cellular automaton is 50 times shorter than that of the CLIP4 machine, execution times for this particular task are in all major respects *identical*.

13.4.2 GLOL

GLOL is the operating language for the Perkin-Elmer research cellular logic machine as well as for the Coulter diff3 GLOPR which is part of the diff3, diff3-50, and diff4 robot microscopes described in Chapter 10. As in CAP4, GLOL permits the user to define a data type which is a binary image array whose elements are automatically indexed without specifying index values. For example, when a GLOL programmer wishes to form the Boolean EXOR of the two binary images A and B with the result directed to C, one simply submits C=A-B at the terminal which, upon carriage return, is immediately interpreted and executed by the operating system. Besides the image array data type, other data types supported in GLOL are scalars (constants) and vectors (generated by sequences of image measurements commands). In GLOL there is also an editor which permits the user to name strings of GLOL commands, called "procedures," and then to execute a procedure by simply submitting EX PROC XYZ, where XYZ is a procedure name.

GLOL leads to extremely compact code, is characterized by a highly interactive program development system, and has a sufficiently limited number of commands so that memorization is easy. As given in more detail below, both branch and loop operations are possible. However, the GOTO program statement is not allowed and no labels are utilized. Thus programming is highly structured and all code is *in-line* permitting extremely easy debugging of even the most complicated command strings. This has proven extremely useful in servicing the hundreds of cellular logic machines in the diff3 and diff4 robot microscopes which are now deployed.

GLOL runs on the Varian 620i, Data General Nova, and the Intel 8086. A detailed description of its major commands are given in the sections which follow, omitting those which have been added by Coulter and are specific to manipulation of the robot microscope optics and mechanics and its peripherals.

13.4.2.1 *Utilities*

DEFINE - (1) Allocates memory locations for a measurement vector, (2) allocates memory locations for a binary image array, (3) designates a procedure name.

EXECUTE - Cause a designated procedure to be carried out.

RUN - Allows a concatenated list of commands and/or procedures to be executed.

DO...STOP - Standard DO-loop terminator (with the

exception that the DO-loop index, increment, and target may be selected from elements of a vector).

IF...ELSE...STOP - Standard branching operation with ... indicating a first and and second list of commands.

OUTPUT - Cause numerical data stored in a designated measurement vector to be delivered to the (1) terminal, (2) magnetic tape, (3) other output media.

READ - Transfer specified image data from magnetic tape to a designated image array in main memory.

WRITE - Cause a designated image array to be stored on magnetic tape.

BEGIN - Rewind magnetic tape storing image data.

EOF - Spool the magnetic tape to the last image recorded on same.

13.4.2.2 *Image Display*

DISPLAY - Cause the contents of a designated array to be displayed on the display screen.

CLEAR - Erase the display screen.

13.4.2.3 *Arithmetic Operators*

GLOL supports the standard arithmetic operations conducted on arrays of numbers using FORTRAN-like instructions. GLOL also permits matrix operations to be executed in a symbolic manner wherein an entire image array (previously identified using DEFINE) is manipulated without indexing. For example, A=B causes each element of the image array designated A to be made equal to each element of the image array designated B. Boolean operators are also supported (AND, OR, EXOR). Finally, the ACQUIRE command is a point operator for use in thresholding a graylevel image and storing same in a binary image array.

13.4.2.4 *Image Transforms*

The major image transform is the Golay Transform as given in Chapters 2 and 10.

13.4.2.5 *Image Measurement*

HISTOGRAM - Compute the probability density function of a graylevel image and store same in a designated image array. Display the result on the display screen.

SMOOTH - Operate cyclically on the designated histogram by pairwise averaging until a given number of modes are achieved, or, alternately, for a designated number of cycles.

COUNT - Sum elements in a designated image array and store the resultant integer as a designated element in a designated vector.

13.4.2.6 *Decision Theoretic*

IDENTIFY - Execute a linear discriminant analysis using pre-assigned weights and a designated measurement vector. The result is classification of the object represented by the measurement vector into one of eight types. This command also may be used to display a table of the number of objects identified as a function of type and to generate a decision matrix.

13.4.2.7 *Example of Use*

In order to provide the reader with an example of a simple GLOL program, the below code performs a skeletonizing operation. It uses subfields rather than the masks of Levialdi for an image acquired from the image scanner and stored in binary image array A.

```
C=0
DO I=1/32
A=M[G(A)A]5-13,1,3
B=M[G(A)A]7-13,,
C=B+C
END LOOP                                   (13.3)
```

This six-command program is equivalent to the 93-step CAP4 program and the 68-step DAP program given above and clearly illustrates the power and simplicity of GLOL as an advanced high-level cellular-logic language.

13.4.3 TASIC

TASIC gets its name from BASIC (Beginners All-Purpose Symbolic Instruction Code) and thus stands for Texture Analyzer System Instruction Code. This chapter ends with a description of TASIC to illustrate still another approach to programming the cellular logic machine. The complete TASIC instruction repertoire includes some 140 commands and sub-commands of which 55 are selected for description in this section. The 55 commands selected cover almost all command categories and are, therefore, illustrative of the complete command repertoire.

As explained by Nawrath and Serra (1979), TASIC is an interpretive language which allows detailed control of the image analysis system and permits, often with a single instruction, the execution of sophisticated image processing functions. In this sense it corresponds to GLOL. TASIC also provides certain macro instructions which are directly allied with specific morphological operators. Furthermore, the user himself may construct sequences of these commands in order to form still more complex operators.

Since TAS is designed with hard-wired circuitry to carry out many elementary morphological transforms directly, the operation of the entire system in an interactive environment is rapid. (It is of interest to note that a lower-level language is available for TAS whose commands are written symbolically on the edge of the display monitor on a touch-sensitive border panel. This permits the user who employs this operational modality to execute pre-determined image processing routines with extraordinary speed without using the keyboard.)

The following sections provide short descriptions of the more significant TASIC commands.

13.4.3.1 *Utilities*

DIM - Dimension either scalars or arrays.
DLS - Commence a DO-loop.
DLE - Terminate a DO-loop.
GTO - Standard GOTO command.
IFC - Branch on specified condition.
GSB - GOTO subroutine.
DST - Set time (day-month-year).
FOR - Specify the data format.
REA - Read the specified data.
WRT - Write the specified data.
OPF - Open a specified file.
CLO - Rewind and close a specified file.

RWD - Rewind (only) a specified file.
MTR - Input/output command for bit planes.

13.4.3.2 *Image Display*

CRT - Display a binary image.
DSG - Display a memory plane.
DCX - Display logical results.
PLO - General data plotting routine.
HST - Plot the histogram.
PPO - Store the current light pen position.
CWT - Light pen tracing routine.
PWT - General light pen control.

13.4.3.3 *Arithmetic Operators*

TAB - General point operator.
CPY - Equate bit planes.
MAB - Equate matrices.
MOV - Equate source and destination.
MSD - General matrix transfer.
ACX - General Boolean Operator.

13.4.3.4 *Geometric Transforms*

PRS - Rotate the field of view.
STA - Translate the field of view.
ITR - Compute contours.
REC - Compute the circumscribing rectangle of an object.
MHS - Set horizontal size of mask.
MHP - Set horizontal position of mask.
MVS - Set vertical size of mask.
MVP - Set vertical position of mask.
MSK - Store current mask.

13.4.3.5 *Image Transforms*

ADA - Filter for halo removal.
GR2 - Gradient filter.
FLK - Non-linear highpass filter.
RNG - Golay transform.
DEF - General cellular logic transform.
SQL - Skeletonizing transform.
BUP - Region labeling operation.
NBO - Count without convergence.

13.4.3.6 *Image Measurement*

HST - Compute histogram.
HDF - Compute histogram moments.
ADF - General histogram measurements.
ASO - Measure integrated absorbance of region(s).
TRA - Measure integrated transmittance of region(s).
ARE - Measure area of region(s).
PER - Measure perimeter of region(s).
PRJ - Measure average projection.
EXT - Compute maximum and minimum within array.

13.4.3.7 *Example of Use*

Because of the sophistication of the TASIC commands the user may readily skeletonize an object (see sections above for comparison) by means of the following command

SQL A (13.5)

13.5 COMPARISON

The four languages described in this chapter were chosen specifically to illustrate the diversity of current languages for programming both cellular logic machines and array automata.

CAP4 is a simple, machine-oriented language whose assembly-like instructions relate directly to the functions computed by the processing element in the CLIP4 array automaton. Here the user has direct control over the step-by-step operations of the machine. In contrast, the user of GLOL and TASIC has available a high-level language where a single command replaces an entire subroutine of CAP4 instructions. DAP-FORTRAN forms a bridge between the standard FORTRAN repertoire and an extension of this repertoire to include commands useful in manipulating data within the DAP array. Here the user is conscious of the specific data manipulations being made as the contents of the processing elements are shifted in various directions and then combined using both arithmetic and Boolean operations.

GLOL, unlike TASIC, has a relatively small command repertoire which concentrates on the manipulation of bilevel images created by thresholding based on histogram analysis. The only image transforms available to the user of GLOL are the Boolean operators and the Golay parallel pattern transform.

The user of TASIC also has available the Golay parallel pattern transform (as well as Boolean logic) but has a far greater repertoire of commands at his disposal. Note that not only are DO-loops permitted but also the GOTO statement is allowed as well as, of course, the use of program labels. Perhaps the deficiency of such a language is that it is complex (155 commands plus their paramaterization lead to over one thousand possibilities!).

The use of CAP4 is centered in the Image Processing Group at University College London and a small number of other laboratories in Great Britain. GLOL has few programmers but is used in hundreds of installations worldwide due to the wide dissemination of the diff-series of robot microscopes. The user community for DAP-FORTRAN is primarily restricted to the users of the few DAP machines which have been installed. These users employ these machines primarily on a service basis using dial-up connections. Note, in particular, that the DAP has no method for displaying the contents of the 64×64 array. This makes interactive use difficult. TASIC is used, of course, exclusively by those groups which have purchased the Leitz TAS system. The use of this system is confined primarily to the European community.

In summary, it is evident that the programming languages used for cellular logic machines and array automata have no universality. This means that thousands of man hours are wasted since programs are not transferrable from one type of machine to another. This is a pity, but, unfortunately, is characteristic of almost all research in this field at present.

14. FABRICATION METHODS

14.1 INTRODUCTION

During the past decade (1973-1983) semiconductor IC (Integrated Circuit) fabrication methods have progressed from LSI (Large Scale Integration) to VLSI (Very Large Scale Integration). The year 1983 saw the introduction of such machines as the VLSI Hewlett-Packard 9000 (HP9000) comprised of some 450 thousand transistors. The HP9000 is a 32-bit CPU using one micrometer line widths which requires anisotropic etching accomplished by using reactive-ion technology. This compares with the early 1970s when LSI was used to fabricate the first 4K memory chip employing eight to ten micrometer line widths. From 1973 to 1983, device complexity has advanced by a factor of two every 18 months.

Until now IC patterns have been etched on photoresist-coated, chromium-on-glass masks. Pattern transfer has been by optical imaging using such instruments as the Perkin-Elmer Microline. Now pattern dimensions are approaching the thickness of the photoresist films and the depth of dopant ion diffusions. This explains why reactive-ion anisotropic-etching techniques have had to be developed. In the next few years, as dimensions decrease still further, optical imaging of masks on the semiconductor substrate will have to be discarded in favor of direct electron beam lithography or, perhaps, xray lithography. In making these submicrometer devices, fabrication technology will approach the limits of current understanding of device behavior. In fact, the number of dopant atoms in the gate regions of MOS-FET structures will be less than 1000 resulting in heretofore unexperienced statistical variability in device characteristics.

Unfortunately, the designer of the cellular automaton has had little or no access to VLSI technology. The chips required for such machines as CLIP4 and MPP are numbered in the thousands and are decidedly in the low-volume market. This means that typical IC manufacturers are totally disinterested in mounting the VSLI design effort required to fabricate chips for such systems. As Trimberger (1983) points out, to design a single, million-transistor chip at current rates of design productivity consumes from forty to eight-hundred designer years. This enormous expenditure of effort prohibits its application to all but the highest production chips. As time passes, the number of designer years may gradually decrease with the introduction of such VLSI design aides as the so-called "silicon compilers."

For cellular automata chips the design costs are as enormous as the market is small. This fact has had a strong impact on the sophistication and complexity of the two cellular automata chips produced to date. This chapter presents a description of the design and fabrication approach employed for both of these chips, i.e., that used in CLIP4 and that used in the MPP. It should be remembered that the designs of both of these chips were undertaken in either the early 1970s (CLIP4) or the mid-1970s (MPP) and, therefore, are based on the technology available during those periods of time and the other constraints mentioned above.

14.2 THE CLIP4 CHIP

The design of the CLIP4 chip was commenced in 1973. At that time inquiries were made of several major IC manufacturers as to the possibility of designing and fabricating the type of chip required. Invariably, the reply was, without guaranteed sales in the range of 10-100 thousand devices per year for several years, no IC manufacturers were interested in becoming involved. The result was to employ independent design houses which were rare creatures, indeed, in 1973. The CLIP4 program at University College London (UCL) was involved with a series of three such design houses from 1973 through 1982.

The chronology for the development of the CLIP4 chip is given in Table 14-1. As this table shows, the interval 1973-1976 was occupied working with Design House A with the final result (December, 1976) that their final design was found to be non-viable. In mid-1977, a contract was placed with Design House B, which, after two years succeeded in producing a first prototype chip. Although meeting some of the design requirements, the prototype chip had some fundamental faults which led to its rejection by UCL. This was followed shortly by the demise of Design House B as a company in July 1978. This misfortune

Table 14-1 CLIP4 Chip Fabrication Chronology

Year	Month	
1973	October	Discussions commence with Design House A.
1975	January	Project funding received. Design House A begins work.
1976	December	Design House A design proves non-viable.
1977	June	Contract placed with Design House B.
1978	June	First prototype device from Design House B with some faults.
1978	July	Design House B discontinues business activities.
1978	September	Contract placed with Design House C to complete design.
1979	May	Prototype produced by Design House C.
1979	July	Correct logical behavior of prototype verified.
1980	January	Total of 1400 chips delivered to reduced specification.
1980	February	Array of 1152 chips assembled and working.
1981	January	Redesign commenced by Design House C to improve reliability.
1981	September	Logical verification of redesigned prototype complete.
1982	Jan-Dec	Delivery of production redesigned devices.

was followed by the writing of a third contract in September 1978 with Design House C. From this there followed a second prototype in May of 1979 whose correct logical behavior was verified that July. By January 1980, 1400 chips were delivered. These chips had somewhat reduced operating speed specifications but were otherwise satisfactory. By February 1980 the first working CLIP4 array of 1152 chips was assembled and in operation.

Design efforts on the CLIP4 chip continued by Design House C into 1981-1982 for the purpose of improving design reliability. New prototypes were made, logical verification performed, and delivery of these improved chips commenced. This lengthy, and at times frustrating, design-test-redesign cycle was, unfortunately, typical of pioneering efforts in LSI custom design in the 1970s.

14.2.1 Design Considerations

The fundamental logical design of the CLIP4 chip has not changed since its inception in 1973. At various stages the implementation of this design has been altered in search of reduced size, increased reliability, etc. Since, at the time of the first design, large, inexpensive DRAMs (Dynamic Random Access Memories) were not commercially available, the DRAMs were included as part of the chip itself. The system design and system software were based on this decision to use on-chip DRAM. This meant that this original design decision had to be perpetuated during the entire chip design process even when other alternatives, i.e., off-chip DRAM, became available. Other parameters which have remained constant throughout the various design stages are given in Table 14-2.

Table 14-2 CLIP4 Chip Specifications

PACKAGE	40-PIN DIP
TECHNOLOGY	METAL-GATE NMOS
TRANSISTORS	3000 TOTAL
	100 PER PROCESSOR 200 PER DRAM 600 CONTROL LOGIC

Table 14-3 Effects of Redesign on CLIP4 Chip Parameters

YEAR	1976	1979	1982
LOGIC TYPE	4-PHASE	4-PHASE	2-PHASE
CHIP SIZE (mm)	6.3 × 6.3	4.5 × 4.5	3.5 × 3.5
CLOCK CYCLE (ns)	–	1000	400
LINE WIDTH (μ)	8	8	5
SUBSTRATE BIAS (v)	-3	-1	0
CLOCK VOLTAGE (v)	+15	+15	+7
CHIPS PRODUCED	–	2000	5000+

As of 1973, the maximum package size available was the 40-pin DIP (Dual Inline Package). This structure determined the trade-off point between requirements of reasonable economy (many processors on the chip left few pins for control) and those of reasonable speed (more control pins and faster operation). Since only MOS technology was suitable as far as power requirements were concerned for a 3000-transistor chip, essentially no choice in technology existed in 1973. (NMOS or iso-planar CMOS would be available alternatives today.) Thus, practical considerations as to availability determined the initial choice of MOS technology and this choice was maintained.

Other parameters changed considerably during the various design phases. These are given in Table 14-3.

14.2.2 Chip Testing

For a custom chip, such as the CLIP4, chip testing for logical errors is a substantial problem. Testing the CLIP4 chip occurs at three stages. First, probe testing of the dice is used. Second, there is testing of each chip after packaging. Third, chips are tested when incorporated in the entire system. The criteria for the first two stages of testing are that the complete system test should take less than one second per chip and that the development time for the test specifications should not be too costly. This requirement means that tests are necessarily a subset of what would be an impossibly large complete test repertoire. This also implies that many chips which pass the first two stages will fail in the final systems test. This, of course, is frustrating. It is, however, typical of the production cycle for a custom circuit prototype where no recursion is possible, i.e., the results of systems testing come too late to be used in modifying device tests. All of these considerations would change, of course, if commercial development were to be undertaken. Also, under commercial development, tests concerning the detection of the crosstalk and pickup which also limit chip performance would be incorporated as well as the above mentioned tests for logical errors.

14.2.3 Design Description

Figure 14-1 shows the floor-plan of the final CLIP4 chip design (Design House C). The chip size is 3.5×3.5 millimeters with the total number of transistors equaling three thousand. As can be seen, there are eight processing elements per chip. DRAMs are located in the central portion of the structure. The Boolean processors and control circuitry are located separately. (For a full description of the logical design of the CLIP4 processing element, see Chapter 11.) There is a 32-bit DRAM (comprised of 200 transistors) per processing element, shared control logic (600 transistors), and an additional 100 transistors for each processing element to implement the Boolean processors, neighborhood gates, registers, logic, etc. Figure 14-2 is a photograph of the resulting physical structure.

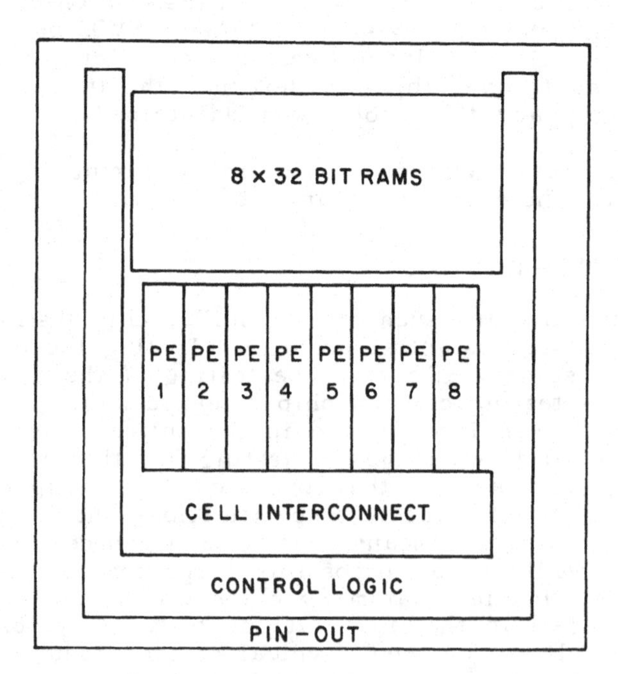

Fig. 14-1 Floor-plan of the CLIP4 chip.

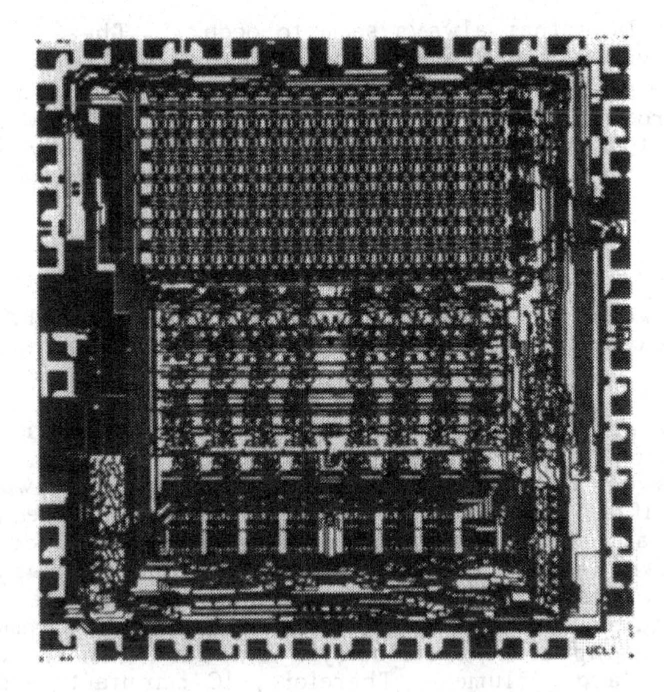

Fig. 14-2 A completed CLIP4 chip.

14.3 THE MPP CHIP

As more fully described in Chapter 11, the MPP has 16,896 processing elements organized in a 128×132 array. Each processing element has six one-bit registers, a shift register with a programmable length of 2 bits to 30 bits, a full adder, and a RAM (Random Access Memory). The shift register and three of the one-bit registers are used to perform bit-serial arithmetic. The remaining one-bit registers are used for logic, routing, masking, and input/output.

Design of the MPP chip started in 1978. At that time, CMOS-SOS appeared to be the appropriate technology. In 1980, the layout of a CMOS-SOS chip was completed and samples were produced and tested. They satisfied the logical specifications but were not able to meet the 10-megahertz clock rate desired. Also, production cost estimates were very high. For these reasons and the fact that other chip technologies were now available, it was decided to redesign the chip.

A design using HCMOS technology and a better floor-plan was started in late 1980. Working samples of the second design were available for testing in 1982 (after an iteration to fix the design flaws that always seem to occur). These samples operated correctly at 10 megahertz. All critical path timing specifications were met with wide margins suggesting a good yield in production. Production runs of this design were started in late 1982 and completed in early 1983 with very good yield.

14.3.1 Design Considerations

The two major influences in the design of the MPP chip were RAM capacity and data routing between processing elements.

Early application studies suggested that a RAM memory of 256 bits per processing element would be sufficient. Later studies showed that most of the application programs would run faster if a larger memory were available. (There doesn't seem to be any limit to the amount of memory that good programmers can use effectively.) The most economical way to give each processing element a large RAM is to couple standard RAM chips with the custom chip containing the remainder of the processing element circuitry. Standard RAM chips are produced in large volumes. Therefore, IC manufacturers use their best designers and special production processes to make large low-cost memories. Neither this talent nor the processes are available for low-volume custom chips.

Another consideration is the fact that memory technology changes at a rapid rate. A custom chip containing a large RAM would soon be obsolete. For these reasons the custom chip only contains the processing element logic. Standard memory chips are used for the RAMs even though this requires a pin per processing element to couple data to the RAM, the choice actually reduced the pin-out of the MPP custom chip since the chip no longer needs to couple to the RAM address bus. The address bus is 16 bits wide to allow up to 65,536 bits of RAM per processing element.

In order to satisfy the requirement of routing data between any pair of neighboring processing elements in 100 nanoseconds, all 16,896 processing elements had to be packaged in one cabinet. Each custom chip had to contain the logic of at least eight processing elements to package the array in a cabinet of reasonable size without using exotic packaging techniques. The logic of eight processing elements plus the on-chip control logic requires about 8000 transistors, which is a reasonable number with HCMOS technology. The best pin-out is obtained when the eight processing elements are organized as a 2×4 array.

14.3.2 Design Description

Figure 14-3 shows the floor-plan of the final MPP chip. The processing elements are laid out in one line (even though they are logically organized as a 2x4 array) to simplify the distribution of control signals. These signals come from the two boxes labelled K LOGIC. Control signal delay is critical so each processing element is long and thin. This topology minimizes the length of the horizontal control lines. Delays in critical data paths (memory transfers and inter-chip routing) were minimized by grouping the required functional blocks towards the top of the chip and placing less critical functions toward the bottom. The shift registers are wider (because they loop back) and were laid out along the bottom edge of the chip. The bi-directional drivers to route data between neighboring chips were laid out on the sides above the K LOGIC. The bi-directional drivers for RAM data and parity bits were laid out along the top edge (DO through D7 plus PAR). The main data bus of each processing element feeds an input of an eight-input three of inclusive-or gates. The output of this tree is fed off the chip through another driver (IOR).

Figure 14-4 is a photograph of the MPP chip. Solid State Scientific Inc. manufactured this chip using HCMOS technology. The design used five-micron design rules. The result was put

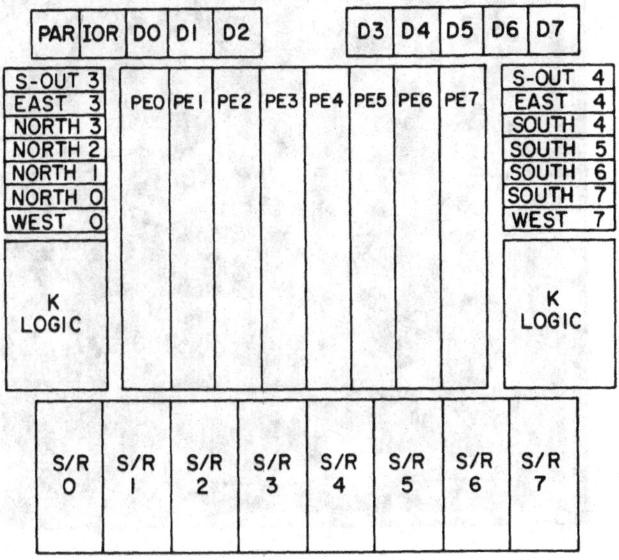

Fig. 14-3 Floor-plan of the MPP chip.

through a 10% shrink to yield a final chip size of 6.0×3.3 millimeters. Two power supplies are used. Seven volts is employed for the internal circuitry; five volts, for the output translators. The chip is packaged in a 52-pin flat pack. It dissipates about 250 milliwatts when operating at 10 megahertz.

Fig. 14-4 Photograph of the MPP chip.

REFERENCES

Ahuja, N., "Dot pattern processing using neighborhoods," IEEE Trans. Pattern Anal. Machine Intell. *PAMI-4*(3), 336-343 (1982).

Akers, S. B., "A rectangular logic array," IEEE Trans. Comput. *C-21*, 848-857 (1972).

Arcelli, C. and Levialdi, S., "Neuron counting in three dimensions; A proposal," Acta Cyber. *5*, 65-68 (1973).

Arcelli, C., Cordella, L., and Levialdi, S., "Parallel thinning of binary pictures," Electron. Lett.*11*, 148-149 (1975).

Banks, E. R., "Universality in cellular automata," Proc. 11th Switch. Automata Th. Conf. (1970), pp. 216-224.

Batcher, K. E., "Design of a massively parallel processor," IEEE Trans. Comput. *C-29*, 836-840 (1980).

Batcher, K. E., "Bit-serial parallel processing system," IEEE Trans. Comput. *C-31*(5), 377-384 (1984).

Blum, H., "A transformation for extracting new descriptors of shape," in *Models for the Perception of Speech and Visual Form* (Wathen-Dunn, W., ed.) MIT Press, Cambridge (1967).

Bracewell, R. N. and Riddle, A. C., "Inversion of fanbeam scans in radio-astronomy," Astrophys. J. *150*(2), 427-434 (1967).

Brenner, J. F., Lester, J. M., and Selles, W. D., "Scene segmentation in automated histopathology: Techniques evolved from cytology automation," Pattern Recog. *13*(1), 65-77 (1981).

Burks, A. W. (ed.), *Essays on Cellular Automata*, Univ. of Illinois Press, Urbana (1970).

Cauer, W. A., German Patent No. 892,772 (Dec. 1950).

Causley, D. and Young, J. Z., "The flying spot microscope - Use in particle analysis," Research *8*, 430-434 (1953).

Clarke, K. A., "Computer assisted tomography with an array processor," Image Proc. Gp., Univ. College London Rpt. 81/6 (1981).

Cook, C. E. and Bernfeld, M., *Radar Signals*, Academic Press, New York (1967).

Danielsson, P. E., "Getting the median faster," Comput. Graph. Image Proc. *17*, 71-78 (1981).

Davies, E. R. and Plummer, A. P. N., "Thinning algorithms: A critique and new methodology," Pattern Recog. *14*, 53-63 (1981).

Deutsch, E. S., "On some preprocessing techniques for character recognition," in *Computer Processing in Communications*, Brooklyn Polytech. Press (1969), pp. 221-234.

Dinneen, G. P., "Programming pattern recognition," Proc. Western Joint Comput. Conf. (1955), pp. 94-100.

Duff, M. J. B., "Parallel computation in pattern recognition," in *Methodologies of Pattern Recognition* (Watanabe, S., ed.) Academic Press, New York (1969).

Duff, M. J. B., "Cellular logic and its significance in pattern recognition," in Proc. 21st AGARD Avionics Panel Technical Symposium, Rome, Italy (1971), pp. 25/1-25/13.

Duff, M. J. B., "Lay planning," Proc. 5th Internat'l. Conf. Pattern Recog. (1980), pp. 300-304.

Duff, M. J. B., Jones, B. M., and Townsend, L. J., "Parallel processing pattern recognition system," Nucl. Instr. Methods *52*, 284-288 (1967).

Duff, M. J. B. and Watson, D. M., "Automatic design of pattern recognition networks," in Proc. Electro-Optics Internat'l Conf., Brighton, Great Britain (1971), pp. 369-377.

Duff, M. J. B. and Watson, D. M., "CLIP3: A cellular logic image processor," in *New Concepts and Technologies in Parallel Information Processing*, (Caianiello, E. R., ed.) Noordhoff, Leyden (1975), pp 75-86.

Duff, M. J. B., Watson, D. M., Fountain, T. J., and Shaw, G. K., "A cellular logic array for image processing," Pat. Recog. *5*, 229-234 (1973).

Duff, M. J. B., Watson, D. M., and Deutsch, E. S., "A parallel computer for array processing," Info. Proc. *74*, 94-97 (1974).

Fisher, R. A., "The use of multiple measurements in taxonomic problems," Ann. Eugenics *7*, 179-188 (1936).

Flanders, P. M., "Fortran extensions of a highly parallel processor," in *Infotech State of the Art Report on Supercomputers*, Vol. 2 (1979), pp. 117-133.

Flanders, P. M., Hunt, D. J., Reddaway, S. F., and Parkinson, D., "Efficient high-speed computing with the distributed array processor," in *High-Speed Computer and Algorithm Organization*, (Kuck, D. J., Lawrie, D. H., and Sameh, A. H., eds.) Academic Press, New York (1977), pp. 113-127.

Fountain, T. J., "A survey of bit-serial array processor circuits," in *Computing Structures for Image Processing* (Duff, M. J. B., ed.) Academic Press, London (1983).

Fu, K. S. and Mui, J. K., "A survey of image segmentation," Pattern Recog. *13*(1), 3-16 (1981).

Gardner, M., "On cellular automata, self-reproduction, the Garden of Eden and the game 'life'," Sci. Amer. *224*(2), 112-117 (1971).

Gerritsen, F. A., "Design and implementation of the Delft Image Processor DIP-1," Ph.D. Thesis, Dept. Elec. Engrg., Delft Univ. of Technol., 1981.

Genchi, H., Mori, K., Watanabe, S., and Katsuragi, S., "Recognition of handwritten numerical characters for automatic letter sorting," Proc. IEE *56*(8), 1292-1301 (1968).

Goetcherian, V., "From binary to gray tone image processing

using fuzzy logic concepts," Pattern Recog. *12*, 7-15 (1980).

Golay, M. J. E., "Apparatus for counting bi-nucleate lymphocytes in blood," U.S. Patent 3,214,574 (1965).

Golay, M. J. E., "Hexagonal parallel pattern transformations," IEEE Trans. Comput. *C-18*, 733-740 (1969).

Golay, M. J. E., "Analysis of images," U.S. Patent 4,060,713 (1977).

Goldstine, H. H., *The Computer from Pascal to von Neumann*, Princeton University Press (1972).

Graham, M. D., "The diff4: A second-generation slide analyzer," in *Computing Structures for Image Processing* (Duff, M. J. B., ed.) Academic Press, London (1983).

Graham, M. D. and Norgren, P. E., "The diff3 analyzer: A parallel/serial Golay image processor," in *Real-Time Medical Image Processing* (Onoe, M., Preston, K., Jr., and Rosenfeld, A., eds.), Plenum Press, New York (1980), pp. 163-182.

Gray, S. B., "The binary image processor and its applications," Rpt. 90365-5C (unpublished), Informational International Inc., Los Angeles (1972).

Gray, S. B., "Local properties of binary images in two dimensions," IEEE Trans. Comput. *C-20*(5), 551-561 (1971).

Hafford, K. J. and Preston, K., Jr., "Three-dimensional skeletonization of elongated solids," Comput. Graph. Image Proc. (1984).

Haralick, R. M., "Automatic remote sensor image processing," in *Digital Picture Analysis*, (Rosenfeld, A. ed.), Springer-Verlag, Heidelberg (1976), pp. 41-63.

Herron, J. M., Farley, J., Preston, K., Jr., and Sellner, H., "A general-purpose high-speed logical transform image processor," IEEE Trans. Comput. *C-31*(8), 795-800 (1982).

Hilditch, C. J., "Linear skeletons from square cupboards," in *Machine Intelligence IV* (Meltzer, B. and Michie, D., eds.) Edinburgh Univ. Press, Edinburgh (1969), pp. 403- 420.

Hilditch, C. J. and Rutovitz, D., "Chromosome recognition," Ann. N.Y. Acad. Sci. *157*, 339-364 (1969).

Hunt, D. J., "The ICL DAP and its application to image processing," in *Languages and Architectures for Image Processing* (Duff, M. J. B. and Levialdi, S., eds.) Academic Press, London (1981).

Huttman, E., German Patent No. 768,068 (March 1980).

Izzo, N. F. and Coles, W., "Blood cell scanner identifies rare cells," Electronics *35*(17), 52-57 (April, 1962).

Justusson, B. I., "Median filtering: Statistical properties," in *Two-Dimensional Signal Processing II* (Huang, T. S., ed.), Springer-Verlag, Heidelberg (1981), pp. 161-196.

Karnaugh, M., "The map method for synthesis of combinatorial logic circuits," Trans. AIEE *72*(PtI), 593-597 (1953).

Kemeny, J. G., "Man viewed as a machine," Sci. Amer. *192*, 58-67 (1955).

Kepler, J., "Harmonices Mundı," in *Omnia Opera*, Vol. 5 (1619), reprinted by Hyder and Zimmer, Frankfurt (1864).

Kirsch, R. A., "Experiments in processing information with a digital computer," Proc. Eastern Joint Comput. Conf. (1957), pp. 221-229.

Kittler, J., Illingworth, J., and Paler, K., "The magnitude accuracy of the template detector," Pat. Recog. *16*(6), 607-613 (1983).

Kolmogoroff, A., "Interpolation und extropolation von stationären zufälligen folgen," Bull. Acad. Sci. U.R.S.S., Ser. Math. 5, 3-14 (1941).

Kosulajeff, P.A., "Sur les problèmes d' interpolation et d'extrapolation des suites stationaires," Compt. Rend. Acad. Sci. U.R.S.S. *30*, 13-17 (1941).

Kruse, B., "A parallel picture processing machine," IEEE Trans. Comput. *C-22*(12), 1075-1087 (1973).

Kruse, B., "System architecture for image analysis," in *Structured Computer Vision* (Tanimoto, S. and Klinger, A., eds.) Academic Press, New York (1980), pp. 169-212.

Kruse, B., Danielsson, P. E., and Gudmundsson, "From PICAP I to PICAP II," in *Special Computer Architectures for Pattern Processing* (Fu, K.S. and Ichikawa, T., eds.) CRC Press, Boca Raton (1982), pp. 127-155.

Lavington, S.H., *Early British Computers*, Digital Press, Bedford (1980).

Lee, C. Y. "Algorithm for path connections and applications," IRE Trans. Electron. Comput. *EC-10*, 346-365 (1961).

Lester, J. M., Williams, H. A., Weintraub, B. A., and Brenner, J. F., "Two graph-searching techniques for boundary finding in white blood cell images," Comput. Biol. Med. *8*, 293-308 (1978).

Levialdi, S., "CLOPAN - A closed-pattern analyzer," Proc. IEE *115*(6), 879-892 (1968).

Levialdi, S., "On shrinking binary picture patterns," Comm. Assoc. Comput. Mach. *15*(1), 7-10 (1972).

Levinson, N., "A heuristic exposition of Wiener's mathematical theory of prediction and filtering," J. Math. Phys. *26*(2) 110-119 (1947).

Lobregt, S., Verbeek, P. W., and Groen, F. C. A., "Three-dimensional skeletonization: Principle and algorithm," IEEE Trans. Pattern Anal. Mach. Intell. *PAMI-2*(1), 75-77 (1980).

Lougheed, R. M. and McCubbrey, D. L., "The Cytocomputer: A practical pipelined image processor," Proc. 7th Ann. Internat'l Sym. Comput. Arch. (1980), pp. 1-7.

Mansberg, H. P. and Segarra, J. M., "Counting of neurons by flying spot microscope," Ann. N.Y. Acad. Sci. *99*(2), 309-322 (1962).

Maruoka, A., "Cellular automata," J. Inst. Electron. Commun. Engrg. Japan *61*(10), 1073-1083 (1978).

McCulloch, W. S. and Pitts, W., "A logical calculus of the ideas immanent in nervous activity," Bull. Math. Biophys. *5*, 115-133 (1943).

McCormick, B. H., "The Illinois pattern recognition computer - ILLIAC III," IEEE Trans. Electron. Comput. *EC-12*(6), 791-813 (1963).

Mendelsohn, M. L., Mayall, B. H., Prewitt, J. M. S., Bostrom, R. C., and Holcomb, N. G., "Digital transformation and computer analysis of microscope images," in *Advances in Optical and Electron Microscopy*, (Cosslett, V. and Barer, R., eds.) Academic Press, London (1968) pp. 77-150.

Minnick, R. C., "A survey of microcellular research," J. Assoc. Comput. Mach. *14*, 203-241 (1967).

Moore, G. A., "Applications of computers to the quantitative analysis of microstructures," U.S. Nat'l Bureau Stds. Rpt. No. 9428 (1966).

Moore, G. A., "Automatic scanning and computer process for the quantitive analysis of micrographs and equivalent subjects," in *Pictorial Pattern Recognition*, Thompson Book Co., Washington (1968), pp. 275-326.

Morgenthaler, D. G. and Rosenfeld, A., "Three-dimensional digital topology: The genus," Rpt. TR-980, Comp. Sci. Ctr., Univ. of Maryland (1980).

Nadler, M., "An analog-digital character recognition system," IRE Trans. Electron. Comput. *EC-12*(6), 814-821 (1963).

Nakagana, Y. and Rosenfeld, A., "A note on the use of local Min and Max operations in digital picture processing," IEEE Trans. Sys. Man. Cyber. *SMC-8*, 632-635 (1978).

Nawrath, R. and Serra, J., "Quantitive image analysis: Theory and instrumentation," Micros. Acta *82*(2), 101-111 (1979).

Nishio, H., "A classified bibliography on cellular automata theory - With focus on recent Japanese references," Proc. IEEE Symp. Uniformly Structured Automata and Logic (1975), pp. 206-214.

Noguchi, S. and Oizumi, "A survey of cellular logic," J. Inst. Electron. Commun. Engrg. Japan *54*(2), 206-220 (1971).

North, D. O., "Analysis of factors which determine signal-to-noise discrimination in radar," Rept. PTR-6c, RCA Laboratories, Princeton (June 1943).

Ohmori, K., Naito, S., Nanya, T., and Nezu, K., "An application of cellular logic for high speed decoding of minimum-redundancy codes," Proc. Amer. Fed. Info. Proc. Conf. *41*, 345-351 (1972).

Post, E. L., "Finite combinatory processes - Formulation I," J. Symbol. Logic *1*, 103-105 (1936).

Potter, J. L., "MPP architecture and programming," in *Multicomputers and Image Processing* (Preston, K., Jr. and Uhr, L., eds.) Academic Press, New York (1982).

Potter, J. L., "Image processing on the Massively Parallel Processor," Computer *16*(1), 62-67 (1983).

Preston, K., Jr., "The CELLSCAN system - A leucocyte pattern analyzer," Proc. Western Joint Comput. Conf. (1961), pp. 175-178.

Preston, K., Jr., "Automatic differentiation of white blood cells," in *Image Processing in Biomedical Science* (Ramsey, D.M., ed.) Univ. Calif. Press, Berkeley (1969), pp. 97-117.

Preston, K., Jr., "Feature extraction by Golay hexagonal pattern transforms," IEEE Trans. Comput. *C-20*(9), 1007-1014 (1971).

Preston, K., Jr., "Use of the Golay Logic Processor in pattern-recognition studies using hexagonal neighborhood logic," in *Computers and Automata* (Fox, J., ed.) Polytechnic Press, New York (1971), pp. 609-623.

Preston, K., Jr., "Application of cellular automata to biomedical image processing," in *Computer Techniques in Biomedicine and Medicine* (Haga, E., ed.) Auerbach, Philadelphia (1973), pp. 295-331.

Preston, K., Jr., "Image manipulative languages: A preliminary survey," in *Pattern Recognition in Practice* (Gelsema, E. S. and Kanal, L. N., eds.), North-Holland, Amsterdam (1980).

Preston, K., Jr., "The crossing number of a three-dimensional dodecamino," J. Combin. Info. Sys. Sci. *5*(4), 281-286 (1980).

Preston, K., Jr., "Comparison of parallel processing machines: A proposal," in *Languages and Architectures for Image Processing* (Duff, M. J. B. and Levialdi, S., eds.), Academic Press, London (1981).

Preston, K., Jr., "Tissue section analysis: Feature selection and image processing," Pattern Recog. *13*(1), 17-36 (1981).

Preston, K., Jr., "Cellular logic algorithms for graylevel image processing," in *Multicomputers and Image Processing* (Preston, K., Jr. and Uhr, L., eds.), Academic Press, New York (1982).

Preston, K., Jr., "Progress in image processing languages," in *Computing Structures for Image Processing* (Duff, M.J.B., ed.), Academic Press, London (1983).

Preston, K., Jr., "Four-dimensional logic transforms," PhysicaD (June, 1984).

Preston, K., Jr., Duff, M. J. B., Levialdi, S., Norgren, P. E., and Toriwaki, J-i., "Basics of cellular logic with some applications in medical image processing," Proc. IEEE *67*(5), 826-857 (1979).

Prewitt, J. M. S. and Mendelsohn, M. L., "The analysis of cell images," Ann. N.Y. Acad. Sci. *128*, 1035-1053 (1966).

Prewitt, J. M. S., "Object enhancement and extraction," in *Picture Processing and Psychopictorics* (Lipkin, B.S. and Rosenfeld, A., eds.) Academic Press, New York (1970).

Reddaway, S. F., "The DAP approach," in *Infotech State of the Art Report on Supercomputers*, Infotech Ltd., Maidenhead (1979), Vol. 2, pp. 309-329.

Reddy, D. R., "Computer architecture for vision," in *Computer Vision and Sensor-Based Robots,"*, Plenum Press, New York (1980).

Reeves, A. P., "Parallel algorithms for real-time image processing," in *Multicomputers and Image Processing* (Preston, K., Jr. and Uhr, L., eds.) Academic Press, New York (1982).

Reeves, A. P. and Bruner, J. D., "Efficient function implementation for bit-serial parallel processors," IEEE Trans. Comput. *C-29*, 841-844 (1980).

Reynolds, D. E., "Automatic generation of image segmentation procedures for cellular array," Ph.D. Thesis, Image Processing Group, University College London (1983).

Rosenfeld, A., "Connectivity in digital pictures," J. Assoc. Comput. Mach. *17*(1), 146-160 (1970).

Rosenfeld, A. and Kak, A. C., *Digital Picture Processing*, Academic Press, New York (1976).

Schrandt, R. G. and Ulam, S. M., "On patterns of growth of figures in two dimensions," N. Amer. Math. Soc. *1*, 642-651 (1960).

Selfridge, O. G., "Pattern Recognition and modern computers," Proc. Western Joint Comput. Conf. (1955), pp. 91-93.

Serra, J., *Image Analysis and Mathematical Morphology*, Academic

Press, London (1982).

Shaefer, D. H., "TSE computer," IEEE Proc. *65*(1), 129-138 (1977).

Shelton, G. L., Jr.,"Pattern recognition preprocessing techniques," U.S. Patent 3,339,179 (1967).

Sherman, H., "A quasi-topological method for the recognition of line patterns," in *Information Processing*, Butterworths, London (1960), p. 232.

Slotnick, D. L., Borck, W. C., and McReynolds, R. C., "The Solomon computer," Proc. Western Joint Comput. Conf. (1962), pp. 87-107.

Slotnick, D. L., "The fastest computer," Sci. Amer. *224*(2), 76-87 (1971).

Smith, A. R., III, "Cellular automata theory," Tech. Rpt. No. 2, Stanford Electronics Laboratories, Stanford University (1969).

Smith, A.R., III, "Cellular automata and formal languages," Proc. 11th Switch. Automata Th. Conf. (1970), pp. 216-224.

Smith, A. R., III, "Simple computation - Universal cellular spaces," J. Assoc. Comput. Mach. *18*, 339-353 (1971).

Sproule, D. O. and Hughes, H. J., British Patent No. 604,429 (July 1948).

Srihari, S. N., "Understanding the bin of parts," Prof. Intern'l Conf. Cyber. Soc., Denver (1979), pp. 44-49.

Srihari, S.N. and Srisuresh, P., "A shrinking algorithm for three-dimensional objects," IEEE Conf. Comput. Vision Pattern Recog., Washington (1983), pp. 392-393.

Stamopoulos, C., "Parallel image processes," IEEE Trans. Comput. *C-24*(4), 424-433 (1975).

Sternberg, S. R., "Cytocomputer real-time pattern recognition," Proc. 8th Auto. Imagery Pattern Recog. Sym. (1978), pp. 205-214.

Sternberg, S. R., "Automatic image processor," U.S. Patent No. 4,167,728 (1979).

Sternberg, S. R., "Parallel architectures for image processing," in *Real/Time Parallel Computers* (Onoe, M., Preston, K., Jr., and Rosenfeld, A., eds.) Plenum Press, New York (1981).

Sternberg, S. R., "Biomedical image processing," Computer *16* (1), 22-34 (1983).

Suenaga, Y., Toriwaki, J-i., and Fukumura, T., "Range filters for processing of continuous-tone pictures and their applications," Sys. Comput. Cntrl. *5*(3), 16-24 (1974).

Taylor, W. K., "An automatic system for obtaining particle size distributions with the aid of the flying-spot microscope," Brit. J. Appl. Phys., Supp. 3, 173-180 (1954).

Thatcher, J. W., "Universality in the von Neumann cellular model," in *Essays on Cellular Automata* (Burks, A.W., ed.) Univ. of Illinois Press, Urbana (1970).

Tojo, A., "Pattern description with a highly parallel information processing unit," Bull. Electrotech. Lab. *31*(8), 930-946 (1967).

Tojo, A., "Distance functions and minimum path connections," Bull. Electrotech. Lab. *32*(9), 1930-1942 (1968).

Tojo, A., Yamaguchi, T., and Aoyama, H., "Pattern description with highly parallel information processing unit. VI-Construction and simulation," Bull. Electrotech. Lab. *33*(5), 479-505 (1970).

Toriwaki, J-i., "Topological properties and topology-transformations of a three-dimensional binary picture," Proc. Internat'l. Pattern Recog. Conf., Munich (1982).

Toriwaki, J., and Fukumura, T., "Extraction of structural information from digitized grey pictures," Comput. Graph. Image Proc. *7*, 30-51 (1978).

Trimberger, S., "Reaching for the million-transistor chip," IEEE Spectrum *20*(11), 100-102 (1983).

Tsao, Y. F. and Fu, K. S., "A parallel thinning algorithm for 3-D pictures," Comput. Graph. Image Proc. *17*, 315-331 (1981).

Turing, A. M., "On computable numbers, with an application to the Entscheidungs-problem," Proc. London Math. Soc., Series 2 *42*, 230-265 (1936).

Ulam, S. M., "On some mathematical problems connected with patterns of growth of figures," Proc. Symposia Appl. Math., Amer. Math. Soc. *14*, 214-224 (1962).

Ullman, J. R., *Pattern Recognition Techniques*, Butterworths, London (1973).

Unger, S. H., "A computer oriented toward spatial problems," Proc. IRE *46*, 1744-1750 (1958).

Unger, S. H., "Pattern recognition and detection," Proc. IRE *47*, 1737-1752 (1959).

Van Vleck, J. H. and Middleton, D., "A theoretical comparison of visual, aural, and meter reception of pulsed signals in the presence of noise," J. Appl. Phys. *17*, 940-971 (1946).

Von Neumann, J., "The general logical theory of automata," in *Cerebral Mechanisms in Behavior - The Hixon Symposium* (Jeffries, L. A., ed.) Wiley, New York (1951).

Voronoi, G., "Nouvelles applications des parametres continus a la theorie des formes quadratiques. Deuxieme memoire: Researches sur les parallel-oedres primitifs," J. Reine Agnew. Math. *134*, 198-287 (1908).

Walton, W. H., "Automatic counting of microscopic particles," Nature *169*, 518-520 (1952).

Weston, P., "Photocell field counts random objects," Electronics *34*, 4647 (Sept. 1961).

Wiener, N., Extrapolation, *Interpolation, and Smoothing of Stationery Time Series*, MIT Press, Cambridge (1949).

Wood, A., Ph.D. Thesis (unpublished), Image Processing Group, University College London (1983).

Woodward, P. M., *Probability and Information Theory, with Applications to Radar*, McGraw Hill, New York and Pergammon Press, London (1953).

Yamada, H. and Amoroso, S. M., "Structural and behavioral equivalences of tessellation automata," Inform. Control *18*, 1-31 (1971).

AUTHOR INDEX

SUBJECT INDEX